ENCYCLOPEDIA OF GARDEN DESIGN

ENCYCLOPEDIA OF GARDEN DESIGN

EDITOR-IN-CHIEF **CHRIS YOUNG**

BE INSPIRED TO PLAN, BUILD, AND PLANT
YOUR PERFECT OUTDOOR SPACE

CONTENTS

8 Foreword

Chris Young, Editor-in-Chief

12 Design inspiration

From formal and family to natural and productive, explore the history and key ingredients of some of the most interesting garden design styles around. Plus, find inspiration for your own plot.

- 14 First questions
- 22 Garden design trends
- 28 Garden styles explained
- 32 Formal gardens
- 46 Informal gardens
- 60 Family gardens
- 74 Natural gardens
- 88 Small spaces
- 102 Productive gardens
- 116 Statement gardens

130 How to design

A comprehensive guide to the principles of garden design and how you can apply them to create a beautiful design of your own.

- 132 Assessing your garden
- 142 First principles
- 180 Creating a plan
- 198 Designing with plants
- 222 Choosing materials

continued >

continued

242 Making a garden

All the practical information and step-by-step guides you need to bring your garden designs to life.

252 Building garden structures
- 254 Laying a path
- 256 Laying a patio
- 258 Building a retaining wall
- 260 Laying decking
- 262 Putting up fence posts
- 264 Laying a gravel border
- 266 Building a pergola
- 268 Making a raised bed
- 270 Making a pond

272 Planting techniques
- 274 How to plant trees
- 276 How to plant shrubs
- 278 How to plant climbers
- 280 Laying a lawn
- 282 Meadow-inspired plantings
- 284 Aftercare and maintenance

286 Plant and materials guide

Expert advice to help you choose the perfect plant for any situation and the right materials for your design.

- 288 Plant guide
- 350 Materials guide

- 368 Resources
- 370 Understanding hardiness ratings
- 371 Designers' details
- 372 Acknowledgments
- 379 Index

Foreword

Have you ever sat – just sat – in your garden, thinking, looking around, taking in the view? Not really looking at anything in particular, but thinking about anything and everything to do with your garden, asking yourself, "What if I planted a tree there?", or "If I moved those paving slabs, what would I put in their place?" Whether you were aware of doing this or not is, in a way, immaterial because what you have been doing is visually making this piece of land your own, and coming up with thoughts and ideas for improving your outside space. Welcome then – whether it be for the first or fiftieth time – to the world of garden design.

The concept of garden design is nothing new: when humans first cultivated land, and enclosed arable crops and livestock, they were delineating usable space to its best advantage. This may not be design as we understand it now (obviously, aesthetics were of no practical value then), but it was about making spatial relationships based on need. This was about designing an environment to suit individual daily, monthly, seasonal, and yearly requirements.

Since that time, the process of creating a garden has evolved according to style, fashion, prowess, skill, aptitude, wealth, travel, experimentation, and history, but it can all be distilled down to that first need. In essence, garden making is all about a human being exerting some level of control over their own surroundings. And, really, that is all garden design is today.

As is set out by my fellow authors in this book, creating a garden can be an intricate and time-consuming process, but the fundamental starting point is to remember that garden design is about creating an outside space that you (or your client) want. Many discussions will ensue after that initial thought – from what style you want, to working out how sustainable your garden can be. But don't let the detail bog you down too much or too early in the process. Of course, detail is essential for a successful garden, but holding on to that vision, that desire, is a key part of the process. This book will help you, not only with the nuts and bolts of garden making, but also to focus the vision and, I hope, help make it become a reality.

Welcome in
Successful garden design is about creating usable, attractive, and well-made spaces that suit the owner's personal needs.

Opposite, clockwise from top left
Eye of the beholder
Sometimes, beautiful design expressions can be created by mirroring shapes, like this sculpture and round-flowered Allium.

Personal space
Good design should reflect the wishes, likes, and dislikes of the garden owner – regardless of the country or climate.

Good form
Successful designs use flower colour, leaf shape, and tree stems to create a balance of colour and form.

Plan your plan
Putting your ideas onto paper, or computer, is an essential step when designing your garden.

Considered style
Successful spaces are created when planting colours and combinations complement the hard landscaping materials.

So why is there still a need for an encyclopedia such as this? In truth, because designing a garden can be something of a lonely experience. Even though we are constantly bombarded with images, suggestions, and information (books, websites, social media, and magazines), it is rare to be able to look in one place for everything – from plant selection to gravel colour, from fence posts to tree heights. The very nature of having so much choice can render the designer/gardener/client more than a little confused as to what they actually want from their garden. The activity of making a garden can also be influenced from so many quarters – everything from plants to hard materials – that a designer needs a refuge of sorts, where questions are answered and problems resolved. I hope this book will be that refuge in this ever-crowded, information-heavy world.

Often, coming up with an overarching vision for what you want your garden to be like is the easiest part of the process. It is translating that vision into a reality that takes the bulk of the time: working out how parts of a garden can sit together, how planting interest throughout the year can be sustained, deciding on hard landscaping materials that will work in all weather conditions, and so on. These are the stimulating – and at times frustrating – aspects of the process, but they make the difference between an unusable piece of land adjoining your property and a beautifully designed garden.

The chapters in this book take you through these very stages of garden design, helping to demystify the unknowns and clarify the unclear. I sincerely hope you enjoy it and, as a result, make the best garden you possibly can.

Chris Young
Editor-in-Chief

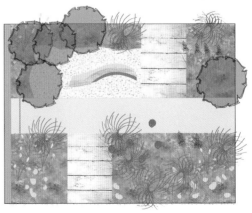

14	First questions
22	Garden design trends
28	Garden styles explained
32	Formal gardens
46	Informal gardens
60	Family gardens
74	Natural gardens
88	Small spaces
102	Productive gardens
116	Statement gardens

DESIGN INSPIRATION

What do you want to do in your garden?

Your garden is an extension of your home and it should provide a place for you to enjoy life to the full. When thinking about any changes that you plan to make to the garden, it is important to consider how you want to use the space, and not just now but in the future as well. This can range from keeping very busy to doing as little as possible. Ask yourself a series of questions about the garden's many roles. Do you want a space for

ENJOY THE PLANTS AND WILDLIFE

The active gardener
Digging, sowing, and planting bring great rewards as plants grow and change throughout the seasons. Colours and textures evolve, and there is something new to see each week. Plants attractive to birds, bees, and butterflies bring borders to life.

ENTERTAIN AND HAVE FUN

The room outside
Gardens are often described as "outdoor rooms" and can be planned as extensions of the house. Ensure continuity with features such as stylish furniture, screens, painted walls, canopies, and planters. An open-air room can be used for entertaining and socializing, while also offering children space for energetic play.

entertaining, a play area while the children are young, or do you simply want a peaceful but beautiful garden in which to relax? Bear in mind that your needs, and those of your family, are likely to change with time, and that it may be more difficult to make significant changes to the garden in the future as it establishes and matures. Ideally, come up with flexible ideas that can be adapted. A range of different requirements might suggest the creation of separate and possibly hidden areas within the same garden.

APPRECIATE THE PICTURE

Simple solution
Gardens for busy people need to be easy to maintain, but they can still be lovely to look at. They require simple design solutions with a strong overall concept and a pleasing layout for long-term appeal, allowing owners to sit back and enjoy the view.

RELAX AND UNWIND

A peaceful space
One of the special joys of having a garden is that you can simply sit, doze, read, or do nothing in the open air, surrounded by the sounds and scents of plants and wildlife. Gardens designed for this purpose can provide the perfect antidote to the stresses and strains of everyday life.

How do you want to feel?

Gardens stimulate emotions: upon entering a space people immediately respond to it. When planning a new design, you may choose to be bombarded with sensory stimulation, a riot of vibrant colour, textural diversity, or striking features to excite and energize. Or you might want a place for quiet reflection and contemplation, or perhaps even a space for therapy and healing, such as a calm, simple garden with evergreen trees

EXCITED AND UPBEAT

The dynamic garden
Exciting, stimulating sensations can be created using vibrant, hot colours, spiky plants, sharp lines, challenging artwork, varied textures, and bold use of lighting. But be warned: strident garden designs can be overpowering.

REJUVENATED

Refreshing space
The presence of water, creating sunlit reflections and offset by natural plantings, can help to evoke a feeling of energy, growth, and rejuvenation. Soft colours and a complementary selection of natural materials enhance the mood. These are places for "recharging your batteries" after a long day at work.

and bushes, and a reflective pool. If you have enough outdoor space to play with, it may be possible to demarcate different areas for different moods by making effective use of screening or tall plants. Creating a new design for a garden provides an opportunity to change or enhance the atmosphere of each area through layout, distribution of paths and spaces, and light touches of detail and decoration. Colour, shape, fragrance, and foliage will also affect the tone, and by using these elements you can help to foster positive moods and emotions.

A SENSE OF WELLBEING

PEACEFUL AND CALM

Restoring health
These gardens should be private, unchallenging spaces, and are often characterized by culinary, therapeutic, and medicinal plants, such as herbs with their appealing scents, or nourishing crops such as fruit trees. They provide a reassuring, relaxed, and restorative environment.

Contemplative moods
Cool colours, simple flowing shapes, delicate scents, and restricted use of materials and planting will create a calm and peaceful mood in the garden. Simple focal elements, waterfalls, and carefully chosen lighting help to enhance these uncluttered spaces.

What will your garden look like?

Garden visits, shows, and plant nurseries, as well as magazines, books, television programmes, social media, and websites, will provide anyone wishing to change their garden with a wealth of inspiration. But remember, the key to successful design is not just collecting ideas and trying to combine them in your design. Rather, it is a process of reviewing and editing a range of ideas, with the aim of developing a coherent overall

FILLED WITH FLOWERS

A TROPICAL RETREAT

A HINT OF HOLIDAYS

Grow your favourite flowers
Your garden can be a horticultural extravaganza, or a setting for favourite plants. These gardens are seasonal and offer change and continuous involvement. Try to work to a clear overall concept in terms of colour, texture, and structure.

Sculpt with plants
Bold-leaved plants bring a sense of the exotic and can be used to create a lush, enclosed garden with a subtropical feel. Choose plants carefully to ensure that they will not get too big and are suited to your site's soil and climate.

Recreate a summer break
Why limit your holiday to a fortnight, when you can pretend to be on a summer trip all year? Adapt ideas seen on your travels: for example, fragrant lavender beds and window boxes brimming with ivy-leaved geraniums for echoes of southern France.

appearance for your garden, whether you are revamping a mature plot or starting with a blank canvas at a new house. A good way of approaching this is to have a clear image of the look you are hoping to achieve and to carefully select elements, features, materials, and plants that combine to produce a unified composition. Make notes, collect pictures, sketch ideas. Some starting points are given below, from the traditional to the modern, to the imaginative and quirky. Use them as a prompt to see which style suits you best.

A SPACE TO REFLECT

Make a sanctuary
A tranquil setting, characterized by straight lines, simple shapes, subtle lighting, and a coherent layout, provides a comfortable space for retreat from modern-day life. Avoid clashing materials and keep planting simple.

CHIC AND MINIMAL

Cut out the clutter
Restrict yourself to no more than three complementary materials and a muted colour palette, but combine them beautifully. A large, dramatic water feature or sculpture adds a dynamic quality to a pared-down design.

FUN AND FUNKY

Show your creative side
Perhaps better suited to show gardens or temporary installations, these quirky gardens are attention-grabbing but require artistic flair and confidence to be successful. Not for the shy or retiring, but they can be great fun while they last.

How much do you want to do?

The amount of time you have to devote to your garden on a daily, weekly, or monthly basis should be a major consideration when thinking about an overall design and its future maintenance. Unless you have a very simple, easy-care garden, with hard landscaping and evergreen planting, the list of tasks normally changes seasonally, with less to do in the cooler winter months. In a high-maintenance garden with mixed flower borders, lawns,

EVERY DAY

Regular upkeep
Most small gardens will not need attention more than two or three times a week at most, although a plot filled with lots of containers will require daily watering in hot, dry spells. Generally, larger gardens with lawns, mixed borders, a diverse range of plants, and productive growing areas will take up more of your time.

ONCE A WEEK

The weekend gardener
This is the most common category, especially for people who only have spare time at weekends. Formal lawns require weekly mowing and edge-trimming in summer, and weeds need to be kept in check throughout the garden.

fruit trees, and a vegetable plot, spring and summer are very busy seasons. Lawn-mowing, hedge-trimming, pruning and feeding fruit trees, sowing and transplanting vegetables, plant propagation, and ongoing cultivation all take time. This may be the garden you want, but be realistic about how much time you can spare to keep it looking good. Working in your garden, watching it mature, and admiring the results are immensely pleasurable, but do plan for maintenance in advance, and budget to bring in help if necessary.

TWICE A MONTH

SIX TIMES A YEAR

Keep it practical
Most shrubs, climbers, and perennial plants require attention at intervals. Seasonal pruning may be required in spring and autumn, borders need weeding and feeding, and flowering plants such as roses should be dead-headed regularly. Lawns are impractical in this category, although meadows are an option.

Minimal maintenance
Gardens requiring infrequent attention will exclude lawns and hedges. Plan for "low" rather than "no" maintenance. Many trees and shrubs need only an annual tidy-up, and hard landscaping just occasional attention, such as sweeping or cleaning.

Garden design trends

Show gardens come and go, but chances are the trends they exhibit will eventually percolate into the mainstream. Just think of the catwalks of the fashion world – much of what may seem crazy or fantastical previewed at those glitzy events will, in time, make its (albeit slightly more subdued) way to the high street and online retailers.

Much is the same for the garden design world. Over the past few decades, there have been myriad innovations in materials and design effects. The crossover from engineering and architecture to garden design – think metal walkways, glass balustrading, outdoor fireplaces, lighting effects, rigid plastics for furniture, and so on – has blurred the styling of inside and outside. The "outdoor room" (epitomized by garden design guru John Brookes from the 1960s) has never been better furnished, styled, and innovated.

So what are the trends affecting garden design as we enjoy this wealth of creative products and materials? It is clear that increasingly all our decisions will be – and should be – underwritten by sustainability and environmental awareness. We should consider everything we do against the (potentially negative) impact it has on the environment. If we are digging up old paving, could we reuse it somewhere else rather than throwing it away? If we live in an area of high rainfall, should we consider more gravel and permeable paving to help slow down the rate of water run-off? How can we see the wider role of gardens as a network of wildlife spaces, connecting across a town or city, and between urban environments and the countryside?

The answer to these questions lies in the multitude of ways we can work with nature to create the gardens we love. And difference is to be celebrated; like the natural world, we need variety. Formal gardens are as relevant as natural gardens; productive gardens are as useful as small spaces; rooftops are as important as country estates. What needs to underline any garden is the principles by which it is made: to be responsible in sourcing products; to use plants that are sufficiently resilient so they can flourish in our changing climate; and to increase chances for wildlife to feed and thrive.

Top and above
Whether plants are in the ground or, as here, growing on a roof, the more greenery and colour you can bring to your outside space, the better.

Making space not only beautiful but also useful throughout the day is part of the modern garden.

Sustainability and the environment

If sustainability and environmental awareness are the basis on which we should create our gardens, in what specific ways should these principles inform our collective design decisions?

Increasing plants

It may seem obvious, but the plants we grow are so much more than an aesthetic decision. They may be medicinal or healing in some way; they might provide rare and important habitats for wildlife; they may extend the season of available food for increasing biodiversity; and they may help cool and soften our urban lives. Think about where you buy your plants from – source them locally if you can, to reduce transport miles – and always buy plants that are grown in peat-free compost. And if you're really unsure about which plants to grow, just plant what you love.

Perennial refuge above right
The stems of perennials look beautiful year-round and offer a refuge for wildlife.

Please the pollinators far right
Insect hotels placed among pollinator-friendly planting make a welcome home for wildlife.

Back door kitchen above
Position a few edible plants in different-sized containers near your back door for easy access.

Edimental

If you've got limited space, you want to make the most of every centimetre of soil. To that end, a new approach is "edimental" – literally, a combination of "edibles" and "ornamental". This doesn't mean hybridizing a new strain of plants! Rather, it's about ensuring the plants you grow provide both an aesthetic benefit and opportunities for eating. So instead of choosing a shrub for the back of a border, why not consider a fruiting currant, which looks great, has flowers, and produces fruit? Or if you want a mid-height focal point plant, grow some rhubarb instead of something purely ornamental. Once you start thinking like this, it makes every planting decision that bit more challenging in terms of what you will choose, but more interesting in the long run.

Mixed use
Growing different types of plants together – trees, vegetables, herbs, and flowers – is a clever use of space.

Edimental approach
Ensuring there's an interesting mix of planting allows edibles to thrive among more colourful companions.

GARDEN DESIGN TRENDS 25

Loving soil

Some gardeners are disrupting much of what we once considered standard lore – by not digging the soil. "No dig" is a global movement, especially when growing vegetables, where adding a thick layer of compost mulch (on top of cardboard if weeds are a problem) reduces the need to disturb the soil by digging. Seedlings and small plants are planted directly in a few centimetres of compost sitting on top of undisturbed soil below, where they can grow freely and easily. Reducing the need to dig helps retain carbon stored in the existing soil; it prevents access to light for latent weed seedlings; and it maintains the wondrous world of soil health and mycorrhizal connections (the association between soil fungi and plants). Plus, if we're honest, it's great for gardeners as it means we don't have to dig as much.

No dig, no-brainer
Crops such as beetroot (above right) can be planted in the compost of a no-dig garden when young and will establish easily. Taller plants, such as kale 'Cavolo Nero', can be interplanted with other crops for efficient use of space (right).

Reducing plastic

It's important to recognize that plastic has been an amazing, versatile material. It has provided so much, and in a garden, it's not really possible to be 100 per cent plastic-free. High-quality plastic, such as for a water butt, hose, or wheelbarrow, can be the more affordable choice and still last half a lifetime. Where consumer changes can be made swiftly is by avoiding single-use plastic – for plant pots, compost bags, carrier bags, and wrapping, for instance. Reuse and recycle plastic where you can, and always consider alternatives to single-use plastic, such as coir, wood, or bamboo, or biodegradable plant pots and trays. Finally, never buy astroturf or plastic grass – as it degrades over time, microplastics are released into water courses and the wider environment.

Paper weight above left
Growing seedlings in paper-based cells is a great way to cut down on plastic use. You can use cardboard toilet roll tubes for larger seedlings.

Loyal service left
Going plastic-free in a garden is almost impossible, but if you do allow plastic, then hard-wearing, durable products such as compost bins and water butts will serve for many years.

Thinking long term

Throughout society, many people are trying to buy high-quality products that look great and last a long time. The same principle applies with garden purchases – whether it be well-made furniture, paving slabs that won't crack or shatter when it's frosty, or metal products that don't rust. Of course, budget implications are a reality for almost everyone, but where it's possible and right to do so, choose products that are well made and will last in the long term.

Plan for rain top
Managing water will be vital in the coming years – planting and water features should be able to cope with varying levels of rain intensity.

Green roof above
Planted roofs allow rainwater to be captured and the percolation rate slowed down before water enters the main drainage system.

Alluring aesthetics

There is no "right" design for a garden – it's simply what you want it to be. We are all, constantly, stimulated with images, videos, and audio that can inform, inspire, and excite us. Through myriad choices and all the noise, stay with the style you want; refine your choices through online sites such as Pinterest (or in good old-fashioned scrapbooks) to stick to the brief; and choose what you love the look of. It may not always be other people's choice, but it will certainly be yours.

Inside out above
The architectural integration of indoors-outdoors continues apace, with glazing technology enhancing the link.

Personal choice below
There is no right or wrong in terms of style – topiary may be a thing of beauty for many, or a maintenance headache for others.

The benefits of nature

Some of the most potent words of our time, "wellbeing" and "mindfulness", are core to a healthier body and mind. For every person, the way to greater wellbeing and mindfulness will be different, but there is clear scientific proof that connection with nature, seasons, and growing plants can all help to improve physical and mental health. The more we take this into consideration, the more we can adapt our outside space to support our (and of course nature's) health.

Connect with nature
Science has proved the connection between nature and wellbeing, and space to enjoy the outside should be integral to all garden design.

The bigger picture

There is little doubt that we are at a tipping point for our changing global climate, with many people feeling overwhelmed and powerless. Yet there is so much we can do. Just taking a moment to think about sustaining wildlife, increasing biodiversity, and considering our individual actions can all contribute to a better environment. Taking time to think of these things will collectively help the bigger picture. Revel in having any outside space and know you can do your bit. And the more we grow plants, look after our soil, and reduce our negative environmental impact, the more gardens will play a vital role in our future.

All-encompassing gardens
Gardens are multi-faceted, offering spaces to sit and relax; homes for wildlife; places to offset climate change; areas for beauty and art; and, of course, spaces to grow beautiful plants.

Garden styles explained

Opposite, clockwise from top left
The power of plants
Prairie-style planting is a dramatic way to create naturalistic swathes of colour over a large area.

Space to play
The open space and minimal planting in this garden offer the flexibility to accommodate many uses.

Eclectic influences
Combining different design elements can create a space that equals more than the sum of its parts.

Leafy mix
Plants with colourful and attractively shaped leaves lend a lush, exotic look to contemporary designs.

Ideas explored
Garden style takes ideas and inspiration from around the world that can be easily adapted and recreated.

Making a statement
Modern materials, strong lines, and understated planting give this design a bold, contemporary edge.

In design terms, style refers to the way in which we express ideas and organize materials, plants, colours, and ornaments to create a composition that can be understood and appreciated. While some garden styles are short-lived fashions, others represent major movements, each with their own aims and motives. In classically inspired formal design, for example, order, repetition, and symmetry are used to create strict visual and spatial balance. This style dates from antiquity, and even when interpreted for modern gardens, the basic design principles still apply. Overlay modernist interpretations on formal or classical designs, and some really exciting stylish and individualistic gardens can be created. By contrast, natural-style gardens, embraced by many as an antidote to today's problems – such as declining biodiversity and climate change – are concerned with sustainability, the garden as a habitat for wildlife, and planting inspired by the example of nature. In these gardens, natural features and plants dominate, with the hand of the gardener well concealed.

External influences

Garden styles commonly draw inspiration from cultural or historic reference points, which give them a particular theme. The aim is to create a stylized interpretation of reality, rather than an accurate representation. Japanese-themed gardens, for instance, often lack the original philosophical and religious meaning but are nonetheless atmospheric. Similarly, the traditional cottage garden is a highly romanticized view of the simpler artisan model.

Broader issues and lifestyle changes have also helped to shape garden design. The increase in foreign travel has given gardeners a taste for the al fresco life (as seen in places like the Mediterranean), and for more exotic planting, which is being used increasingly in city gardens where warm microclimates allow a broader range of plants to thrive. Meanwhile, concerns about the environment are driving the use of sustainable materials, considering the long-term impact of our actions and ensuring that gardening practices take into account the needs of wildlife.

Functional space

The idea of the working garden has long been a recurrent feature of garden history, where the focus has involved growing food for the table. While the current trend for healthy eating has put home produce at the heart of many gardens once more, the functional requirements of gardens today are far broader, and reflect individual lifestyles more closely. Hence, families commonly require space for leisure, play, and socializing, while other gardeners seek refuge from daily pressures in a calm space, ideal for rest and relaxation.

The way ahead

As population densities increase, the urban garden is coming under ever greater pressure, diminishing in size but increasing in value. A century ago, a one-acre plot would have been considered quite small, but now people fill balconies, roof terraces, and postage stamp-sized gardens with vibrant ideas, creating a new idiom in direct contrast to much larger and expansive country gardens in which abundant space is the key characteristic.

Just as the form and function of gardens are changing, new styles are also being developed. Statement gardens often celebrate the human-made, creating dramatic and sometimes thought-provoking gardens that can be humorous or whimsical, philosophical and profound, short-lived or permanent. Designers of these conceptual or non-conformist spaces have thrown out the rule books to make statement gardens for a future generation. The cultural connection in many of these designs is strong, with some offering social commentary or presenting a reflection of modern society. Other designers mix up styles to create a fusion of the old and new, perhaps weaving cottage-style planting into a sharp modernist ground plan, or employing the latest materials, sculptures, and technology in a formal, symmetrical layout.

As styles and references merge, so innovative ideas, fresh possibilities, and new idioms arise. Where once garden style was seen as conservative and predictable, it has now been rejuvenated and celebrates change. In addition, new links with architecture and art are being forged, and garden design is now considered a dynamic and socially relevant discipline.

Opposite, clockwise from top left
Cottage dream
Generously filled borders and a bountiful approach to planting are typical of the informal, cottage-garden style.

Wildlife habitats
Even small garden ponds and boggy areas provide an excellent habitat for a wide range of wildlife.

Urban living
Ever-decreasing outdoor space is forcing gardeners and designers to develop creative new solutions.

Productive patch
Attractive vegetables and herbs integrate easily into most garden styles, even where space is limited.

Blue-sky thinking
Modern garden designers are constantly pushing the boundaries to bring new materials, textures, and combinations together.

Formal rules
A parterre illustrates the symmetry and geometry of the formal garden style.

Formal gardens

Designed as expressions of humankind's dominance over nature, the features and natural elements in formal gardens are contained in an imposed geometry and structure. This idea is rooted in classical architecture and design, and many of the best historic examples of this type of garden can be seen in France and Italy.

A successful formal garden has a balanced design, often achieved through symmetry and a clearly recognizable ground plan or pattern. Organized around a central axis or pathway, formal plans may focus on a key view through the garden from the house. In larger gardens, there may be space for several axial routes that cross the central path, and sometimes reach out into the wider landscape. Sculpture, water, or decorative paving are also used to punctuate the areas where these routes intersect. Modern formal gardens, especially in an urban environment, may not use symmetry and can be more introspective, but careful balance, clean lines, and a clear, strong ground plan remain important features. The geometry of the formal garden is clear and easily identifiable, but scale and balanced proportions are key considerations.

The material palette tends to be kept to a minimum, with gravel and paving most frequent. However, decorative elements such as cobbles and brick are still popular, as are more modern materials such as Corten steel, concrete, limestone, and composite decking. Water is employed either as a reflective surface or used for jets and fountains.

Lawns and hedges are traditionally key planting features in formal gardens, the latter helping to define space, edge borders, create parterres, and form knot gardens. Modern interpretations also often make use of clipped and shaped plants, such as pleached or multi-stem trees, to add height and frame focal points. They also often continue to employ a restrained and limited planting palette. Foliage is usually of key importance, and flowering plants are repeated and used sparingly amid evergreens, which provide year-round structure and texture.

Top and above
Dynamic water features provide movement.

Symmetry about a central axis attracts attention to focal points – such as sculpture or water features – in a formal garden.

What is formal style?

Formal garden design relates directly to the classical architecture of Greece and Italy, although a modern take on formal style is somewhat different. Ordered gardens originally provided a setting for the villas of the wealthy or powerful across Europe, echoing the symmetry of their grand houses. Known as "power gardening", it was seen as the ultimate in garden-making, embodying a sense of control. Although famous formal gardens, such as Versailles, are vast, this traditional interpretation of the style can be applied to gardens of any size, even tiny urban spaces, where ordered, balanced designs work very well. Symmetry about a central axis is crucial to emphasize the focus of the garden. Planting and construction are geometric and simple, with lawn, clipped hedges, and avenues forcing planting into order, and balustrades, steps, terraces, and wide gravel pathways all conspiring to unify the garden space.

Traditional formal gardens in detail

Classical formality demands an axis, or central line, which is the basis of the garden plan. This could be a pathway or lawn, or even a central planting bed. Generally, the axis focuses on a dominant feature, such as a sculpture, statue, or ornament.

If space allows, cross-axes can be created; some larger gardens have multiple axial routes that create views along and across the garden. A dramatic sense of scale and proportion is essential as planting and paving are often kept simple – one reason why many modernists appreciate this style and have reimagined it.

The space should be divided into halves or quarters. Larger gardens can be partitioned further, but divisions should be sizeable to maximize the impact of long vistas, or the repetition of topiary or trees. Parterres, water pools, and expanses of lawn are typical of classical formality; examples by contemporary designers may also feature decorative borders that soften the garden's structure. By contrast, modernist interpretations challenge this approach, introducing asymmetry and even avoiding ornamentation.

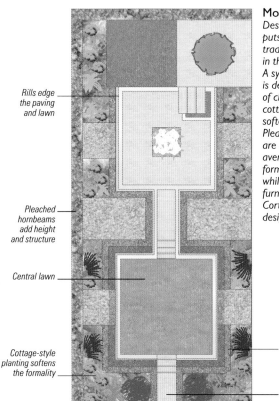

Modern twist
Designer Charlie Albone puts a modern spin on traditional formal style in this elegant garden. A symmetrical floor plan is defined with layers of clipped box, while cottage-style planting softens the rigid lines. Pleached hornbeams are a nod to the classic avenues of trees seen in formal country gardens, while the contemporary furniture and pavilion, and Corten steel rills, bring the design up to date.

- Rills edge the paving and lawn
- Pleached hornbeams add height and structure
- Central lawn
- Cottage-style planting softens the formality
- Clipped box walls define the symmetry
- Central axis

FORMAL GARDENS

Vaux le Vicomte by André Le Nôtre.

DESIGN INFLUENCES

Although some of the earliest Islamic gardens were formal in layout, often divided by rills into quarters, classical and Renaissance influences have come to define this style. The doyen of the formal garden is André Le Nôtre, one of a long line of gardeners turned designers who found fame in France under the reign of Louis XIV. The gardens he designed at Versailles and Vaux le Vicomte are his most famous legacies. The false perspectives, level changes, and reflective pools of both gardens are typical of Le Nôtre's approach to design, which won him the affection of the King.

Hedges, vast lawns, water features, and parterres of box and cut turf, often decorated with coloured gravel, as seen in Le Nôtre's work, set the tone for all formal gardens that followed, with views and perspectives manipulated for the best theatrical effect.

Key design elements

1 Symmetry
The symmetrical balance of a traditional formal design can be achieved at any scale. Here, an olive tree and a parterre form a focal point in a circle that intersects the pebbled and paved central path.

2 Statuary
Gods and mythological creatures were the original subjects of statuary in formal gardens. In modern designs, contemporary figurative subjects and abstract works function well as focal points.

3 Topiary
Clipped hedging, typically box or yew for evergreen structure, is used to define space. Topiary provides architectural definition, and dwarf box hedges are used to form patterns in traditional parterres.

4 Ornament
Large, ornate urns, often on plinths or balustrades, provide focal points or punctuation. Modern formal gardens use the same technique, although elaborate decoration is reduced.

5 Natural stone
Paving provides an architectural element for pathways and terraces. Sawn and honed natural stone slabs can create regular patterns, or they can be used to edge lawns and gravel paths.

Contemporary formal style

Formal style encompasses a far broader range of gardens than may at first seem obvious. The term has often been used to recall a range of historic examples, but it can also accurately describe contemporary applications, including many modernist gardens. In these, planting is an architectural element, a part of the composition but not the main reason for the garden's creation. Clipped hedges, specimen trees, and large blocks of planting provide simple, sculptural surfaces or screens, which complement the horizontal expanses of timber, stone, concrete, or water. The aim is for a composition as a whole, forming a beautiful, elegant space. The formal style adapts well to different scales and locations, from urban courtyard to rural retreat; the designs may feature symmetry or be refreshingly asymmetric but they generally emphasize leisure and outdoor living.

Contemporary formal style in detail

Formal designs suit gardens of any size, but while traditional iterations may employ decorative embellishments such as statuary and topiary, modern examples rely more heavily on scale and proportion to create drama, providing gardens with open, uncluttered spaces that offer the perfect setting for outdoor living.

Most formal gardens are based on some form of geometric layout, with the horizontal lines of rectangles providing a sense of movement. These dynamic lines contrast with the verticals of trees, hedges, or walls, and slice through space to unite different sections of the garden. A modernist take may employ asymmetrical design; traditional versions often rejoice in symmetry.

Materials are selected for their surface qualities – decking, polished concrete, limestone, and gravel produce expansive surfaces, often punctuated by water or specimen trees. These materials require stunning high-quality finishes and architectural precision. Fine lawns, clipped hedges, and simple planting are common to both traditional and modernist formal gardens, but contemporary designers sometimes use a more complex palette.

Tonal accents
Pale tones, accented by green, make a formal space look clean and ordered.

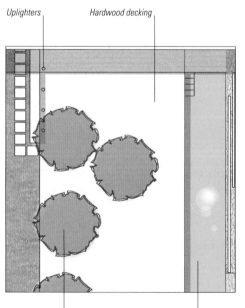

Uplighters   Hardwood decking

Inside out
Here the main terrace of Casa Mirindiba in Brazil (right), designed by Marcio Kogan, extends into the garden to create a sheltered space, part interior and part exterior in character. The long, narrow swimming pool reflects the stone wall, and lighting picks out surfaces and tree canopies to create interest after dark.

Minimalist tree planting

Reflective swimming pool, or "lap" pool

FORMAT GARDENS

VILLANDRY RECREATED

Key to French 17th-century formal gardens were the concepts of order and harmony as inspired by Italian Renaissance gardens a century before. Symmetry and geometry were important, with beds kept in perfect order by vast teams of gardeners. Designs generally featured an elevated terrace offering views over the gardens for best effect. Among the best-known examples of French formal gardens today are those at the Château de Villandry, in the Loire, France. These gardens are largely a lavish early 20th-century recreation of the original style, and include water gardens, flower gardens, and an ornamental potager (kitchen garden). It features a grid of nine squares, each with a different internal layout, but planted with vegetables of contrasting colour set within low box hedging.

Formal potager at Château de Villandry in France.

Key design elements

1 Geometric layout
A central axis is common in traditional formal gardens, but in other designs is rarely a dominant feature. Rectangles of lawn, water, paving, or planting interlock more intuitively to create sharply defined but irregular patterns.

2 Quality materials
The clean lines of steel, concrete, glass, and timber emphasize the precision of the manufacturing process. Paving joints are minimized, and subtle lighting is used to enhance the surfaces.

3 Limited plant palette
The range of different plants is limited and often planted in large blocks or masses. Grasses and perennials are interplanted to catch the light and create movement.

4 Architectural furniture
Garden furniture is traditionally classically styled, but contemporary classics, such as the sculpturally inspired Barcelona chair, set the tone for elegant recliners, simple tables, and matching benches.

5 Reflective water
Reflective pools create unruffled surfaces and bring light into the garden. Modern technology now allows water pools to brim or overflow, maximizing the expanse and impact of the reflective surface.

Interpreting the style

Although the rules of formality are simple and clear, it is still a remarkably versatile style. The overall layout can be completely symmetrical and axial, or you can select just a few formal elements. One axis can be more dominant than another, for example, or a series of balanced, rectangular beds can be veiled by soft, romantic planting. You can also experiment with the style and opt for a traditional look or bring formality right up to date.

Contrasting elements top
An overflowing bowl creates a focus at the centre of this parterre in an enclosed corner of the Alhambra, bringing a dynamic quality to the formal planting.

Contemporary order above
A simple rectangular lawn, elegant pleached hornbeams, and a pale paved surface create restrained formality. The three plinths and subtle lighting lend focus.

Urban formality above right
Limestone paving creates a crisp, formal edge to this lawn, offering clear definition. Pleached lime trees provide increased privacy in this urban space.

Ornamental hedging right
A parterre-style panel of box cartouches makes a decorative statement of light, shade, and texture. The pattern will read particularly well from the first floor.

FORMAL GARDENS

"Set the geometric rules of formality, then decide which ones to break"

Aquatic symmetry far left
Pools and a connecting rill form the focus of this formal arrangement, with the sculpture and fountain on the central axis. The planting is then arranged symmetrically.

Sculpted greenery left
Here, the tightly clipped topiary supports the axial layout. The mossy path itself breaks the rigid formality, with lawn softening the edges of the rustic paving slabs.

Softer planting below left
Steel edging evokes a sense of formality in this grid-pattern garden, and is in stark contrast to the soft, light-catching grasses and perennials that fill the borders.

GARDENS TO VISIT

VAUX LE VICOMTE, Seine-et-Marne, France
Designed by Le Nôtre using false perspectives and axial layout. vaux-le-vicomte.com

VERSAILLES, Yvelines, France
André Le Nôtre's best-known garden. chateauversailles.fr

VILLA GAMBERAIA, Settignano, Italy
Garden of allées and formal compartments that radiate around the house. villagamberaia.com

ALHAMBRA & GENERALIFE, Granada, Spain
Evidence of the Islamic influence on formal design in Europe, with water as a central theme. alhambra.org

DUMBARTON OAKS, Washington DC, US
Originally designed as a series of formal spaces and vistas, but with some naturalistic planting. doaks.org

Formal garden plans

In these traditional and modern interpretations of the style, important elements remain consistent. Geometric lines and the use of quality materials are common to all. Planting in Declan Buckley's garden is fulsome, contrasting with Andy Sturgeon's minimalist approach. The colour palette in all is restrained, with greenery offset by crisp, clean surfaces, exemplified by Matt Keightley's garden. Size need be no barrier, as Charlotte Rowe's design proves.

Layered planting

In designer Declan Buckley's own garden, a rich tapestry of layered planting sits alongside the bold formal geometry of paving and a pool; the use of reflective water increases textural impact. There is a great sense of contrast here, between the open, light terrace and the narrow pathways.

Key ingredients
1. *Phyllostachys nigra*
2. *Euonymus japonicus*
3. *Fatsia japonica*
4. *Pseudosasa japonica*
5. *Geranium palmatum*
6. *Astelia chathamica*
7. *Buxus sempervirens*
8. *Cycas revoluta*

Declan says:
"After years spent growing plants in pots on a roof terrace, it was a relief to have a garden to plant them in. The site is a long rectangle, overlooked by five-storey houses, so bold and layered architectural planting helps to screen the site and provides privacy. Conversely, the end wall of my own house is solid glazing, which gives me a dramatic view across the pool and into the luxuriant planting.

"London's warmer temperatures allow more tender and unusual species to thrive, and plants were chosen for their texture and form – flower and colour came second. A strong, simple framework softened by foliage is key to all my projects."

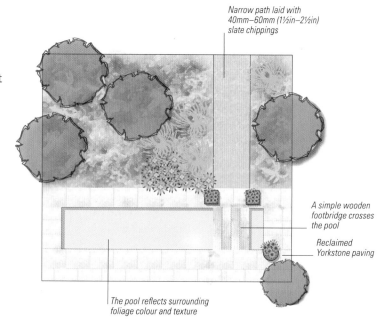

Narrow path laid with 40mm–60mm (1½in–2½in) slate chippings

A simple wooden footbridge crosses the pool

Reclaimed Yorkstone paving

The pool reflects surrounding foliage colour and texture

FORMAL GARDENS 41

Traditional and contemporary unite

In this modern, urban take on the formal style, designer Matt Keightley cleverly makes a nod to the traditional with a design that manages to be both symmetrical and asymmetric. There are decorative flourishes and lavish planting, but with restrained use of materials and colour.

Key ingredients
1 Pleached hedge
2 Ferns, aspidistra, and other foliage plants
3 Lawn
4 Rendered screen
5 Clipped rosemary domes
6 Limestone detailing

Elegant modernism
This elegant formal garden very much lives up to modernist ideals with its open expanse of lawn, clean uncluttered lines, and use of high-quality materials, where planting is restrained, foliage-led, and simply part of the designer's palette.

At first glance, the layout appears symmetrical, but look closer and many elements such as the sculptural tree and seating area with its screen are carefully positioned off-centre, while more traditional elements such as gravel, pleached trees, and clipped evergreens recall features from formal gardens of the past.

The lawn, framed by gravel and placed centrally, is key. It emphasizes the garden's width, making it feel more spacious while providing an uncluttered space at the garden's heart. At the same time, its shape with the four rosemary domes also suggests a hint of Italian formality within a contemporary space.

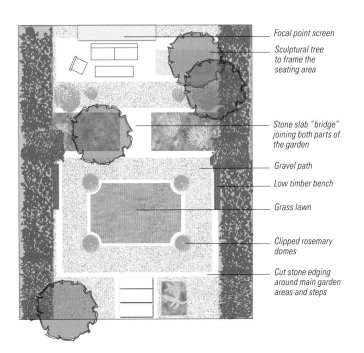

- Focal point screen
- Sculptural tree to frame the seating area
- Stone slab "bridge" joining both parts of the garden
- Gravel path
- Low timber bench
- Grass lawn
- Clipped rosemary domes
- Cut stone edging around main garden areas and steps

Classic lines

In this traditionally formal small space, designed by Charlotte Rowe, the simplicity of design works well: the beds retain a mix of just a few species. The urn and *Ligustrum* topiary add height and a sense of scale to the scheme, while the *Hydrangea* provides an elegant focus to the central axis.

Key ingredients
1. *Ligustrum delavayanum*
2. *Hydrangea macrophylla*
3. *Artemisia* 'Powis Castle'
4. *Geranium sanguineum*

Charlotte says:
"My simple, understated design for this front garden in Kensington had to fit in with the regulations of the local conservation area. I used Yorkstone and bricks to match similar detailing on the house façade and evergreen screening for privacy but kept the overall design simple and understated.

"I'm often influenced by modernist designers, such as Luis Barragán and Dan Kiley, so it was interesting to retain a sense of precision here in such a classical format. I think of the hard landscaping materials as the bone structure of the garden, which the planting can soften and enhance."

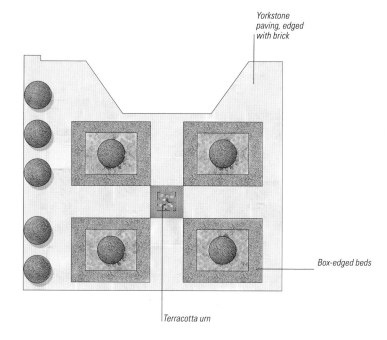

Yorkstone paving, edged with brick

Box-edged beds

Terracotta urn

FORMAL GARDENS

Up on the roof

In a restricted city space, this rooftop garden by Andy Sturgeon makes excellent use of the great outdoors. The low-maintenance design creates an extra room in which to entertain, with materials providing the focus and simple planting offering shelter and privacy.

Key ingredients
1 *Fargesia rufa*
2 Iroko bench
3 *Astelia chathamica*
4 Gas-fired flambeaux

Andy says:
"This space suited the client, who was young and enjoyed entertaining friends, but wasn't particularly interested in gardening.

"The water became the focus of the garden. It is very shallow, to reduce the weight on the roof, but highly reflective to excite and entrance; combining it with fire proved a particularly complex detail.

"I normally design larger spaces that are not so minimal, but my approach to this project suited the client and the rooftop location, and I enjoyed responding to the challenge. More specifically, the client wanted to be able to sit outside in all weather and seasons, hence the canopy and the water, fire, and bench combination.

"I call upon a wide range of inspirations, from shop-window treatments to contemporary art, and find this input particularly useful in urban situations."

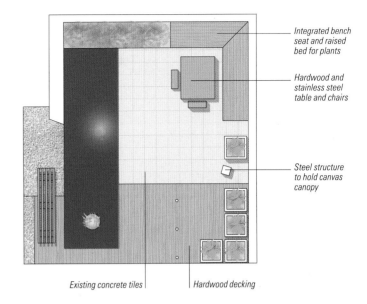

Integrated bench seat and raised bed for plants
Hardwood and stainless steel table and chairs
Steel structure to hold canvas canopy
Existing concrete tiles
Hardwood decking

CASE STUDY
BUILDING BLOCKS

The simple, clean lines of this garden betray an exacting design that has formal, contemporary detailing at its heart. Crisp blocks of planting, paving, and water are set out on an asymmetrical floor plan, reflecting strong design principles.

Visual play
Contrasts of texture and form are used to great effect in the design. Smooth paving, a reflective water feature lined with flat pebbles, trim beech hedging, and multi-stemmed *Osmanthus* trees, conspire to create bold visual effects.

Ordered space
The space in the garden is set out in rectilinear blocks of paving, planting, and water. Some of the areas are open, others are enclosed, hiding and then revealing aspects of the design as the visitor walks through the different spaces.

Designer **Marcus Barnett**

FORMAL GARDENS **45**

Asymmetrical plan
The stone wall panels were inspired by a Piet Mondrian painting. Combining calm, clean lines and strict geometry with an asymmetrical plan, they perfectly represent modernist design principles and are brought to life in this formal garden.

Floral contrasts
The different flower shapes provide refined contrasts and accents. The yellow daisy-like flowers of *Doronicum* stand tall above tiny *Euphorbia polychroma* bracts, and contrast in colour and form with red tulips and blue cornflowers.

Clear colours
The colour palette shows a typically formal restraint. Shades of green predominate, allowing the primary colours of red, blue, and yellow to shine through.

Informal gardens

This style comprises some of the most popular and often-seen gardens. It includes cottage gardens, country gardens, and plant-filled gardens where the main emphasis is on what is growing, rather than how it is displayed. Often, they are the gardens of plant enthusiasts. These are gardens that have evolved gradually over the years with the impression of little planning, but grown instead through plant purchases and horticultural endeavour.

Celebrated for their abundant planting and carefree nature, cottage gardens are traditionally simple and regular in layout, with a path to the door and maybe beds on either side. They were first used as productive spaces in rural locations, where the focus was on food rather than flowers.

Traditional cottage gardens were also championed by the famous garden designer Gertrude Jekyll, who used them to form the basis of the Arts and Crafts planting schemes she devised at the turn of the 19th century. This style is followed in many country gardens and can be regarded as a more sophisticated extension of cottage garden style, with a simple, often formal layout but with generous informal planting.

The scale of informal gardens is generally intimate, sometimes even restrictive to movement, as dense planting is allowed to spill across pathways. Self-seeding is encouraged, as are plants that can colonize gaps in paving. Hedges are frequently used to divide the garden, often forming a series of enclosed spaces with different planting schemes and atmospheres in country gardens. The combination of often diverse, usually soft, and riotous planting with formal clipped hedges and decorative topiary results in one of the most successful contrasts in this design style. Away from the house, in larger gardens, there may be room for meadow planting and native hedges that create a wilder impression.

The most appropriate hard materials for use in informal gardens are natural stone or brick, with weathered or rescued materials favoured for their aged and subtle appearance. Gravel is also used for pathways, partly because it allows easy self-seeding, and simple post-and-rail or picket fences also suit this naturalistic design style.

Top and above
*Jewel-like aubretia cascades over a weathered stone wall.
Decorative produce adds colour to a working garden.*

What is informal style?

The profusion and sheer romance of an informally styled design, particularly when applied to cottage and country gardens, wins the hearts of many designers globally. This is mainly due to the luxuriant planting with its use of colour and texture, and the sheer range of plant species involved in this quintessentially relaxed English style. Informal planting is, in fact, far from random but uses thematic or coordinated flower and foliage colour within small compartments or "rooms", as seen to great effect in the gardens at Sissinghurst Castle or Hidcote Manor.

Cottage and country garden style

The layout of cottage and country informal gardens is often simple and geometric, although many diverge from this pattern into more idiosyncratic twists and turns. Pathways are often narrow, with plants partially obscuring a clear way through. This romantic planting softens the appearance of a garden, blurring the layout, and brings you into close contact with scent, foliage textures, and spectacular blazes of colour. Many modern informal gardens employ elements of this style, perhaps the exuberant planting set within a less ordered layout, and the different areas less formally delineated.

Paved areas are often constructed from brick, gravel, setts, or cobbles, which allow mosses, lichens, or creeping plants to colonize joints and surfaces. Larger gardens may employ stone slabs. Simple seats, old well heads, tanks, pumps, and "found" materials make interesting focal points and create a serendipitous quality. Arbours or arches decorate thresholds between the garden spaces, often divided with hedging, walls, or trellis.

Lawns provide a visual rest from the colour and profusion, while fruit and vegetable beds retain the simple geometry of the earliest cottage gardens, with stepping stones, brick, or compacted earth paths providing access.

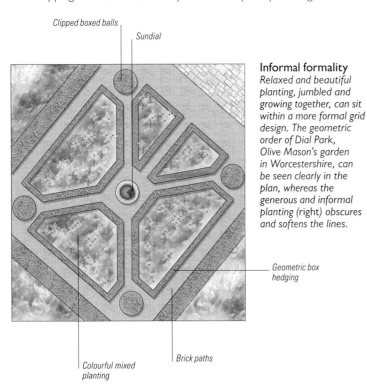

Clipped boxed balls
Sundial
Colourful mixed planting
Brick paths
Geometric box hedging

Informal formality
Relaxed and beautiful planting, jumbled and growing together, can sit within a more formal grid design. The geometric order of Dial Park, Olive Mason's garden in Worcestershire, can be seen clearly in the plan, whereas the generous and informal planting (right) obscures and softens the lines.

Summer colour in a garden for all seasons

With its wide range of foliage textures, tumbling climbers, colourful perennials and perfumed flowers, Olive Mason's garden is planted for year-round interest. In spring, green and white foliage predominates, interspersed with subtle drifts of daffodils, tulips, hyacinths, and forget-me-nots. The colours intensify in early summer (*above*) to warm pinks and mauves, with roses, geraniums, delphiniums, clematis and centaureas. As summer progresses into autumn, the palette deepens to the cerise, deep blues and purples of asters, phlox, dahlias and aconites, and in winter everything is cut back to reveal the simple pattern of the box hedges, enhanced by a bark mulch spread over the bare beds.

INFORMAL GARDENS

Key design elements

1 Profuse planting
Informal gardens with complex planting may require intensive maintenance. The art lies in the skilful association of planting partners, and the selective editing of species that become too dominant.

2 Rustic furniture
The patina of timber garden furniture changes organically over time; plants can be encouraged to weave through it to create an impression of apparently natural, but actually cultivated, recolonization.

3 Rose arbours
These make pretty shelters for seating, and can also be used to link different areas. Here the intense colour and delicate scent of a pink rose help to awaken the senses on a walk through the garden.

4 Weathered paths
Brick, stone sett, and gravel pathways provide textured surfaces as a foil to the complex planting on either side, allowing plants to seed and soften the boundary between path and border.

5 Vegetables and herbs
Productive borders are often seen in informal gardens, with cut flowers and herbs used in association. This attractive mix softens the functional appearance of these areas, and may also help to control pests.

DESIGN INFLUENCES

The modern interpretation of the cottage garden – and informal gardens more generally – is based to a great extent upon the work of Gertrude Jekyll and her architect partner, Edwin Lutyens. They created many outstanding designs in the 1890s under the auspices of the Arts and Crafts Movement. Jekyll used local cottage gardens around Surrey as the inspiration for her planting schemes, teamed with elements from her Mediterranean travels and colour theories developed during her fine art training.

Jekyll and Lutyens designed and planted enormous borders in a luxuriant and romantic style, which brought timeless cottage-garden qualities to the country estates of some of the wealthiest Edwardian families. Their approach set the agenda for the English garden over the next century.

Munstead Wood designed by Gertrude Jekyll.

DESIGN INSPIRATION

Interpreting the style

Almost anything goes with informal gardens – some have a simple geometric layout, others a more flowing, organic ground plan; this lack of a prescriptive approach is the point. Many have a traditional, rather rustic feel, the hand of the gardener well disguised; they are gardens of content rather than style, where plants come first. Often they are tended by enthusiasts. Different garden areas are usually about creating sites with microclimates where particular plants flourish.

Autumn glow top
A rustic fire pit means this quiet corner can be enjoyed even as autumn leaves begin to fall.

Profusion on the menu above
Planting flourishes in the microclimate by an old wall, with profuse hydrangeas and grasses as well as planted pots blurring paving edges. A weathered dining table is a temporary home to a collection of cacti, and a wall-mounted mirror adds depth to the space.

Naturally relaxed left
Relaxed planting is a key part of the informal style, with evocations of natural habitats common. Sometimes the informal feel can be enjoyed over a larger area: here, a sown prairie mix of Rudbeckia and Echinacea brings a naturalistic feel.

INFORMAL GARDENS

Lose the lawn left
Lawns are not an essential component of the informal style – gravel set within brick circles provides a low-maintenance alternative, softened by typically exuberant planting. A metal bowl makes a charming pool and focal point.

Waterside planting below
Ponds provide keen gardeners with an opportunity to grow moisture-loving plants, and so naturally styled creations are often part of an informally styled garden.

Urban informality above
Even in a tiny city garden, a touch of the country can be effective: lavish planting of rustic favourites such as hydrangea and soft grasses spills from the raised beds and a submerged seating area.

Exotic adaptation left
In tropical designs, informality can be the best way forward, with large, luxurious, and fast-growing plants that spill over each other.

Planting is key

In modern schemes, there has been a move from traditional mixed borders to a more limited planting palette, such as the architectural hedges and monocultures typical of Jacques Wirtz's designs, or drifts of colour evident in the work of Piet Oudolf or James Hitchmough. These designers rely on the movement and light-capturing qualities of grasses and seedheads, which provide a long season of interest.

Painting with flowers top
Christopher Lloyd experimented with vivid colour in his garden at Great Dixter in Sussex. He combined clashing pinks and reds, at the time flouting conventional colour theory.

Arbour in a storm of colour above
Classic features of informal gardens, arbours and pergolas allow climbing plants to move in from boundary fences and walls and become centre-stage performers.

Autumn glory right
The mahogany seedheads of Phlomis stand out against the green, silver, and bronze mounds of grasses and perennials in these stunning deep borders.

INFORMAL GARDENS

Exuberant border left
Splashes of colour illuminate this haze of planting and emerge skywards, adding vertical interest. Transparent veils of grasses and perennials create the romance.

Reflections on a theme below
At Gresgarth Hall, climbers cascade from walls, and poolside terrace plantings provide a feel of wild profusion, while immaculate lawns and hedges offer a formal counterpoint.

Catching the light right
These graceful borders, planted with a mix of golden feathery grasses and eye-catching red Sedum, encircle this sunny seating area with movement and light.

Virtuoso planting below right
In this garden, Piet Oudolf mixed broad masses of colour with drifts of grasses to create a soft meadow effect. The wave-clipped yew hedges provide a contrast in architectural form.

"The luxury of space and abundant planting create the magic"

GARDENS TO VISIT

BORDE HILL, West Sussex
Combines many different garden and planting styles, including water gardens.
bordehill.co.uk

GREAT DIXTER, East Sussex
Inspiring garden that uses colour creatively.
greatdixter.co.uk

HESTERCOMBE, Somerset
A garden by Edwin Lutyens and Gertrude Jekyll, plus an 18th-century landscape garden.
hestercombe.com

KIFTSGATE COURT, Gloucestershire
An outstanding 20th-century garden.
kiftsgate.co.uk

ROUSHAM PARK HOUSE, Oxfordshire
William Kent's early 18th-century masterpiece. rousham.org

SCAMPSTON HALL, Yorkshire
Includes Piet Oudolf's dazzling walled garden.
scampston.co.uk

Informal garden plans

Abundant planting and a mass of forms, textures, and colours define an informal garden, with hard landscaping taking a back seat. The lively soft planting in Gabriella Pape and Isabelle Van Groeningen's design comes in many colours, while Jinny Blom celebrates pinks and reds in a limited, warm palette. Despite their abundant appearances, the modern examples designed by Nigel Dunnett and James Barton are held together with well-defined lines and shapes.

Sea of plants and flowers

This garden was designed by Gabriella Pape and Isabelle Van Groeningen for the RHS Chelsea Flower Show as an homage to Karl Foerster, a great nurseryman who experimented with perennial plants. It creates the sensation of swimming through the foliage and flowers.

Key ingredients
1 *Digitalis purpurea* 'Alba'
2 *Hakonechloa macra* 'Aureola'
3 *Hosta* 'Sum and Substance'
4 *Veronica* 'Shirley Blue'
5 *Paeonia lactiflora* 'Duchesse de Nemours'
6 *Aquilegia chrysantha*
7 *Hosta* 'Royal Standard'
8 *Achillea* 'Moonshine'

Isabelle says:
"This layout was based on Karl Foerster's own garden in Potsdam, Germany, so it's not typical of our work. The planting, however, is. Influenced by the English style, it incorporates colourful matrix planting, and drifts of plants and flowers are reminiscent of Edwardian woodland gardens. These themes recur a lot in our work.

"Our influences are varied and we often bounce ideas off each other to develop design solutions. English garden designers, such as Vita Sackville-West, Geoffrey Jellicoe, and Charles Wade, are a major influence. We also create gardens and their planting around existing elements."

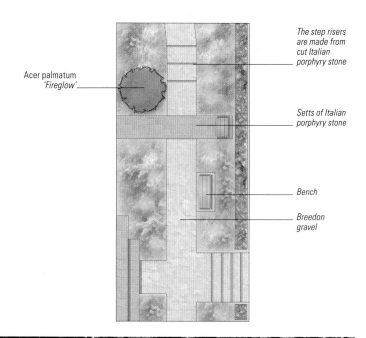

The step risers are made from cut Italian porphyry stone

Acer palmatum 'Fireglow'

Setts of Italian porphyry stone

Bench

Breedon gravel

INFORMAL GARDENS

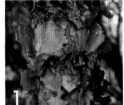     

Restrained palette

Modernist treatments, such as simple, clean paving, provide a cool contrast to the hot-hued palette of plants that tumble and explode around this garden by Jinny Blom. In true cottage style, the seemingly haphazard, densely packed planting pockets soften and relax the more ordered layout. The use of gravel allows plants to self-seed, creating additional random patterns of spontaneous growth. Grasses, seedheads, and bulbs create veils of foliage and texture.

Key ingredients
1 Betula nigra
2 Akebia quinata
3 Geranium PATRICIA ('Brempat')
4 Allium sphaerocephalon
5 Verbena bonariensis
6 Panicum virgatum 'Heavy Metal'

Jinny says:
"This view is just one part of a multi-levelled garden – the different parts of which are connected by walkways and steps, so that, overall, the design flows nicely. The clients were a young family, and, as a result, the design needed to be robust, allowing the children to play freely.

"We agreed a strategy of hardwearing, virtually indestructible materials that would be softened with romantic planting. This seems to have paid off, as the garden has matured well. We have recently added yew hedging in order to create a visual anchor in winter.

"I was inspired by the work of Italian architect Carlo Scarpa. In terms of flow and visual stimuli, his work was very important in creating the design."

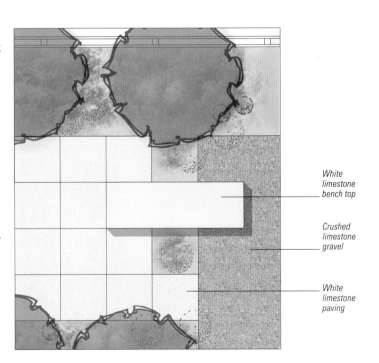

White limestone bench top

Crushed limestone gravel

White limestone paving

Practising what you preach

Nigel Dunnett is a professor at the University of Sheffield and a landscape designer. He is renowned for his research into sustainable planting and urban drainage systems, and this small informal garden, which sits on a north-facing slope, puts many of his findings into practice.

Key ingredients
1 *Euphorbia palustris*
2 *Geranium sylvaticum*
3 *Lonicera periclymenum* 'Serotina'
4 Green roof
5 *Euonymus alatus* 'Compactus'
6 *Astilbe chinensis* var. *taquetii* 'Purpurlanze'
7 *Caltha palustris*
8 *Acorus calamus*

Nigel says:
"I wanted to create a woodland glade, with closely planted birch forming a light canopy and linking with the surrounding countryside. Clipped hornbeam hedges provide enclosure and structure alongside softer successional planting.

"Perennials form a dense ground cover, almost eliminating the need for weeding. The planting is 50 per cent natives and 50 per cent cultivated garden plants – together they give almost year-round colour. The pond is filled with run-off from the paved surfaces and helps to manage the drainage in the garden, which has been a huge success."

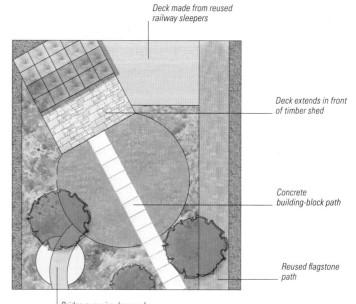

Deck made from reused railway sleepers
Deck extends in front of timber shed
Concrete building-block path
Reused flagstone path
Bridge over circular pond

INFORMAL GARDENS

  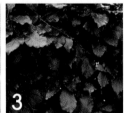   

Sustainable informality

Dr James Barton and his wife developed the informal design of their sustainable garden in Westphalia, Germany, over a number of years. The garden is modest in size, yet includes a rich range of planting – ornamental and native species, selected for interest and their ability to thrive as good neighbours, are intermingled. A system of pathways provides easy access to them.

Key ingredients
1 *Nymphaea alba*
2 *Iris sibirica*
3 *Fagus sylvatica*
4 *Angelica archangelica*
5 *Carpinus betulus*
6 *Lychnis flos-cuculi*

James says:
"In its early days, this was a family garden, but since our children left home it has evolved into something else.

"We develop areas as we gain new ideas, but the basic layout of the garden, as a series of "rooms", remains the same.

We have structured the spaces with beech and box hedges, or with fences, and we have also created a range of small, informal seating areas to provide different views through the garden. In the main, we use perennials and shrubs, with some annuals added as necessary to provide splashes of colour.

"For inspiration, we have visited many open gardens, primarily in the Netherlands and southern England. However, we were originally inspired by a visit to a small private garden in Germany, the owner of which was the president of a local society, the Gesellschaft der Staudenfreunde, of perennial plant enthusiasts."

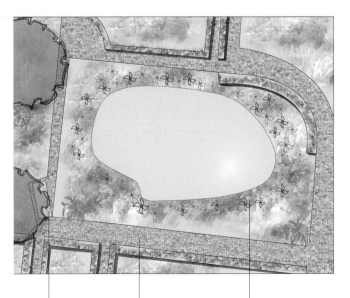

Timber bench, made from recycled wood

Paths made from granite, reclaimed when local streets were repaved

Dense planting around pool offers habitats for wildlife

DESIGN INSPIRATION

CASE STUDY
MODERN INFORMALITY

This garden proves the concept that a plant-filled, country garden can be re-imagined in a contemporary way. While planting is generous, the overall palette is restrained, floral colours limited, and green (from the foliage) the dominant hue.

Planting first
Lushly planted evergreens, including climbing *Trachelospermum jasminoides* – with its seductively perfumed white summer flowers – ensure the living element of this garden is the first thing you notice and admire.

Planting the joins
Paving cracks have been planted with *Soleirolia soleirolii*, which helps separate different zones and materials; together with generous planting, this softens the overall hard landscaping and gives an impression of establishment and age.

Designer **MATT KEIGHTLEY**

INFORMAL GARDENS 59

Layered planting

A multi-stemmed *Acer* is underplanted with a dome of yew (*Taxus baccata*) and shade-loving white *Digitalis*. Removing lower tree branches like this allows a wider range of plants to flourish in the same space.

Added detail

Sometimes a simple placement of a pot plant can bring a lovely touch to a space. This *Origanum* has a shaped pot that ties in with the colour of the outdoor furniture, and the flowers stand out against the green of the garden.

Impromptu seating

In a small space, outdoor furniture can create a cluttered look; if you are entertaining, lightweight, movable chairs can be a solution, creating a beautiful garden that doubles as a relaxing place for a catch-up with friends.

Family gardens

As leisure time increased in the middle of the 20th century, the concept of a garden shifted from being a formal area to be admired from afar – or simply a place for growing fruit and vegetables – to a space that provides a focus for family life. Specific areas devoted to relaxation, children's play, and dining have become increasingly popular, and today, these spaces form the template for many family designs.

Family gardens are often a blend of styles. Their layouts can be rectangular or curved, with flexible designs for children's areas that will accommodate their changing needs as they grow. Play equipment sometimes helps to introduce colour into the design, while planting areas that attract a range of wildlife can also provide entertainment for young ones.

The safety of babies and young children is a top priority in these gardens. If water is to be included, jets and cascades where the main water reservoir is underground are better than open water features. However, naturalistic ponds are perfect for older children, who will enjoy the aquatic creatures and wildlife these features attract.

Decking (wooden and recycled composite) and natural stone are popular materials for dining and seating areas, with bark chippings or other soft yet resilient materials providing practical surfaces for play spaces. In larger gardens, the transition between the children's and adults' areas can easily be managed with separate, designated spaces, but in smaller plots, the design may need to be more adaptable, perhaps using play equipment that can be cleared away as night falls. Lighting can also help to create a different ambience for adults to enjoy after dark.

Planting in a family garden needs to be robust and easy to maintain; it should also be free from toxic plants and sharp thorns. Open space is important for children's games; hardwearing turf is the best choice for lawns.

Top and above
Natural surroundings can be adapted to create play areas.
A swimming pool provides hours of fun for older children.

What is a family garden?

A family garden can be almost any style that has been adapted to provide a flexible space for games, room for entertainment and play, and an area for dining. The smallest of gardens can accommodate a sandpit or swing, while larger plots have space for separate adult- and child-friendly zones.

Family gardens in detail

The concept of the outdoor room celebrates family life. Terraces need to be large enough to accommodate a dining table and chairs, with space for a barbecue or even an outdoor kitchen.

For play, there are two schools of thought: structured play relies upon equipment, but children have different needs as they grow, so flexibility is important. For example, a small sandpit located close to the house allows parents to watch their young children more easily; then, as they grow and move down the garden to seek more adventure, swings, slides, and climbing frames can be introduced.

Unstructured play provides a rich and interesting environment in which children can be encouraged to take some risks – building dens, pond-dipping, climbing trees, and watching wildlife. This requires a more subtle approach to design and one in which parents cannot be too precious about their gardening exploits, giving preference to the needs of their inquisitive children.

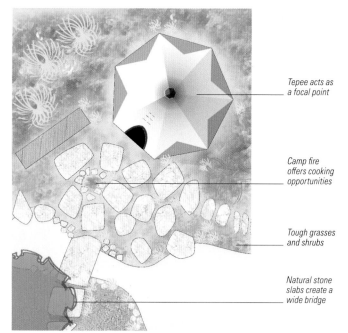

Tepee acts as a focal point

Camp fire offers cooking opportunities

Tough grasses and shrubs

Natural stone slabs create a wide bridge

Natural playground
Designed by Chuck Stopherd of Hidden Garden, this garden (right) for older children offers valuable opportunities for outdoor play. The tepee, fire pit, and pool, hidden behind trees, provide a natural setting for children to take risks and explore their environment.

A 1950s family garden designed for play.

DESIGN INFLUENCES

The opening up of the garden as a family facility is a relatively recent occurrence, although outdoor dining *en famille* has always been a tradition in Mediterranean countries. Thomas Church's book, *Gardens Are For People*, first published in 1955, changed perceptions of the garden and signalled a move away from intensive gardening and towards the development of the outdoor room. Later, John Brookes developed these ideas in his designs and 1969 book *Room Outside: A New Approach To Garden Design*. Today, gardens are places of enjoyment, education, and fun for families to share.

Key design elements

1 Play equipment
The children's area can feature large items of play equipment, such as a swing or climbing frame. If space is limited, some items may still be included by adapting a pergola or similar structure.

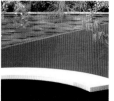

2 Colourful materials
Splashes of bright, primary colours are an essential ingredient in a family garden. These can be introduced via planting, equipment, or hard landscaping.

3 Dens and tents
Part of the children's area could include a den: a place of their own where they can extend their imagination through play. It may be sited within view of the house or tucked away in a corner.

4 Tough plants
Plants have to be versatile and tough to withstand rough treatment from children and pets. Closely planted, often with some evergreens and seasonal colour, they must also be easy to look after.

5 Wildlife features
Ponds with sloping sides to allow creatures access, boxes for birds, habitats to give shelter to hedgehogs, and plants to attract bees and butterflies, are all ideal for family gardens.

6 Easy-care seating
Seating needs to be suitable for children and adults. Furniture that can be left uncovered all year and requires the minimum of care and maintenance is the most practical.

Interpreting the style

A family garden is about sharing your space. The dining area is the social hub around which the design revolves, and can be created with a paved or decked terrace that links into a lawn or into more structured play areas with integrated or temporary play equipment. Swimming pools or natural ponds make reflective centrepieces for gardens where older children play.

Versatile space above
A large-scale chessboard is both a design feature and a challenging family game, making the most of a quiet retreat surrounded by textured foliage planting.

Star attraction below right
Central to the design of this contemporary garden, the turquoise pool is both functional and decorative. Safety covers or security fences may be introduced if necessary.

Secret hideaway far right
In a secret corner of this densely planted garden, a den of willow and brushwood becomes the focus of adventure and discovery, providing an escape from the adult world.

FAMILY GARDENS   65

Adventure playground far left
A play house that can only be accessed via a footbridge – fun for kids, but perhaps too precarious for adults – allows children to escape, and control who visits.

Wildlife haven left
This large reflective pond and the reed margins provide a range of wildlife habitats that can be observed from the various vantage points located around the banks.

Safe play area below
This built-in sandpit is close enough to the house to be monitored, but planting creates the illusion of another world. A cover will provide protection from the weather.

"Helping to bring families closer together is perhaps the garden's most important role"

Colourful entertaining above
This vibrant area is part of a modern design, and combines cooking, dining, and relaxation, offering a fun area where the whole family can decamp to escape the confines of the house.

Tree-top retreat right
A tree house takes pride of place here, acting as both a retreat for children and a decorative focal point. It also offers a hideaway for adults when the children are in bed.

FAMILY GARDENS TO VISIT

ALNWICK GARDEN, Northumberland
Created with children in mind, with water features and a gigantic tree house.
alnwickgarden.com

CAMLEY STREET NATURAL PARK,
King's Cross, London
Ponds and meadows, and hands-on activities.
wildlondon.org.uk

CAMDEN CHILDREN'S GARDEN,
Camden, New Jersey, US
Four-acre interactive garden for families.
camdenchildrensgarden.org

MILLENNIUM PARK, Chicago, US
Offers a programme of interactive family events and workshops.
millenniumpark.org

Flexible space

Perhaps the most hard-working of all spaces, a garden designed for families must offer flexible, stimulating space for all. During the day, it may need to be an outdoor playroom, sports field, nature reserve, or simply a restful retreat. In evenings and at weekends, the area may become an outdoor lounge, kitchen, or dining area for adults to relax with family or friends, even after dark. It must be a safe yet bright and beautiful place that appeals year-round, tailored to the needs of the family as it grows.

Luscious lawn right
A lawn provides open space for fun, games, and even makeshift picnic space, as well as somewhere cool to lie in the sun.

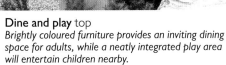

Dine and play top
Brightly coloured furniture provides an inviting dining space for adults, while a neatly integrated play area will entertain children nearby.

Courtyard calm above
Stone and gravel create flexible, functional surfaces, while a covered seating area ensures rain showers or hot summer sun need not disturb outdoor relaxation in this urban oasis.

Multi-use space above
Deceptively simple, this multi-faceted garden provides seating and dining options indoors and out, with open space for games and a wildlife pool all set within verdant yet varied planting.

FAMILY GARDENS 67

"Gardens designed for families must offer flexible, stimulating space for all"

Safe habitat left
Children are often fascinated by nature, so including elements that attract wildlife is an important part of family garden design. This simple timber structure provides habitat for overwintering insects and creates a nature reserve exploration space for children.

Discreet space top
A trampoline can be successfully and discreetly integrated into a lawn – it need not dominate even in small gardens. In later years, the space might become a circular pool.

Splashes of colour above
Brilliant colour dominates this sun-filled, compact space, the terracotta painted wall clashing with the bougainvillea overhead, which offers some shade for outdoor dining.

Family garden plans

These varied examples offer different takes on a family garden: two focus on relaxation, providing comfortable, restful, even contemplative spaces to get away from the daily hustle and bustle, while one provides space for vegetable growing alongside dedicated lawn and meadow areas. The last garden features a smart deck for entertaining and a children's play area.

Plant-filled secluded space

In this sunken garden by Catherine MacDonald, a seating area immersed in soft, tactile planting creates a perfect retreat for a spot of peace. Warm flowers and foliage tints stand out from the grey-painted woodwork.

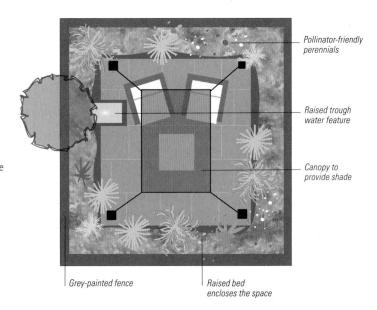

- Pollinator-friendly perennials
- Raised trough water feature
- Canopy to provide shade
- Grey-painted fence
- Raised bed encloses the space

Key ingredients
1 *Stipa arundinacea* and *Selinum wallichianum*
2 Chairs with cushions that tone in with planting
3 Blue *Eryngium*
4 Tall peach *Eremurus* and terracotta *Achillea*

Tranquil refuge
Lavish colour-themed planting surrounds this charming seating area, perhaps used for quiet contemplation or a few moments of tranquillity. Through summer, soft-to-touch, pollinator-friendly perennials that are also easy on the eye, such as Achillea and Selinum, flower in a relaxed naturalistic blend that almost gives the feel of being immersed in a beautiful meadow. Painting woodwork a uniform grey provides a cohesive feel, while the raised beds around the paved area means any gardening can be done easily at a convenient height. Cushion accents help to bring out some of the darker ochres and oranges.

Living art

The fish-filled pond is a meditative focal point in Maggie Judycki's contemplative home garden. Rocks, ornaments, and planting are carefully arranged around it and a split bamboo fence filters light in horizontal patterns across its surface. The leaves of a *Sassafras* and a *Betula* merge and rustle above.

Key ingredients
1. *Acer rubrum*
2. *Sassafras albidum*
3. Bamboo fence
4. *Hosta* 'Francee'
5. *Cotoneaster salicifolius* 'Gnom'
6. Japanese bathing stool
7. *Sarcococca hookeriana* var. *humilis*
8. *Betula utilis* var. *jacquemontii*

Maggie says:
"This is my own garden, and it's been a work in progress for many years. I started out as a stone sculptor, which has helped me to use and understand hard materials. I tend to start with them and soften the surfaces with planting.

"Sitting places are important to me too. A favourite is the Japanese bathing stool, ideal for contemplation when I'm feeding the koi carp. Living art and the movement it creates is also fascinating — we can see the pool from the house, and it's a constantly changing view. The garden is typical of my work in that I customize the space for each client."

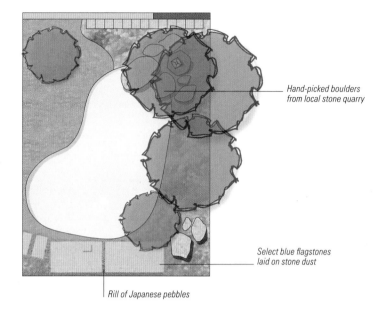

Hand-picked boulders from local stone quarry

Select blue flagstones laid on stone dust

Rill of Japanese pebbles

A gardener's retreat

This family garden by Jane Brockbank offers three distinct areas. A border-edged lawn is closest to the patio and house. In the centre lies a productive space with greenhouse and raised wooden beds, where fruit and vegetables can be easily cultivated with minimal bending. Beyond is a relaxed meadow area with grasses and pollinator-friendly plants.

Key ingredients
1 Petersen brick and gravel path
2 Greenhouse
3 *Cirsium rivulare* 'Atropurpureum'
4 Raised growing beds
5 Meadow
6 Lawn

Stylish family garden
Garden designer Jane Brockbank wanted the productive section of this family garden to be celebrated, so she opted to position the raised beds for fruit and vegetable growing in the centre of the garden, separating the naturally styled area at the bottom of the garden from the part closer to the house, with its more formal patio and lawn. The relatively simple linking path uses Petersen bricks, which are used in the structure of the house, providing visual cohesion. The wilder area, with its long grass and naturalistic planting, was inspired by the flora of the East Anglian fens.

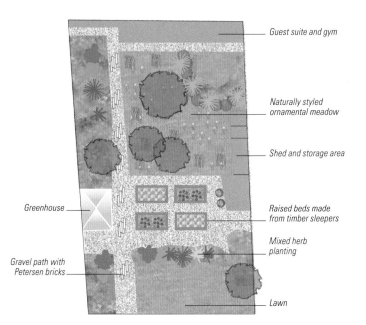

- Guest suite and gym
- Naturally styled ornamental meadow
- Shed and storage area
- Greenhouse
- Raised beds made from timber sleepers
- Mixed herb planting
- Gravel path with Petersen bricks
- Lawn

FAMILY GARDENS

Corner piece

The sophisticated look of this family garden by Claire Mee was achieved with an elegant decked terrace for dining, while the pergola at the end of the plot gives the children a play area, complete with swing. The spaces are divided by a grove of olive trees, which offer privacy and add height. The tree canopies have been lifted to leave clear stems that create dramatic shadows; light also reflects on the silvery foliage.

Key ingredients
1 *Olea europaea*
2 Hardwood decking
3 *Allium hollandicum* 'Purple Sensation'
4 *Sisyrinchium striatum*
5 Bark chippings
6 *Origanum vulgare* 'Aureum'

Claire says:
"This urban garden occupies a corner plot, so it's an unusual shape. My ideas for the design were developed from the house's architecture, and from the interior design and decor. I'm often influenced by the interiors of hotels, restaurants, and bars, which use different materials so well.

"Wide windows look down the length of the garden, and we used clear-stemmed olives to provide privacy without blocking this view. Elsewhere, I like the contrast between the softer planting and the architectural specimens. The client also wanted a terrace outside the French doors to match the floor-level in the house, and I designed a large timber deck to make this link (legally, a paved surface would have to be lower to avoid the damp course)."

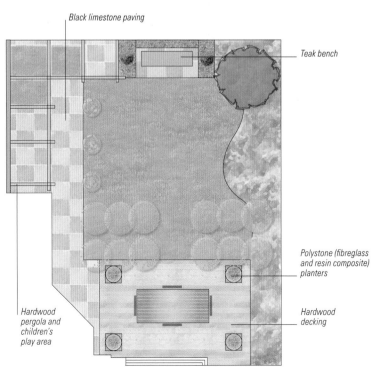

Black limestone paving
Teak bench
Polystone (fibreglass and resin composite) planters
Hardwood decking
Hardwood pergola and children's play area

CASE STUDY

FAMILY VALUES

Family gardens should be places of fun, where children have freedom to explore and play safely. Successful designs cater for both young and older users, providing features to entertain little ones, and areas for adults to relax and enjoy the scenery.

Shady canopies

The white birch stems echo the white blooms, while contrasting with the understorey of green foliage. The trees punctuate the design with their bright vertical trunks, and their canopies also offer essential shade, helping to protect youngsters from sunshine.

Soft to touch

Easy-care plants that are soft to touch are ideal for family spaces. Here, shade-loving perennials, shrubs, and evergreen ferns create a leafy blanket, while star jasmine clads the walls, its tiny blooms scenting the air.

Designers **Nick Buss and Clare Olof**

FAMILY GARDENS

Colourful journey
The curved path is colourful and confident, creating a visually exciting journey and a focal point through the duo-tone planting. The small brick setts also lend detail and texture, and complement the tiled box stool.

Bubbling tubes
A great way to introduce water safely into a family garden is with these eye-catching "bubble tubes" filled with clear and dyed water. The sound and movement will fascinate children, while also producing a soothing, calming effect.

Hide and seek
The hollowed tree trunk and woven willow playhouse (*far left*) bring an element of fairy tale to the design, to fire the imagination and provide places to play and hide. Such naturalist structures blend tonally with the planting and wider design.

Natural gardens

Gardens with a natural style are nothing new, with influential designers from the 18th to the 21st century striving to emulate the natural world in a variety of ways. Today, this style focuses primarily on sustainability, with designers incorporating plants and materials that do not diminish the world's dwindling resources. A natural garden will typically include recycled and renewable materials and a diverse mix of plants that offer food and habitats to wildlife.

Introduced in the late 20th century, the New Perennial movement – as espoused by plantsmen such as Piet Oudolf – increased interest in naturalistic gardening styles and has influenced many contemporary designers today. This style emulates the appearance of wild plant communities and combines hardy perennials with grasses, matching plants with their sites so that they flourish with need for little maintenance. More recently, British, Dutch, and German research into sustainable plant communities has also set new design trends.

There is a popular idea that natural gardens must be rustic in character, but this need not be the case. Many modern, elegant designs include local or renewable materials, such as timber from certified plantations, and sophisticated recycled materials.

Most owners of natural gardens adopt an organic approach to controlling pests and diseases, keeping them at bay through use of biological controls and balanced ecosystems, rather than chemical pesticides. Habitats that support local species and help to increase biodiversity are key to these designs, but natural gardens do not rely exclusively on native species. Non-invasive exotic plants that attract beneficial insects and wildlife are also highly useful, offering extra colour and year-round interest.

Extensive prairie and meadow planting is often used in large natural gardens, recreating the appearance of natural habitats yet including at least some cultivated plants. Wild flowers and bee-friendly species can easily be included in smaller spaces, too, providing a range of different habitats even in tiny gardens. Features such as ponds and green roofs are also key elements.

Top and above
Dots of repeated colour in long grass give a natural but tonally consistent feel.
A sympathetically designed swimming pond will attract wildlife.

What is natural style?

When designing a natural-style garden, it's important to consider it resembling a balanced ecosystem, with reduced or minimal levels of intervention – it is this approach that sets the style apart from more traditional gardens. Ecological principles play an important role in creating natural garden habitats in which, although managed, naturalistic plant communities thrive and competition is well balanced. Plants should be carefully matched to the prevailing soil and climatic conditions.

Natural gardens in detail

The materials used in a natural garden should be assessed against a series of criteria. Recycled products are a good idea as they reduce the exploitation of new resources, but sometimes they have a higher carbon footprint, whereas sourcing new timber from managed, renewable, and, preferably, local plantations may be a better option.

Other factors to consider include the permeability or drainage of hard-landscaped surfaces. These should ideally be either porous, in order to top up groundwater, or designed to allow water to run off into a collection unit or water butt, thereby reducing the strain on supplies.

In a natural garden, planting is key, and a healthy variety of wildlife habitats essential. Choose plants that thrive in the prevailing conditions and complement each other, which in turn will help to reduce the incidence of pests and diseases. Soil improvers should come from your own compost heap to boost sustainability.

Rosemary Weisse's garden at Westpark in Munich.

DESIGN INFLUENCES

The change from purely ornamental planting to the creation of successful plant communities started when William Robinson (1838–1935) advocated the integration of native and exotic species, which he called "wilderness planting". The development of American prairie planting, championed by Jens Jensen in the 1920s and 30s, responded to Robinson's ideas, and was later taken up in Europe by the New Perennial movement. Large drifts of grasses and perennials, like those seen in the schemes of Rosemary Weisse in Munich, are typical of this approach (see also pp.282–83). In the UK, the Department of Landscape at the University of Sheffield has produced significant research into sustainable prairie and meadow planting.

Key design elements

1 Green roofs
Green roof systems manage rainwater run-off and provide insulation. Convert existing roofs using pre-planted sedum mats. New structures can accommodate more elaborate habitats.

2 Encouraging wildlife
Increased diversity in natural gardens is key and achieved by creating effective habitats for wildlife. The more habitats there are, such as old logs, bee hotels, and insect-friendly planting, the greater the diversity.

NATURAL GARDENS

Wildlife haven

Designed as a natural-style, sustainable garden by Stephen Hall (*left*), this beautiful design shows how precious resources, such as water and wildlife, can be supported and protected. The garden includes a range of diverse habitats, including a pile of decaying logs and tree stumps to provide homes for rare beetles, small mammals, and overwintering amphibians, such as frogs and toads. The traditional-style building is built entirely from sustainably sourced cedar, and features a green roof planted with sedum species. Research shows that green roofs help to insulate buildings and keep them cool when temperatures rise, reducing the need for heating and air-conditioning. They also attract beneficial insects when the plants are in flower.

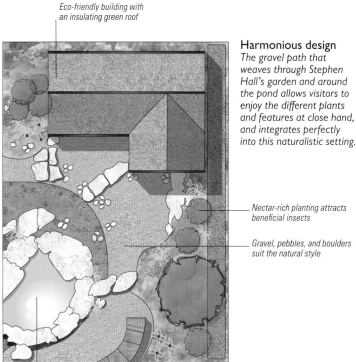

Eco-friendly building with an insulating green roof

Harmonious design
The gravel path that weaves through Stephen Hall's garden and around the pond allows visitors to enjoy the different plants and features at close hand, and integrates perfectly into this naturalistic setting.

Nectar-rich planting attracts beneficial insects

Gravel, pebbles, and boulders suit the natural style

Wildlife pool attracts insects, birds, and small mammals

3 Rainwater harvesting
However small, water butts are an excellent way to catch and store rainwater. If you need something with a larger capacity, underground storage and pump mechanisms are available.

4 Rustic garden furniture
Wherever possible, support your local economy by commissioning a craftsman close to home to make your furniture. All products should be made from responsibly sourced, natural materials.

5 Recycling features
The recycling of organic waste through composting is vital. Several compost bins may be required in order to maintain and rotate supply. Think carefully about their location, as they need regular access.

6 Naturalistic ponds
Wildlife ponds with sloping sides that allow easy access, and margins planted to provide cover, offer a natural habitat for aquatic creatures, as well as birds and insects, such as dragonflies.

Features of natural style

Natural gardens have a distinctive style, where the hand of the gardener is as disguised as possible; "natural" design can be seen just in the planting style or be apparent throughout the garden. Close to the house, there can be elements of formality or practicality, perhaps a terrace or seating area, and often with a contemporary edge. In the outlying garden, informal spaces dominate, including natural features such as ponds or areas of meadow-style planting, and occasional artificial features such as sculpture. These areas are linked with paths, and visitors are led by focal points, viewing areas, and resting places. Garden boundaries often blur with the natural landscape beyond.

Boundaries and perspective

The boundaries of these gardens have to be carefully considered – fence panels and brick walls usually look out of keeping with the natural theme. In large country gardens, windbreak planting is frequently the first element to be introduced, but this can obscure views. Compromises have to be made, often producing limited or narrow vistas into the wider landscape, yet this restriction forces designers to evaluate views and different perspectives carefully, which can increase the drama. Hedges are a great solution, especially for smaller urban natural gardens; they also provide soft structure, dividing up internal space or serving as screens. Hurdles made of woven hazel or dry-stone walls are other small-scale rustic alternatives. Simple wooden post and rail, or smarter metal estate fencing, is in keeping for larger gardens.

Hard paving materials are generally used near the house, with gravel routes through the garden. Planting schemes have to be appropriate for the large scale with a relaxed feel and an impression of natural plant communities. Mown lawns may still be welcomed, but meadows with mown paths or prairie planting provide more natural texture and seasonal colour.

William Robinson's natural style.

DESIGN INFLUENCES

In the 1870s, the English designer William Robinson revolutionized attitudes to gardening with softer, more naturalistic planting schemes that combined exotics and native species. Through his writing and the gardens he developed at Gravetye Manor, he influenced prominent designers such as Gertrude Jekyll, Vita Sackville-West, and Beatrix Farrand.

Later gardens by Thomas Church and Dan Kiley relied more on the manipulation of space and links to the wider landscape. They tended to use existing or native planting, creating harmonious designs with a much simpler palette.

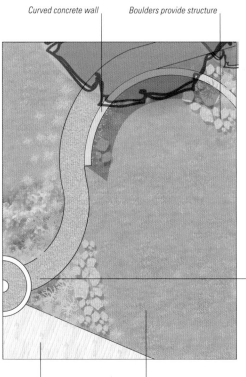

Curved concrete wall
Boulders provide structure
Decked walkway adds texture
Lawn links planting and paths
Bound gravel path meanders through the garden

Natural contrast
Here, Andy Sturgeon uses concrete, decking, and bound gravel paths to create a fluid transition between level changes in this contemporary yet naturally styled garden in south-east England (right). Large boulders stand out against the soft sweeps of planting beyond, and grasses provide movement and light in the deep planting beds. The wide boundary hedges screen views of neighbouring properties.

Key design elements

1 Naturalistic planting
Extensive borders provide the opportunity for meadow-style plantings that merge with the landscape. Areas below trees can be planted with woodland perennials, all visually appealing and great for wildlife.

2 Large pools and streams
Natural springs may provide the basis for ponds and streams, but they can be introduced artificially to create reflective surfaces and wildlife habitats, or for new planting opportunities.

3 Views into the landscape
The garden experience can be dramatically enriched by linking it to the landscape. Long, narrow views, which open up to a wide natural panorama beyond, produce spectacular effects.

4 Lawns and grass
Mown lawns provide a functional surface and a perfect contrast to long meadow grass with naturalized bulbs or wildflowers. Keep grass pathways as wide and open as possible to minimize wear.

5 Hedging and screens
Hedges are a perfect natural way to define space and control views. Yew produces a dark, dense backdrop that is perfect for colourful borders; mixed hedges work well on a larger scale.

6 Natural materials
Local stone that weathers to produce varied surface textures, such as Yorkstone, is often seen in traditional country gardens. A more contemporary feel is possible with limited use of concrete and decking.

Interpreting the style

Simple, quality materials used for hard landscaping in most natural gardens play second fiddle to planting, which generally dominates garden layout. Many natural designs have a rather contemporary feel, built features used sparingly with clean, unfussy lines that will help complement the impression of natural profusion, or add a sculptural quality that provides a focal point. Natural style follows an informal pattern in evocations of wild plant communities, often using a mix of cultivated plants, wild flowers, and grasses for a relaxed, carefree feel.

Summer peak top
Achillea, Rudbeckia, Verbena, and grasses peak in high summer, while natural wood and an unusual gravel and stone path sets the contemporary tone.

Green makes the scene above
Hard landscaping elements sink beautifully into lush, verdant planting, yet inject order, practicality, and an interesting diagonal element.

Bird haven above
Feeders and bird tables add dynamic interest, attracting wildlife, especially during harsh winters, when food sources may be scarce.

NATURAL GARDENS

"Planting dominates in natural gardens, while landscaping helps define space"

Stepping up above
Simple yet gorgeous, rustic steps are framed by meadow, edged with box balls that lead the eye to a seat and round "window" in the copper beech hedge beyond.

Hatching an idea left
Arresting, yet natural in form and material, an inventive egg seat hangs above naturally influenced planting.

Provençal landscape top
The wide joints in the pale limestone paths create patterns and allow thymes to colonize. Lavender blues are virtually the only flower colours.

Curated simplicity

Natural gardens appear deceptively simple – some as if they are barely gardened at all. But that is entirely in keeping with the style. In fact, these gardens need as much planning and ongoing care as many others if they are to keep their wild yet beautiful appearance. Human-made plant communities, even if based on natural combinations, constantly change and adapt without input; trees and shrubs, for example, grow and cast shade where there was once sun. Keeping a balanced order of plants and conditions requires consideration.

Wild islands top
Adding a touch of natural style is easy: simply letting lawn grass grow long soon creates a new atmosphere, reinforced by establishing a more tidy boundary.

Contemporary meadow above
A striking modern home clad in wood stands above prairie planting of Rudbeckia and grasses. A terrace is next to the house, but otherwise the effect appears entirely natural.

Poppy appeal left
Cornfield annuals including poppies, cornflowers, and daisies have a surprisingly long season of interest and create a relaxed, rural feel.

NATURAL GARDENS

Rustic pavilion left
Sporting a green roof and made of natural timber posts, this waterside seating area will soon develop a natural patina and fit beautifully with the wild-inspired planting.

Mini nature reserve below left
Sowing an area with wild flowers such as ox-eye daisies (Leucanthemum vulgare) and ragged robin (Lychnis flos-cuculi) creates an impressive effect that rivals traditional borders. An insect hotel reinforces the nature-friendly feel.

Rosy future below
Highly cultivated plants such as roses can be enhanced when used with drifts of wildflowers; if anything, they take on an even more romantic feel.

Grassy effects bottom
A basket-weave path meanders through a border of fine textures, which include the repeated arching rosettes of Hakonechloa macra 'Aureola' – a grass that takes on warm orange tones in autumn.

Natural garden plans

While natural style is prescriptive, it need not be restrictive. Jo Thompson's garden with its woodland feel proves that even a small, relatively shaded space can be pressed into action. In Jane Brockbank's Cambridge garden, a more open, expansive area is set aside for meadow-like planting that is distinctly on the wild side.

A woodland corner

In Jo Thompson's garden, an offset circular bench seat creates a welcoming focal point for this small space, providing both a practical place to sit while also being a retaining wall to the planting. Birch trees provide dappled shade for lush ferns and hostas below, while a limited use of richly hued flowers provides sparkling highlights amid the foliage.

Key ingredients
1 *Betula pendula* (silver birch) trunk
2 Shade-loving perennials, including geraniums, *Alchemilla mollis*, and *Thalictrum*
3 Grasses soften the seating area
4 Colour injected by orange geum

Jo Thompson's garden:
This tiny yet versatile garden pulls off the natural style brilliantly. Minimal landscaping provides maximum function; the plants may be pushed to the edges, but they are still allowed to dominate the space and provide year-round interest as well as shelter and a great sense of privacy. The layered woodland planting is well reimagined using a mix of wild and cultivated plants, masking the brick boundary walls but never feeling heavy or overpowering thanks partly to the lightness of the birch tree canopies. A large wind chime finishes it all off with a sculptural and sensory touch.

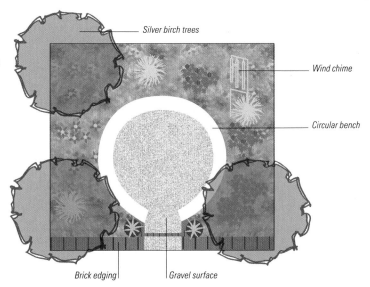

Silver birch trees
Wind chime
Circular bench
Brick edging
Gravel surface

NATURAL GARDENS

Walk on the wild side

Jane Brockbank's natural-style area of a larger family garden is a great example of clever garden design, with several distinct elements, including path, meadow, and a naturalistically planted border. They all merge seamlessly together and help to mask the garden room that forms the area's backdrop.

Key ingredients
1. *Digitalis purpurea* with *Astrantia*
2. *Alnus glutinosa* (common alder)
3. Naturalistic border planting with *Cirsium rivulare* 'Atropurpureum'
4. Meadow grass and wild flowers
5. Blurred edges of the brick and gravel path
6. Orchard trees and ox-eye daisies (*Leucanthemum vulgare*)

Jane Brockbank's garden:
Soft, billowing perennials, planted in a naturalistic style, fill a border on the left side, the extensive use of ornamental grasses linking it effortlessly to the meadow area on the right, with its wild flowers, birch, and orchard trees. The apparently random Petersen bricks within the path create a certain order and rhythm, leading visitors through the planting to the garden room at the bottom. Contemporary charred-wood cladding and the mimimalism of the design help to make the structure unobtrusive, acting as a foil for the meadow and perhaps suggestive of walking deeper into a shady wood.

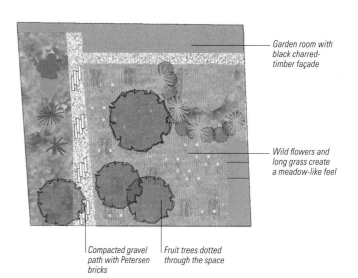

- Garden room with black charred-timber façade
- Wild flowers and long grass create a meadow-like feel
- Compacted gravel path with Petersen bricks
- Fruit trees dotted through the space

DESIGN INSPIRATION

CASE STUDY
SUN-KISSED RETREAT

In this Mediterranean-influenced, natural-style design, tough, drought-tolerant plants soften classic arid garden design elements such as a sun-drenched open central space and a trickling water rill. The planting takes centre stage.

Precious water

Water is present in almost every Mediterranean garden, and here the spouts pour into a cool, refreshing rill, adding movement and sound to the design. The rendered wall links tonally with the informal stone paving that divides the space.

Tapestry of colour

The naturally inspired planting scheme cleverly combines a tapestry of different colours and textures, using heat- and drought-tolerant perennial plants, including silvery artemisia, achillea, red *Dianthus cruentus*, and white *Centranthus*.

Designer **Cleve West**

NATURAL GARDENS 87

Cracked terrain

The rocky terrain of the Mediterranean coast is echoed in the irregular stone paving. Mortar joints between the stones allow rain to slowly percolate into the ground, ensuring that any available moisture is not lost.

Form and shade

The pagoda tree (*Styphnolobium japonicum*) in the centre and yew hedges beyond provide much-needed shade and natural structure to anchor the design. They also help to convey a sense of enclosure, creating a private area for relaxation.

Ancient origins

Sculpted columns, made from textured concrete and terracotta, are included to evoke the ruins of an ancient temple. They act like a stage set, contextualizing the design and giving the garden a feeling of permanence.

Small spaces

Increasingly, today's new gardens are getting smaller, especially in urban and suburban areas, where pressure on land is greater than ever. At the same time, there is generally a far wider appreciation of the importance of outdoor space to wellbeing and health. Put simply, gardens matter, regardless of size. The days when people assumed small spaces were not worth the effort or investment are long gone, for while these areas may be limited in scale, they are great in potential.

Small spaces can be some of the most interesting, exciting, and vibrant places to be in. They are also some of the easiest spaces to personalize and transform quickly, but they do need some planning and (later) upkeep. In these limited spaces, things really have to earn their keep. Detail is vitally important, as is quality of finish.

As far as style goes, your choice is vast: a simple yet formal layout can be effective, but you might fancy a tiny productive plot or an uninhibited cottage garden theme. Perhaps you want to plant a mini jungle or transform the space into a versatile outdoor room for the whole family. Do you want a modernist feel with clean lines? Or would you prefer something busy with different elements that are always changing? Maybe you just want a tranquil sanctuary. With a small space, anything goes.

When it comes to materials and plants, remember that you will be seeing this space all the time, so everything needs to be well thought through. Get the basics right: make sure all the hard landscaping is cohesive and of a high quality; details such as paving joins and carpentry will jar if they are less than perfect. Choose your plants with great care – go for ones that look good for as long as possible or have several seasons of interest.

You may find the idea of a permanent design too restrictive when space is limited. It's possible, of course, to make the area far more flexible than is practical in a larger garden. Planting can be in containers that may be moved; the configuration of decoration and furniture can be changed, perhaps to suit the seasons or a particular occasion. Whatever you choose, get it right and enjoy a space you'll want to spend all your time in.

Top and above
Vertical surfaces, such as this water feature, make the most of restricted space.
Growing plants in containers helps to create beautiful and flexible spaces.

What is a small space garden?

Today's gardens have to work hard, providing space for planting, relaxation, play, and entertaining. As the high price of land – especially in urban areas – has squeezed the size of gardens, new ideas for small spaces have emerged. Approaches vary, but most are treated either as functional spaces or as green oases – both offer a private escape or retreat from hectic daily life. In the former, hard surfaces dominate, creating a stage for multiple uses. Architectural treatments to boundary walls, furniture, and water features create elegant "rooms", often lit after dark to create extensions to the home. In the latter, planting dominates, often taking over areas that could have been used for entertainment or play. This intensive planting approach benefits the keen gardener, who may even use the space as a productive allotment.

Small gardens in detail

The use of simple, clear geometry is often the best approach to a small garden. Planting similarly needs careful thought: the trend has been for fewer species that work harder seasonally, providing architectural or sculptural interest, but using large, dramatic plants in a small space can also be effective.

In many city gardens, sliding or folding doors create a seamless transition between interior and exterior "rooms", extending the living area. Paved or decked surfaces help to increase functional space; materials are often selected to match interior finishes, further unifying indoors and outdoors. Pergolas or pleached trees offer privacy in overlooked spaces, while dense planting can achieve the same effect in more naturalistic gardens.

Sculpture provides a focal point, often combined with water used in jets or cascades rather than pools. Built-in seating fits architecturally, but can limit the flexibility of the garden. Stylish furniture and identical containers in a row add drama and rhythm.

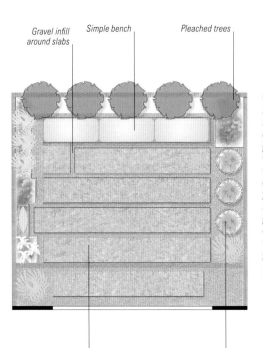

City garden
Here, garden designer Philip Nixon has created a simple but decorative plan with timber-clad walls complementing the furniture, and folding doors that lead out from the house (right). Planting is a mix of perennials, grasses, and evergreens, with the addition of tall pleached hornbeams, which provide valuable screening.

SMALL SPACES

A John Brookes design for a London garden.

DESIGN INFLUENCES

Evocative of country gardens, early city designs were often heavily planted and complex in layout. Today, they have become much simpler.

In 1839, J.C. Loudon – the Scottish botanist, garden designer, and garden magazine editor – responded to increasing urbanization and the diminishing size of city gardens in his book *The Suburban Gardener and Villa Companion*. In it, he classified different design approaches to the small urban garden, including low-maintenance designs.

More than a century later, John Brookes published a series of successful books that, like Loudon before him, addressed designs for smaller plots and explored the idea of the "outdoor room".

More recently, the Japanese have led the way in designing tiny outdoor spaces. In their densely populated cities, balconies or light wells are often the only areas available for planting.

Key design elements

1 Dramatic containers
Clay, stone, or steel plant containers are often repeated for a cohesive effect in a small space. Fill with clipped plants or – for a softer, more informal look – perennials and grasses.

2 Sculptural furniture
Artfully designed furniture – such as bespoke built-in benches, coordinated tables and chairs, or recliners – gives a small garden focus and answers a functional need.

3 Lighting
With the introduction of low-voltage and LED systems, lighting has become more sophisticated. Use it to emphasize a small garden's contours and plants.

4 Pleached trees
In overlooked city gardens, pleached trees (which look like hedges on stilts) provide privacy while using little floor space. Use lime, hornbeam, or holm oak.

5 Stylish materials
Make the most of limited space with a mix of materials to maximize texture and interest. Both natural and human-made materials, such as concrete, glass, and steel, are popular.

Interpreting the style

Gardens of new homes are getting ever smaller as pressure on land bites into available space. The way we use outdoor areas, be it a balcony or courtyard, is changing too, with many home owners now using them simply as an extra room and place to entertain or relax. Almost anything goes in small spaces – no one style or look sums them up, but they can be highly individual, beautifully detailed places. Key when planning a small space is to have a clear theme, know what sort of space you want to create, and avoid too many different materials. Aim for quality rather than quantity.

Perfect for pots above left
Small spaces need to be flexible, and planting in containers means you can move greenery about to suit the occasion, perhaps forming a living screen or even making room for outdoor dining.

Look to the east left
Small spaces adapt brilliantly to Japanese style – just a few elements are needed, with careful placement and attention to detail being everything.

Urban sanctuary top
Beautifully detailed, this small private space includes the reflective, calming element of water, and looks clean and uncluttered while being generously planted.

Rooftop views above
Roof gardens offer challenges and great opportunities – just think of the views. Planting needs to be in containers and windproof, and using climbers adds height without bulk.

SMALL SPACES

Sitting pretty left
Acting as a living sculpture and providing summer shade, a multi-stem Prunus serrula complements a pair of well-chosen chairs for a simple yet refined look.

Room for a pool below
Water can so easily be included in the smallest spaces, providing a cooling, calming focal point that draws the eye with its light-reflective surface.

Time for a dip bottom
The swimming pool in this garden may take up much of the space, but its elegant, unusual design is well balanced by lush planting, the pool's shallow steps inviting you into the water.

"Almost anything goes in small spaces – no one style sums them up"

Hard-working planting

In small spaces, hard surfaces for outdoor living can often dominate, so planting must work harder to compensate. Choose simple, bold, architectural combinations, which are stylish and easily maintained. Lighting, strategically placed, will flatter the space in the evening. For densely planted areas, keep paving simple, using strong textural foliage and colour as a foil to the built-up environment.

Feast for the senses top
Clean, simple lines give a stylish feel to this outdoor dining area; verdant raised beds and a matching planter allow diners to feel immersed in the planting.

A place to entertain above
Raised beds also provide informal seats for relaxing around the fireplace. The mix of ornamental grasses and Allium creates a diffuse screen between two areas.

Just add water above
A barrel serving as a small pond brings aquatic plants such as waterlilies within reach of those with even the smallest garden, balcony, or terrace.

SMALL SPACES

"As space diminishes, the urban garden becomes an increasingly precious resource"

Perfectly contained
Bold planting can impress in small spaces. Tall terracotta pots containing floodlit Osmanthus *flank a trough cascading with* Pennisetum *and* Miscanthus *grasses.*

Hidden gem above
A suspended canopy adds style and privacy to a seating area. Planting is minimal and restricted to containers, tonally linking to the cushions on the benches.

Soothing retreat left
Vertical or wall planting optimizes the restricted space, while retaining a softening effect. A textured panel of basalt provides sound as water trickles over the surface.

A modernist approach

Small gardens lend themselves perfectly to a modernist approach, which creates spaces free from clutter or fuss. A clearly defined geometric layout is key, so that proportions of the main features can be appreciated. Keep material and plant palettes to a minimum, and pay particular attention to the finer details. Fixings can be hidden to create smooth flowing surfaces.

Rooftop retreat
Consistent use of wood for flooring, furniture, and the large container with its soft planting provides a cohesive feel, while a sedum roof adds an unexpected diagonal element.

Bamboo screen
Decking creates a warm, tactile, lightweight surface, which is ideal for city or roof gardens. Here the planting is contained in simple box planters that screen this private space.

"Small spaces are often most successful when a restrained approach is applied"

SMALL SPACES

Complementary colours
Texture, colour, and shape combine to create this small garden. The ochre tones of the brickwork contrast with the warm terracotta-rendered surfaces, while clipped evergreens, grasses, and irises offer natural forms.

Textural composition
Contrasting surfaces of honed limestone, precise dry stone walls, and reflective steel-edged water create the drama here, softened by the dense planting of irises and Stipa beyond.

Classic structure
The rectangular pool, deck, and path of this compact garden are complemented here by blocks of dwarf hedging and an untamed yet unfussy leafy backdrop.

Geometrical design
The architecture of this garden space is the dominant theme, with the rectangular pool based on the dimensions of the picture window. Repeated cordylines arranged along the balcony above create a sculptural splash.

GARDENS TO VISIT

RHS CHELSEA FLOWER SHOW, London
The five-day show contains a specific section of smaller gardens designed for urban situations. rhs.org.uk/chelsea

INTERNATIONAL GARDEN FESTIVAL, Chaumont-sur-Loire, France
Includes numerous contemporary styled smaller gardens. domaine-chaumont.fr

GARDENS OF APPELTERN, Netherlands
A whole range of gardens, including urban style. appeltern.nl

NATIONAL GARDENS SCHEME
Charity providing access to more than 3,500 private gardens. ngs.org.uk

PALEY PARK, 53rd Street, New York, US
One of New York's famous pocket parks. tclf.org/landscapes/paley-park

Small space garden plans

Many small space gardens embrace a simple, geometric layout and a balanced design. The emphasis is on sculptural planting and quality materials, but approaches differ depending on the client's needs. In Vladimir Djurovic's simple, minimalist space, the focus is on clean, uncluttered lines. Also of modernist style, Andrew Wilson's garden is based on a grid, with interest achieved through changes in level and powerful planting. A rather more rustic feel can be seen in Nigel Dunnett's naturalistic design.

Maximizing space

Planting is restricted in this elegant garden by Vladimir Djurovic, where surface and texture are the highlights. The clever lighting design draws attention to the low bench seats made from the same material as the paving, and to the apparently floating fire cowl, which becomes a giant focal point for the terrace.

Key ingredients
1 Red cedarwood table
2 Acer palmatum
3 Lighting
4 Natural stone-honed finish

Vladimir says:
"This garden was developed as a holiday retreat. The space available for the garden was quite restricted, and a major part of the design process was dedicated to creating a sense or illusion of space.

"The brief was quite demanding: the client loves to live outdoors when in residence, and the garden needed to reflect this – with spaces for cooking and dining, relaxing, entertaining large groups of people, and so on.

"The restricted topography and the fact that the house is arranged on split levels also made the connection and sequencing of space more difficult.

"The result is typical of my work – I aim to produce memorable spaces, no matter what their scale. I am inspired by nature, and like to feel that my work brings people closer to the natural world."

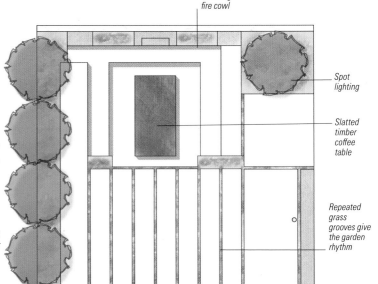

Grid lock

The owners of this property asked Andrew Wilson to maximize available space and inject a semi-industrial quality to complement a new, polished dark green fibreglass house extension with long curtain walls made of glass.

Key ingredients
1 *Betula pendula*
2 *Stipa gigantea*
3 *Deschampsia cespitosa* 'Bronzeschleier'
4 *Yucca aloifolia*
5 *Ligustrum delavayanum*

Andrew says:
"The long, low roof of the new building extension was echoed in the horizontals of the paving, low walls, and steps. The trees, mainly pine and birch, provide towering verticals that produce the classic contrast central to most modernist compositions.

"The garden is paved in coloured, poured concrete that appears to float out across a reflecting infinity-edge pool. Darker rendered walls provide subtle screening and a backdrop for uplighting to create an ambient glow after dark."

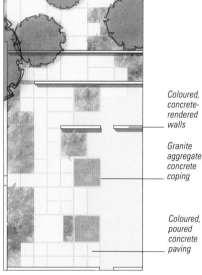

Coloured, concrete-rendered walls

Granite aggregate concrete coping

Coloured, poured concrete paving

Back to nature

This hard-working small garden designed by Nigel Dunnett centres around a converted shipping container, which serves as a garden office. The overall feel is of a naturalistic space with relaxed yet colourful planting, reflective pools, secluded seating area, and sculptural stone walls.

Key ingredients
1 *Betula utilis* var. *jacquemontii*
2 Green roof
3 Insect hotels
4 Dry stone wall with insect hotels
5 *Geum* 'Prinses Juliana'

Integrated design
This small garden manages to pull off quite a feat, neatly incorporating several distinct uses within its limited area. The relaxed planting helps to bind everything together. Birch trees enclose and shelter a private seating area and help blur garden boundaries, while the green roof blends the blue-painted office into the garden. The sculptural bug hotels double as a wildlife-friendly focal point, while low stone walls with integral bug hotels divide up the space. Naturalistic perennial plantings provide long-lasting colour and attract pollinators.

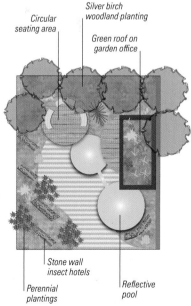

Silver birch woodland planting

Circular seating area

Green roof on garden office

Stone wall insect hotels

Perennial plantings

Reflective pool

DESIGN INSPIRATION

CASE STUDY
ROOFTOP RETREAT

A bijou rooftop terrace is transformed into a restful, plant-filled space. The angular, raised layout reflects the surrounding cityscape and strikes a modernist note; glass panels allow expansive views, providing a feel of space, even though the floor plan is modest.

Towers of flowers

Blue-flowered *Echium pininana* makes a spectacular accent plant, echoing the shapes of tower blocks in the broader landscape. It is a biennial but, after flowering, plants set seed and young plants germinate freely, thriving in shallow soil.

Right place

Plants such as red-flowered *Grevillea rosmarinifolia* and silvery *Artemisia* 'Powis Castle' are well adapted to the conditions, tolerating shallow soil and winds, while providing a long, low-maintenance season of interest.

Designer **Alasdair Cameron**

SMALL SPACES

A touch of privacy
Some smaller trees do well on a small terrace and help add refuge from the open rooftop. Deciduous *Acer negundo* is a good choice; after leaf fall, winter winds will not bother it, and it can be kept bushy and compact with regular hard pruning.

Keeping it light
Cane and wood furniture are light enough to be moved under cover when not in use. The stylish form of these chairs injects a sculptural element amid the planting, and provides a sense of style in keeping with this modern design.

Floral sunlight
With sprays of bright yellow, sweetly scented flowers, *Spartium junceum* is a perfect rooftop shrub, enjoying full sun and thriving in exposed sites.

Productive gardens

Productive food plants have been a part of garden design since the earliest of times, but it was the Victorians who elevated productive gardening to a fine art. Large walled gardens of wealthy estate owners offered a remarkable range of exotic fruit, fresh vegetables, and cut flowers year round, often grown, at least in winter, under glass. They were not, however, the first to mix fruit, vegetables, and flowers in the same area. Medieval abbey gardens were typically divided into small herb and vegetable beds with some decorative planting, and Renaissance gardens in France featured ornamental produce in elegant parterres, known as "potagers". This term is still used today to describe an attractive productive garden.

The Dig for Victory campaign during World War II generated a huge enthusiasm for home-grown produce in the UK, but this waned as wealth increased after the conflict. Today, our increasing desire for good-quality, healthy food, and concerns about the carbon footprint of imported goods, are fuelling a revival of the kitchen garden, albeit on a smaller scale.

Most productive gardens tend to be orderly, with geometric beds separated by paths for ease of access and maintenance. The plants are typically planted in rows (drills), which allow easy picking and maintenance, and mean plants can be protected from pests and inclement weather. Crop rotation – growing plants in different places from year to year to reduce pests and disease build-up – is usually practised.

However, designs today also include tiny spaces, where fruit and vegetables are grown informally in pots on a patio or balcony, or even in a window box. Materials for surfaces focus on the utilitarian – concrete slabs, brick paths, or compacted earth are all practical options and suit the look.

Planting varies seasonally, with fruit trees and bushes providing the permanent structure. Low hedges may also be included, often to contain herbs that tend to flop and spread, while rainwater, required for irrigation, can be captured in butts or other recycling vessels.

Top and above
A scarecrow protects valuable crops.

A raised bed provides easy access for harvesting produce.

What is productive style?

In large productive gardens, the layout and surfaces tend to be functional, creating a sense of ordered abundance, while in smaller spaces, the design is often more relaxed, with planters used to squeeze in as many crops as possible. Traditional designs were influenced by early monastic or physic gardens, which were divided into geometric beds filled with herbs and vegetables, punctuated by taller focal plants, such as bay trees or standard roses, in the centre. These simple design plans are used in contempory edible gardens, too, with bed sizes often shrunk to fit smaller urban plots. Functional paths – made of brick, stone, or gravel – allow space to tend the fruits and vegetables easily, while colourful rows of crops, fruitful containers, and decorative interplanting create garden designs that provide a feast for the eyes as well as the table.

Productive gardens in detail

As the 20th century came to a close, productive planting was pushed to the end of the main garden to give flowers, shrubs, and trees pride of place. Today, this approach is changing, as more people realize that growing food close to home is not only fun, but also allows you to enjoy fruit and vegetables that are either not available in the shops or, like raspberries or blueberries, expensive to buy.

Productive gardens need to be planned carefully to make them easy to manage. When planting in the ground, different crops should be planted in different beds each year to prevent the build-up of soil-borne pests and diseases. In small gardens and on patios or terraces, compact crops, such as tomatoes, chilli peppers, aubergines, and leafy salad crops can be grown successfully in pots or larger planters. Cold frames, greenhouses, and sunny windowsills indoors allow you to extend the growing season, while bee-friendly plants, such as lavender and open-flowered dahlias, inject colour and bring in pollinators to guarantee a good crop.

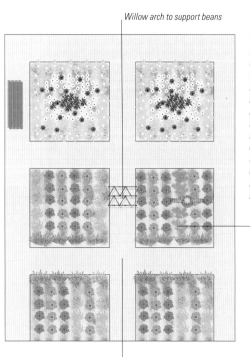

Willow arch to support beans

Wide paths for access and structure

Colourful potager
Here, the ordered character (left) of the vegetable garden, with its rows of crops and strong rectilinear pattern, makes a beautiful impression (right). Tall supports for runner beans and clipped hedging are used to enclose the space, and red dahlias and lavender add extra splashes of colour.

Square beds with a colourful mix of flowers and leafy crops

PRODUCTIVE GARDENS

Traditional walled kitchen garden.

DESIGN INFLUENCES

While many modern productive gardens are a mix of styles, some still echo the regimental formality of the walled kitchen gardens of the great English country houses. Victorian aristocrats showed off their wealth by serving exotic hothouse produce to guests, but the main function of the garden was to provide fresh food for the whole household.

Crops were set out in orderly lines in geometric beds edged with box and separated by paths made of gravel or beaten earth, or ash produced by the glasshouse boilers. Tender fruit trees were trained along south-facing walls that radiated heat to give them additional protection, while soft fruit bushes were grown under netted frames to prevent birds from eating the harvest.

Large, heated greenhouses were often built into the structure of the wall, allowing early cropping and the cultivation of tender produce, such as peaches and apricots.

Key design elements

1 Raised beds
Raised beds were first introduced to improve drainage, but they also provide a sense of order. An increased height of up to 1m (3ft) allows those with a disability to tend their gardens more easily.

2 Wide paths
Pathways should be at least 1m (3ft) wide in order to make the garden easy to navigate. Hard surfaces, such as brick, concrete or stone slabs, or gravel, are ideal since they withstand heavy everyday use.

3 Rustic obelisks
Ornamental features are always put to good use. Trellis and wooden or metal obelisks create height and rhythm in the garden, but also provide support for climbers, such as runner beans or sweet peas.

4 Planting in rows
Crops planted in rows can be easily recorded, cared for, and harvested, and the spaces between rows provide access for weeding. This geometric layout gives these beds their strong character.

5 Practical containers
Pots can be used to grow a wide range of edibles in small gardens and on patios and terraces. Large containers hold more compost and water and require less maintenance than smaller types.

DESIGN INSPIRATION

Interpreting the style

With the ongoing and, for some, renewed interest in growing fruit and vegetables, the idea of having a productive garden has seldom been so popular. Even with small new gardens, you can enjoy great harvests through the year while having a versatile, beautiful, and practical space. Raised beds and a simple, formal layout allow easy access, while structures for climbing vegetables add height. Grow herbs and flowers for cutting, and these hard-working sites can be as attractive to pollinators as they are to visitors.

Four square
A quartet of large wooden containers serve as raised beds for a range of crops, as well as support for a tall, arched pergola, up which runner beans or climbing squash may later scramble.

Edible colour top
People are often surprised how attractive productive gardens can be, whether it's by growing vegetables such as rainbow chard with their vibrant and tasty leaf stalks, coloured leaf kale, or the edible flowers of nasturtiums.

Productive profusion above
The formal layout of this productive space is emphasized with the painted boards and central plinth with planter, as well as the rows of flowering chives repeated along the edge of each bed.

Finial touch right
Raising beds by just a few centimetres improves drainage, retains improved soil, and provides a formal layout. Corner finials enhance that feel and stop hosepipes damaging produce.

Rotate the flavours left
Separate raised beds allow for crop rotation. Growing the same produce in different areas each year helps ensure plant health, minimizing the build-up of pests and diseases.

Buckets of taste below
Some crops, such as chillies, perform well in containers; standing pots within galvanized buckets helps retain moisture in summer and looks great.

Ensuring bumper crops above
This corner of a walled garden is filled with colour and tasty crops. Walls shelter trained fruit trees and cages for soft fruit, while elegant arches support developing runner beans. Orange and yellow calendula attract pollinators, and their individual petals are edible.

Practical and edible right
Strawberry plants do well in raised beds as fruits are lifted clear of the soil, away from pests and moisture that can cause rot. They also ripen more quickly when not shaded by foliage.

108 DESIGN INSPIRATION

Formal or relaxed approach

When planning a fruit and vegetable garden, you can opt for a formal design with regular pathways, or go for a more relaxed approach, using a series of planters and pots. Low hedges or raised beds give coherence to border edges in larger gardens, and beans, corn, and fruit trees provide height. Introduce colour with flowers that attract beneficial insects, or choose those you can eat.

Fruitful balcony
Pots of tomato cordons are tucked into a tiny sunny balcony, which provides a warm microclimate for these tender crops. Tomatoes are ideal, since the plants produce lots of fruit yet take up very little floor space, allowing an area for seating.

"Homegrown produce is one of the joys of a gardening life"

PRODUCTIVE GARDENS

Edible windowbox bouquet left
Strawberries have been planted along with edible flowers, including nasturtiums and pot marigolds, in this contemporary windowbox. The marigolds have a citrus flavour and nasturtiums taste peppery.

Urban kitchen garden below
This small city courtyard has been transformed into a tiny allotment, with baskets of crops and a cleverly designed dining table that doubles as a planter for salad leaves, herbs, and flowers.

Salad in a planter right
Suitable for use in restricted spaces, this stained timber planter contains a mix of salad crops and herbs. Tomatoes or strawberries would also be appropriate.

Eye-catching gourds below
Productive planting can be included in the design of a main garden. Here, gourds are used as a decorative climber, giving privacy to the seating area. Pink dahlias provide late summer colour below.

Lettuce and herb mix
Raised timber planters offer easily accessible beds for herbs and salad leaves. The rough woven rope edging on those shown here helps to combat attacks by slugs and snails.

GARDENS TO VISIT

BROGDALE, Kent
Home of the National Fruit Collection.
brogdalecollections.org

LOST GARDENS OF HELIGAN, Cornwall
Walled garden with many traditional cultivars. heligan.com

WEST DEAN, West Sussex
Beautifully restored Edwardian kitchen garden. westdean.org.uk/gardens

RHS GARDEN WISLEY, Surrey
Includes herb, fruit, and vegetable gardens.
rhs.org.uk/wisley

CHATEAU DE VILLANDRY, France
Formal Renaissance kitchen garden.
chateauvillandry.fr

Productive garden plans

In a productive garden, function generally wins over style, but the two are not mutually exclusive. These three gardens are packed with delicious edible plants, yet each, in its own way, looks great. Borders at Rockcliffe Gardens, Gloucestershire, burst with produce; Bunny Guinness's vegetable garden gives a nod to formality with its timber raised bed; and an allotment society has mixed herbs, flowers, and vegetables in a small space.

Traditional classic

This pleasing vegetable plot includes classic elements that add up to a great productive space. A pleached hedge shelters the site from winds, while gravel paths allow good access to borders with a wheelbarrow when needed.

Key ingredients
1 Greenhouse
2 Crops planted in rows
3 Antique copper tub planted with mint
4 Terracotta forcer for sea kale
5 Cosmos flowers for cutting
6 Rustic frame for annual climbing plants

Rockcliffe Gardens:
Direct sowing of crops in rows (or drills) is a practical and visually appealing technique that allows for thinning and adequate spacing. Alternatively, young plants can be raised in the greenhouse on site and planted out. Simple ornamental detail comes from rustic frames, ideal for annual climbers such as sweet peas, positioned at each corner, and the old copper tub planted with mint. Flowers for cutting make this ordered space pleasing to the eye.

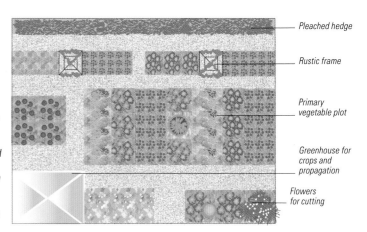

Pleached hedge
Rustic frame
Primary vegetable plot
Greenhouse for crops and propagation
Flowers for cutting

Raising veg

The geometric layout of this garden by Bunny Guinness includes the sort of well-equipped detailing needed in a hard-working space. The raised beds of vegetables are easy to reach and maintain.

Key ingredients
1 *Phaseolus coccineus* (runner beans)
2 *Allium cepa* (garden onions)
3 *Daucus carota* subsp. *sativus* (carrots)
4 *Beta vulgaris* subsp. *vulgaris* (red chard)
5 *Vitis vinifera* (vine)

Bunny says:
"This garden was originally dominated by an overgrown Leylandii hedge. Once this was removed, the space really opened up and a backdrop of native plants was revealed, which help to soften my design.

"The space works hard, which is typical of my approach. The owner is a barbecue enthusiast, so I created a space for entertaining, with a barbecue and built-in sink, and a small greenhouse.

"My influences often come from the architects I work with, and new or interesting ideas I see on my travels."

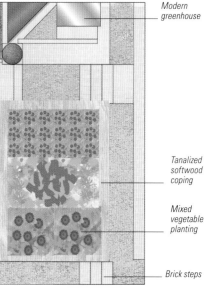

Modern greenhouse
Tanalized softwood coping
Mixed vegetable planting
Brick steps

Garden allotment

This garden was designed by the Manchester Allotment Society for the RHS Flower Show Tatton Park and aims to show how easy it is to integrate a few crops into the average domestic garden.

Key ingredients
1 Beehive-style composter
2 Wild flowers
3 *Ocimum basilicum* (basil) and other herbs
4 *Solanum melongena* (aubergine)
5 *Cucurbita pepo* (pumpkin)

Packed with a variety of herbs, including basil, fennel, sage, and parsley, the crops are squeezed into raised wooden beds and small patches of soil in between. French marigolds (*Tagetes*) are woven through the herb plants, providing colour and helping to deter flying pests.

Tender crops, such as aubergines and tomatoes, are also included. They can be grown outside in a sheltered sunny garden, and ripen towards the end of summer. A few pumpkin plants scramble up supports at the back of the plot.

The white beehive composter creates a decorative yet practical focal point, and wild flowers help to lure pollinating insects to the fruiting vegetables.

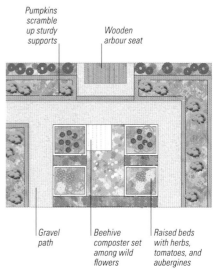

Pumpkins scramble up sturdy supports
Wooden arbour seat
Gravel path
Beehive composter set among wild flowers
Raised beds with herbs, tomatoes, and aubergines

Relaxed productivity

This small garden has been put to good use growing vegetables, fruit, and herbs, as well as ornamentals. Raised beds help maintain order and allow for easy cultivation; a colourful herb bed softens the gravel path on the sunny side, while raspberries thrive in semi-shade.

Key ingredients
1 Gold-leaved marjoram
2 Thyme
3 Nasturtiums
4 Dwarf French beans
5 Raspberries

Thriving veg garden
Generous planting blends productive and ornamental in the form of a herb bed that also features flowering perennials. All this growth spills over raised edges, releasing pleasant aromas when brushed against and attracting pollinating insects.

Dwarf French beans yet to crop fill a central bed — a piece of trellis protects newly planted seedlings of a follow-on crop from cats or unintended trampling. A simple shed serves as a practical focal point, enlivened by a few pots, while the trellis-covered boundary fence allows climbers to scramble and overhanging plants to be easily tied back.

Verbena on the left and right of the garden beds offers height and dots of colour, and helps frame the side of the garden. Nasturtiums growing in the foreground are an attractive feature, while being entirely edible, flowers as well as leaves.

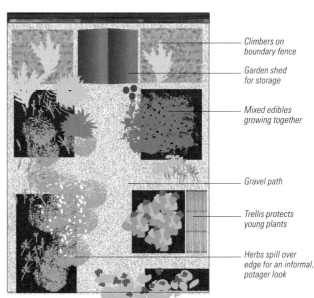

- Climbers on boundary fence
- Garden shed for storage
- Mixed edibles growing together
- Gravel path
- Trellis protects young plants
- Herbs spill over edge for an informal, potager look

PRODUCTIVE GARDENS

Vertical allotment

The advent of vertical growing systems has allowed even the smallest of spaces to be transformed into productive and beautiful places. Many are ideal for growing herbs and salad crops as these are compact and quick to mature.

Key ingredients
1. Runner beans
2. Flowering thyme
3. Chives
4. Red leaf lettuce
5. Crispy green leaf lettuce
6. Alpine strawberry

Versatile space
A vertical growing system has been erected on a sunny, sheltered wall, making the most of a small space. This treatment would be perfect for an apartment with a balcony, small terrace, or roof garden, as the materials used are light and easily installed. Adding automatic irrigation reduces day-to-day care, especially in summer, ideal for busy urban dwellers.

Integrated perfectly with the building's wooden cladding, troughs of compost cascade with veg and herbs — here, lettuce can be cropped whole or individual leaves harvested as required. Herbs and strawberries add flowers, scent, and fruit. A single container planted with two or three runner beans will provide colourful flowers and a modest crop of tasty beans by the back door.

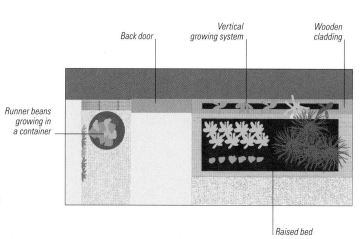

DESIGN INSPIRATION

CASE STUDY
EDIBLE EDEN

Productive gardens can be any shape or size, and even in this small plot, the designer has squeezed a wide range of edibles into raised beds and narrow borders, mingled with flowers that attract bees and other pollinators to create a beautiful, bountiful space.

Elegant yields
Rustic materials and a mix of vegetables, herbs, and flowers reference cottage style. Every bed is crammed with edibles, from beetroot and lettuces to beans scrambling up tepees, but the overall look is decorative and orderly.

Practical paving
The red brick pathway marries well with the traditional styling. Both practical and decorative, it lends an old-fashioned look, while allowing plenty of space for wheelbarrows and a hard surface from which to cultivate and harvest the produce.

Designer **Nick Williams-Ellis**

PRODUCTIVE GARDENS   115

Herb focal point
A clipped bay tree edged with a skirt of culinary herbs – including rosemary, parsley, and thyme – provides a beautiful, aromatic focal point in the centre of the garden, and a readily accessible source of fresh herbs for the kitchen.

Crops in close-up
The wide edging on the raised beds doubles as both work surface and informal seating from which to admire the garden. It also allows crops to be inspected at close quarters so that damage from pests and diseases can be spotted quickly.

Potted extras
In small gardens, compact crops can be grown in pots and containers to increase the growing space. These patio tomatoes have been bred for such a purpose and produce high yields of sweet fruits on small bushy plants.

Statement gardens

Influenced primarily by art or other creative industries, statement gardens often break design conventions and free up designers to make their own set of rules. Conceptual gardens, which are often based on an idea or theme, fit into this category, and examples can be seen at various festivals around the world. Statement gardens can also include any contemporary garden that does not fit neatly into a more conventional style.

Many of these designs celebrate new technologies and employ human-made materials, such as concrete, steel, rubber, fabric, glass, and Perspex, to create impact and visual interest. Lighting is also used to great effect in many of these gardens.

Planting is not always intrinsic to a successful statement garden, but can support the overall message conveyed by the design. When used, planting is often included for its sculptural qualities, and may also emphasize colour, texture, and movement. For some designers, ideas are inspired by ecology or the environment, and their gardens may feature plants that showcase a particular place or habitat.

Design concepts can be applied on a whim, but the best results are achieved where there is a relationship between the garden, its location, and the personality of its owner, or its history and cultural significance.

Key figures in cutting-edge design include the landscape architects Martha Schwartz and Kathryn Gustafson, who have both created ground-breaking gardens. Land art has also been influential in the evolution of this statement style. Examples include the works of Richard Long and Andy Goldsworthy; both designers are renowned for their natural sculptures, which form part of the landscape and intensify visitors' experience of a place.

Top and above
Architectural foliage and flowers provide focal points.
Manufactured materials are mixed with natural elements.

What is statement style?

This style is a mix, sometimes accidental, but often deliberate and boundary-pushing, drawing from a wide range of genres. Short-lived and more experimental, garden shows offer a platform for more conceptual eclectic creations and allow designers the freedom to innovate. Colour, sculpture, and garden art provide focal points and interest, planting often focuses on architectural specimens, and lighting adds to the drama.

Statement gardens in detail

Faced walls are typical of this style, as they provide backdrops or surfaces on which art and sculpture can be displayed. Colour, usually intense and bold, is also important, creating a vibrant, occasionally jarring atmosphere. A wide range of materials can be associated with the style, and in some gardens, the combinations can be quite complex. Designers often use a mixture of human-made and natural surfaces, such as concrete and timber, or stone and steel, and by keeping the overall plan simple, these textural contrasts are more clearly appreciated.

Furniture is frequently used to express particular architectural or stylistic references, or it may also introduce colour. Sculptural plants add scale and drama, and are sometimes repeated to amplify ideas. In addition, colourful and textural planting is a common feature, with containers used to reinforce stylistic concepts.

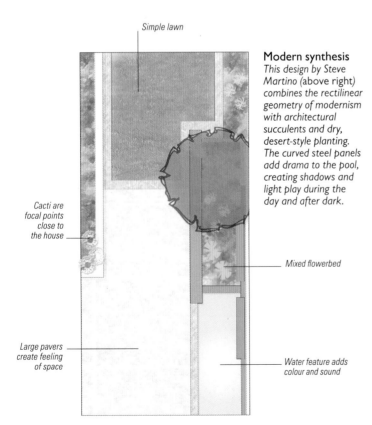

Modern synthesis
This design by Steve Martino (above right) combines the rectilinear geometry of modernism with architectural succulents and dry, desert-style planting. The curved steel panels add drama to the pool, creating shadows and light play during the day and after dark.

- Simple lawn
- Cacti are focal points close to the house
- Mixed flowerbed
- Large pavers create feeling of space
- Water feature adds colour and sound

DESIGN INFLUENCES

This style borrows from a range of ideas with energy and confidence. Travel, a shrinking world, and the internet have opened up access to a wide range of plants, materials, and influences – from jungle planting to Japanese gravel, modernism to Mediterranean, and formal to conceptual style. This gazebo by Michael Schultz and Will Goodman uses Japanese elements with Art Deco and postmodernist overtones. The personality of the resulting designs may not please the purists, but statement garden style is all about breaking the rules.

The Hurst garden by Schultz Goodman.

Key design elements

1 Modern materials Statement garden designs often include materials that are not traditionally associated with gardens, such as glass, steel, and Perspex, with planting softening the lines.

2 Sculptural plants Although a wide variety of plants are used in statement gardens, many have sculptural qualities – grasses, *Yucca*, or *Astelia* are typical, and palms are used for height.

3 Water cascades and fountains Cascades, fountains, and water blades – controlled by an app to produce complex displays – provide movement, atmosphere, and sound.

4 Lighting Light effects are key style devices, picking up architectural details, specimen plants, and decorative topiary. The development of lighting technology and LEDs produces spectacular results and can also inject additional colour.

5 Eclectic floor plan The mixing of styles can produce interesting and complex layouts, with modern designs mixed with drought gravel planting, or formality combined with the asymmetry of Japanese gardens.

6 Vibrant colours Bold colours are often used in surface finishes to make connections between plants and hard materials. Faced walls, ceramics, paving, and lighting can all contribute colour and drama while creating an exciting ambience.

Interpreting the style

Designing statement gardens is a liberating and fun experience, where rules can be rewritten. Colour can be a controlling element, with rich or strident tones making clear connections between materials and planting. Also try using irregular shapes and mix solid materials with transparent glass or Perspex to create a bold, unique design.

Bedrock of design above
Like a geological phenomenon, these angled layers of red sandstone rise out of a pond and are juxtaposed with dry, Mediterranean planting combinations with glaucous foliage.

Solid seating left
A touch of the interior design is brought to this outdoor terrace, with its concrete seating and coffee table. Planting softens the effect in places, and cushions would make the furniture more comfortable.

Playing with the elements right
In this garden of sculptural Jura limestone, fog is used as a device to create a sense of mystery. Its wisps veil and reveal the stone forms in turn.

Golden brown far right
This gravelled courtyard space is unified by striking colour and strong shadows. The simple grove of Mexican fan palms (Washingtonia robusta) creates a brilliant connection with the modern Mexican architecture too.

STATEMENT GARDENS 121

True blue
The Blue Stick garden was inspired by Meconopsis betonicifolia (blue poppy). Two sides of each stick are blue and two red, creating different effects.

"Cutting-edge designs mix up conventional ideas and bend the rules of garden-making"

Water and earth above
Water gently cascades over this ledge, cantilevered from a rendered wall, and into the trough below, creating an oasis in this desert garden. The warm earth-tones echo the sandy soil and glow in the sun.

Blocks and undulations left
White concrete cubes are counterpoints to the turf that ripples across this garden. They create a sculpted quality that offsets the stark walls of the house.

GARDENS TO VISIT

RHS HAMPTON COURT PALACE GARDEN FESTIVAL, UK
Show with a section of conceptual gardens.
rhs.org.uk/hamptoncourt

GARDEN OF AUSTRALIAN DREAMS,
Canberra, Australia
Richard Weller and Vladimir Sitta's garden.
nma.gov.au

FESTIVAL OF GARDENS
Chaumont-sur-Loire, France
domaine-chaumont.fr

CORNERSTONE, Sonoma, California, USA
Regularly changing showcase of innovative design. cornerstonesonoma.com/gardens

Practical beauty

With statement gardens, almost anything goes – if you can think it and build it, there really are no rules. To make them successful in the long term, as well as being beautiful, stimulating, and eclectic spaces, they must also succeed in terms of practicality and durability. Basic features such as planting, seating, boundaries, and paving will probably be present but reconfigured and reinvented, while sculpture and other forms of abstract art are likely to figure heavily in the design.

Cube squared top
Geometry is key to this small space; blocks of strident orange contrast with wooden surfaces, while bold plants such as Aralia soften the edges of cuboid forms.

Timbered path above
An unusual but inviting timber path winds between repeated domes of yew and crosses a rill filled with water. An unusual use of materials is common in statement gardens.

Life behind bars top
Railings form both a screen and a deck over white gravel in this uncompromising composition softened by minimal planting.

Art without boundaries above
A simple fence becomes a work of art using laser-cut Corten steel in a design featuring an Eastern motif of bamboo.

STATEMENT GARDENS   123

Riveting composition above
Lush woodland planting contrasts directly with artistically created rusty, jagged, riveted girders, in a design perhaps providing a nod to a lost industrial past.

Softened by planting left
A sculpture brings a dynamic touch to a verdant corner as it emerges from cool foliage.

Statement pieces above
Arresting artwork is key in statement gardens – here Aeon, a dazzling sculpture, symbolizes the force keeping our planet in equilibrium.

Statement garden plans

Statement gardens vary greatly, but often offer up the unexpected or present usual elements of a garden in a novel, even innovative way. In three of these gardens – Robert Myers' coastal garden, Sarah Eberle's forest edge garden, and Antony Watkins' tropical garden – planting takes a dominant role, but materials and functional form rule in the space by Colm Joseph.

Coastal retreat

The drought-tolerant planting scheme in this garden by Robert Myers is designed to evoke the landscape of the Mediterranean coast. Pine and tamarisk trees offer cool shade at the back with low plants at the front.

Key ingredients
1. *Sesleria nitida*
2. *Centranthus ruber* and *C. ruber* 'Albus'
3. *Jasione montana*
4. *Spartium*, *Crambe*, and *Centranthus*

Relaxed coastal scene
In this statement garden, planting takes centre stage in a naturalistic if abbreviated interpretation of coastal plant communities. The composition is pleasing and will satisfy plant enthusiasts and lovers of modernistic gardens alike. A stone sculpture that doubles as a seating bench provides an arresting focal point. Rendered walls painted lilac complement the silvery tones of drought-tolerant foliage, while stronger colour is provided by the flowers of red valerian, verbascum, salvias, and jasione.

Materials are pale and reflective, mimicking the sun-drenched landscape that is being evoked, and while the overall feel is soft and relaxed, elements such as the stone path that bisects the planting enforce a sense of a human-made structure.

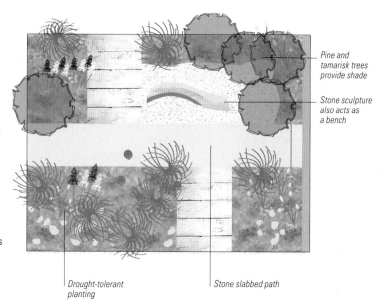

Pine and tamarisk trees provide shade
Stone sculpture also acts as a bench
Drought-tolerant planting
Stone slabbed path

STATEMENT GARDENS

Contemporary courtyard

Colm Joseph designed this tiny Cambridge courtyard to blend perfectly with a new extension, which includes a corridor linking the house to a garden office. The result is a garden with plenty of visual interest. A bold yet restful space, it helps to unite different parts of the home, and is a great place to entertain and relax with its central water feature, simple yet effective planting, and use of quality materials.

Key ingredients
1. Corten steel kitchen window louvres
2. Multi-stem *Malus* 'Rudolph'
3. Woodland-style underplanting
4. Corten steel fins
5. Corten steel water trough
6. Central water feature
7. Weatherboarded garden office
8. Poured concrete hardstanding

Making less more
This courtyard has to look great year-round. The key is the way interest is focused in the centre via a statement water feature. Restrained use of materials links garden to house, be it Corten steel or poured concrete, which is used indoors and out as flooring, and in structural elements such as garden furniture.

Boundaries have been simplified and softened with natural materials such weatherboarding and pleached beech (Fagus). Planting is relaxed; a multi-stem crab apple (Malus) is underplanted with geraniums and ferns, while carpeting Pratia angulata 'Treadwellii' softens edges.

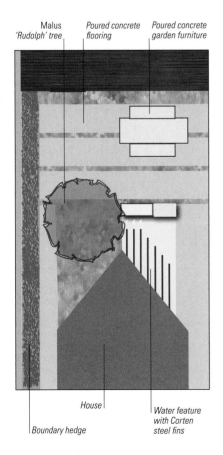

Inspired by the forest edge

Sarah Eberle's forest-edge statement garden is devised around the concept of sustainability and inspired in part by natural rock strata. The planting is lush and cool, and includes many native and naturalized plants in a reimagining of an Irish woodland. A dramatic waterfall forms an arresting centrepiece, cascading in front of a verdant green wall into a stone-edged angular pool.

Key ingredients
1. Lime tree
2. Poolside planting
3. Wooden cladding mimicking vertical rock strata
4. Water cascade
5. Shade-casting tree canopy
6. Path and pond edging

Lush retreat
Planting is pushed to the forefront of the concept here; the garden offers a retreat from the wider world, lush tree canopies providing seclusion and privacy; lower-level understorey planting softening the design's angular, rather edgy layout. The plant choice is restrained, with foliage favoured over flowers – showy blooms are not needed while ferns, conifers, and bold-leaved plants such as Rodgersia and Aralia help provide structure and textural contrast. They show up well against the buff-coloured path, which also helps to lighten the woodland feel.

Innovative use of sustainably sourced wood to mimic rock formations disguises a central structure, which also supports a green wall, while water spills from a trio of conceptual rusted metal girders.

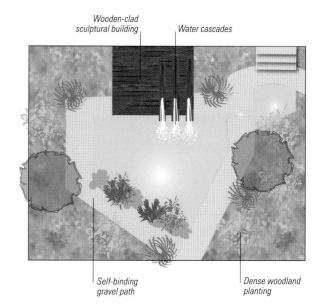

Wooden-clad sculptural building · Water cascades · Self-binding gravel path · Dense woodland planting

STATEMENT GARDENS

Backyard oasis

Mature palms and other exotics give this bold, plant-led urban garden designed by Antony Watkins the flavour of a relaxed subtropical retreat. Incorporating seating areas, raised beds overflowing with diverse planting, and a secluded garden room feature, the space is cleverly divided up using different paving materials, architectural planting, and screening.

Key ingredients
1 *Trachycarpus wagnerianus*
2 Purple-leaved *Canna*
3 Seating area
4 *Chamaerops humilis*
5 *Brahea armata*
6 *Trachelospermum jasminoides*

Tropical hideaway
Providing the feel of a holiday destination, this statement plant enthusiast's garden is filled with unusual and often borderline hardy plants that thrive in a near-frost-free city microclimate.

While planting takes priority here, the layout displays skilled use of space with distinct zoning dividing up the area into sections that serve different functions. Close to the house a tiled patio makes a spot for outdoor dining, while further down the site is a more relaxed seating area amid palms, olives, and cannas.

Raised beds and grouped pots allow for easy cultivation of often brightly coloured planting, while winding paths and bold foliage mean that the full extent of the garden is concealed. At the bottom of the garden is a garden room with a roofed kitchen area for entertaining – it's the perfect summer getaway.

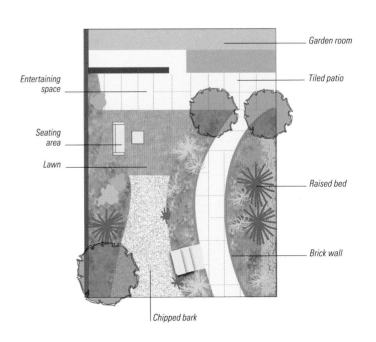

DESIGN INSPIRATION

CASE STUDY
DESIGN FUSION

Fusing a range of styles, from Mediterranean to modern, this statement garden weaves Jurassic-period inspirations into a harmonious design, with large metal structures – inspired by the bony back plates of a stegosaurus – defining the space.

Natural structure

Evergreen trees, including the holm oak (*Quercus ilex*) and strawberry tree (*Arbutus unedo*), lend structure and a sense of permanence to the garden, while other hardy exotic trees with finely cut foliage soften the look.

Prehistoric slabs

The seemingly random floor plan and irregular-shaped paving stones throw out the design rule book. They help to evoke a rugged landscape that references the Earth's tectonic plates as they collide to form new geological features.

Designer **Andy Sturgeon**

STATEMENT GARDENS 129

Bridging the gap
The pathway in the garden steps up to form a bridge across the water, giving the impression that the water has been here for a long time and the paving is a new addition. In other areas, stone slab-like benches suggest ancient rock formations.

Steel screens
Bronze-coated steel slabs stand proud, cutting dramatic shapes that resemble a dinosaur's back plates and providing a focus along the perimeter of the garden. They also present a foil to the fire pit and create hidden areas that heighten the intrigue.

Artful planting
The planting seems informal, even "shaggy" in parts, but this belies a considered approach. Sculptural plants, such as *Corokia* × *virgata* with its tangle of black stems, jostle with colourful perennials, including the fiery orange kangaroo paw (*Anigozanthos*).

132 Assessing your garden
142 First principles
180 Creating a plan
198 Designing with plants
222 Choosing materials

HOW TO DESIGN

Assessing your garden

Top and above
Assess the soil and feed with compost if necessary.

Choosing the right plants for your site is an important first step.

If your plot isn't a blank canvas, take the time to look carefully at what is already in place before you begin work on a redesign. If you have just moved into a property, it is worth waiting to see what plants emerge and how the garden looks at different times of the year. When planning a makeover of an old garden, cost may be a factor, and you may want to retain and incorporate favourite features.

Get to know your garden soil, too, and notice how much sunshine and rainfall the plot receives. This will tell you what plants will thrive in your particular growing conditions, and help you to avoid costly mistakes. Improving drainage by digging in grit, or adding plenty of compost to poor soil, will also broaden your choice of suitable plants.

The drawbacks of a sloping garden can be turned to an advantage by the use of terraces, steps, raised platforms, or suspended decking. Introducing these elements can revitalize a tired garden, giving it a new lease of life. The same is true of an area that stays constantly damp: transform it into a bog garden or pool and enjoy the pleasures of a wide variety of moisture-loving plants and the ensuing wildlife they attract.

Privacy is important, but it is wise to consider your neighbours' needs before making any major changes to a boundary. A tall, vigorous conifer hedge may shield you from view, but does it also cast a long shadow over their patio for most of the day? Legal obligations may come into play, too, so check first before you finalize your design or begin construction around a shared boundary.

Perhaps the most important piece of advice is to take your time before launching into a garden redesign and new landscaping. And if bare or ugly patches are inevitable while work is carried out, remember that strategically placed containers make a quick and effective screen.

Assessing your soil and aspect

Find out as much as you can about your site before you plan a garden. If you ignore the local environment and specific soil and drainage conditions, you could waste money on unsuitable plants, or discover that your planned seating area is in a wind tunnel, or that the lawn turns into a lake in winter.

Identifying and improving soil

Garden soils range from sticky clays to free-draining sands. Clay soil is prone to waterlogging in winter and dries hard in summer, while sandy soil warms up early in spring but is a challenge to keep moist in summer. Clays can be very productive and rich in nutrients if manure and grit are dug in, but sands are typically poor in nutrients and, without adding manure or garden compost mulches, won't retain moisture or nutrients. The ideal "loam" soil contains a mix of clay and sand plus organic matter. Loams are dark and fertile because of the organic content, form a crumb-like structure when forked over, and have good moisture retention. Test your soil (*above right*) before designing planting areas; loams when rolled hold together to form a ball but crumble under pressure.

Testing clay soil
As clay content increases, you can form it into a ball or sausage, then a ring.

Testing sandy soil
This soil crumbles under light pressure, won't form a ball, and feels gritty.

Grit improves drainage
Large quantities of coarse grit worked into the top layer of soil (to fork depth) improve the drainage of heavy clay, but drains may also be necessary on waterlogged soils.

Well-rotted manure benefits all soils
Manure causes fine clay particles to clump together, improving soil structure and drainage. It also helps sandy soil retain water and nutrients, but use it only as a mulch.

Testing acidity

The soil pH is a measure of acidity and alkalinity – 7 is neutral, below 7 is acid, above 7 is alkaline. Acid soils suit ericaceous plants, while many Mediterranean herbs, shrubs, and alpines will grow happily in alkaline, lime-rich conditions. You can pick up clues about your soil by looking around the neighbourhood to see what plants are thriving. Soil type can also vary around a garden due to local anomalies, so carry out several pH tests using an electronic meter or simple chemical testing kit (*right*).

Determining your soil type
Take samples from around the garden, using a test kit to check acidity/alkalinity.

Checking the aspect

The direction your garden faces has a marked effect on how much sun it receives and how exposed it is to wind. To work out your garden's aspect, stand with your back to the house and use a compass to check the direction you are facing.

Typically, south- and west-facing plots are warm and sunny while north- and east-facing gardens are cooler and shadier (*right*). Filtering the gales on an exposed site reduces wind-chill and limits damage to structures and plants. As altitude and distance from the sea increase, temperature and exposure can be adversely affected, whereas urban areas produce and hold heat, keeping gardens artificially warm.

Windy sites
Exposure can restrict your choice of plants as well as your enjoyment of the garden. Provide shelter with deciduous hedging, which will help reduce wind speeds without creating turbulence, or use other permeable windbreaks (see also p.227).

Frost pockets
On sloping sites, cold air rolls down to the lowest point and pools there if its path is blocked. Less hardy plants here can suffer frost damage.

ASSESSING YOUR GARDEN

MORNING

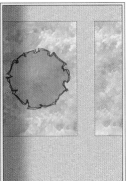

South-facing garden
Gentle sunshine across the garden from the east first thing creates pleasant conditions for summer breakfasts on a patio on the west side of the house.

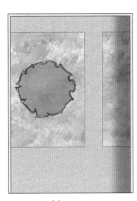

North-facing garden
Morning sun from the east soon disappears behind the house. Plant camellias, and other plants sensitive to morning sun after frost, on the shady east side.

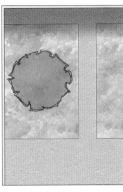

East-facing garden
Enjoy breakfast on a patio by the house, but avoid planting wall shrubs here that are sensitive to morning sun after frost. Cold east winds can scorch tender foliage.

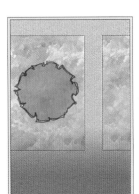

West-facing garden
The area near the house is shaded for most of the morning and a cool retreat in hot weather, but for early sun, design a seating area at the end of the garden.

MIDDAY

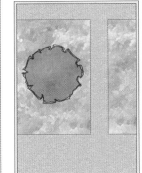

South-facing garden
In the height of summer, walls reflect the sun's heat and the whole garden is exposed to the sun, so you and your plants will bake without additional shade.

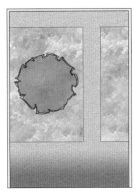

North-facing garden
The area next to the house is completely shaded, but the top end of a longer garden could be in full sun — perfect for a seating area and some sun-loving plants.

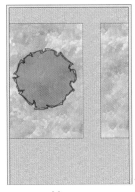

East-facing garden
Sun filters across the garden from the south but disappears behind the house in the afternoon. Cool after midday, this is a good aspect for a shady conservatory.

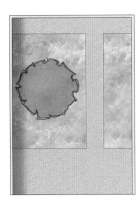

West-facing garden
Most of the garden is in sun at midday, especially in summer. Tender wall shrubs thrive on the house and north and west boundaries. A patio to the south offers shade.

EVENING

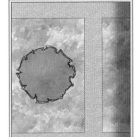

South-facing garden
Heat radiated from walls keeps the patio warm into the night. Most areas of the garden are ideal for frost-tender plants since the garden is warm all day.

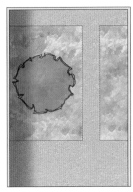

North-facing garden
Gentle light from the west offers an ideal aspect for woodland plantings. A patio on the east side of the garden will capture evening sunlight in summer.

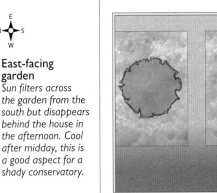

East-facing garden
The area by the house is shady. It can feel chilly sitting out because walls haven't absorbed heat during the day; make a patio at the far end of the garden for evening sun.

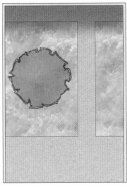

West-facing garden
A dining area by the house allows people to bask in late evening sun, but you may need some shade. Walls absorb sufficient heat to keep the area warm on summer nights.

Managing slopes and drainage

Predicting how water moves around, and how it can be redirected, is the basis of drainage design. As a general rule, all surfaces should be on an incline and water must flow away from buildings. In most cases, the water runs off hard surfaces, such as terraces or steps, into the soil where it is absorbed. However, sites on hills or with heavy, compacted soil can present drainage problems, and you may need to seek specialist help to avoid waterlogged conditions or flooding.

Drainage issues

All waterproof surfaces (roofs and paved areas) prevent water from draining naturally; the water must be channelled to flow into municipal drains, or to run into soakaways or, if in small quantities, directly on to planting beds. The type of soil in a garden will affect drainage, with heavy soils (clays and silts) causing more problems than free-draining types (sands, gravels, and sandy loams).

On a steep site, water will flow quickly, seeking a low point and, eventually, an underground pipe, open ditch, or stream. Particular attention needs to be paid to water moving over bare soil or sparsely vegetated surfaces where it will cause gullies and erosion. However, if the landscape is undulating or contained, water will gather in the dips and in larger wet areas, such as bogs or ponds, and will need an overflow.

If you have a difficult site, determine the upper level of the groundwater (water table) as it may affect where you decide to position your drains or soakaway.

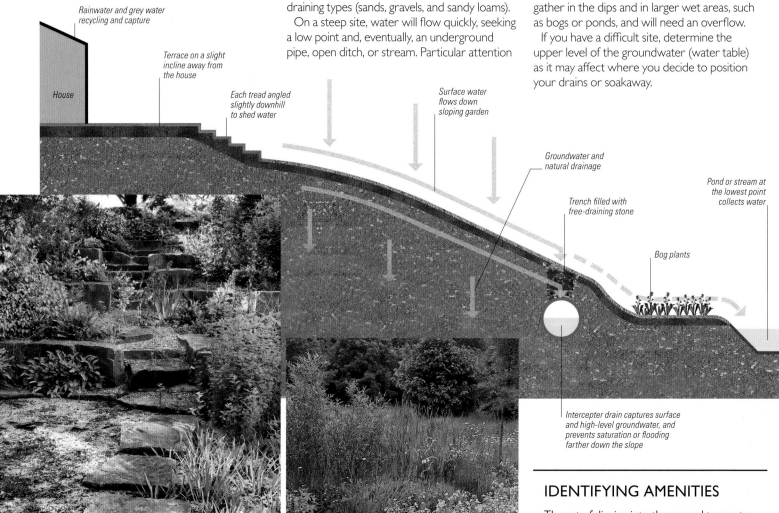

Sloping garden
All rainwater falling on this garden will eventually find its way into the ground or to the pond, which is located at the lowest point. An overflow may be needed to channel any excess water into an underground drain or soakaway.

Moisture-loving plants
Groundwater may be a problem, but it is also an opportunity. A naturally high water table or a butyl-lined bog garden can make an ideal place for growing a range of beautiful moisture-loving plants.

IDENTIFYING AMENITIES

The act of digging into the ground to create ponds, alter slopes, or install drains can hit underground services (such as water and gas pipes and electrical cables) or existing drains and sewers. Never excavate the site unless you know what is directly below, and do not presume that amenities are in the exact locations shown on local council plans. Take your time to identify problems and employ a specialist surveyor if you are in any doubt.

ASSESSING YOUR GARDEN 137

Rainwater collection
This recycled barrel holds enough rainwater to cover a short period of dry weather and makes an attractive addition to the overall appearance of a garden.

Reduce flooding risks

Where drainage is not managed carefully, it can cause flooding, both in your garden and in the local neighbourhood, if storm drains are unable to cope with the excess. In the UK, there are regulations about paving over front gardens, so check before any redesign. To prevent flooding, install a Sustainable Urban Drainage System (SuDS) by creating areas where water can collect and then be absorbed slowly into the ground following heavy rain. Planted areas absorb large quantities of water, helping to mitigate flooding. You can also include small depressions that act as temporary ponds, filled with plants that thrive in wet and drier conditions. The aim is to retain all the water that falls on the garden in the garden. Also install water butts and use the captured rainwater on your plants.

Flow diagram
Where waterlogging is not severe, excess surface water can be directed into a drainage ditch or pond. If the water table is high, you will need to install an underground drainage system, preferably using a specialist contractor.

Garden pool
An informal pool can be used to capture excess water and will serve as a perfect habitat for wetland and aquatic plants and animals.

Design considerations

If your garden is on a sloping site, you will need to create flat, usable surfaces. Often this requires construction work, so when drawing up plans, consider budget and time constraints, the overall size and shape of the proposed spaces, and possible access for earth-moving machines. More complex solutions may be required for steeper sites and slopes that are less stable, or where especially large level areas are required.

DECKING AND PLATFORMS

To construct flat platforms or walkways on a slope with minimal disturbance to existing ground levels, it is best to use timber. Decking is especially useful where access for earth-moving is difficult, when slopes are too steep to alter, and on undulating surfaces around wetlands. However, it is short-lived compared to other landform solutions.

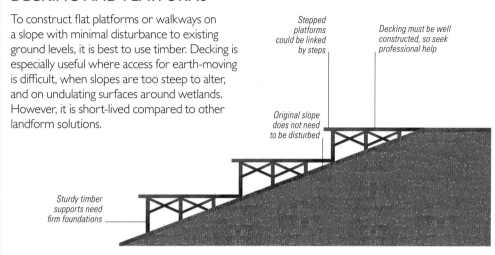

TERRACING

Small-scale terracing can be used to make horizontal planting beds on a slope. A series of retaining walls, set one above the other, provide structure, then soil is cut away from the slope for backfilling. Work can be done by hand or with a mechanical digger. Any large-scale terracing will require the advice of professional designers and engineers.

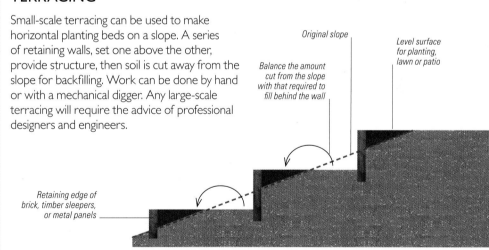

CREATING GENTLE SLOPES

Undulating land can be landscaped into gentle slopes or flatter areas. Excess soil or hardcore may be generated, or more required to achieve the desired levels, and, in both cases, this may increase the cost. Any changes will destroy existing vegetation and cannot be carried out beneath the canopies of trees that you want to retain.

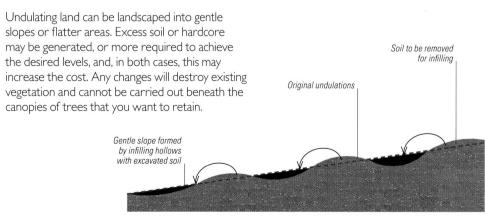

Assessing your garden options

When thinking about a new design for your garden, first ask yourself which elements you like and want to keep, and which you dislike. Next, consider your budget – does it allow you to add a new feature, adapt the existing garden, or will you decide to go for a wholesale makeover, with a new planting design and landscaping? It is often possible to rejuvenate a tired or mature garden, even with a limited budget, by taking a fresh approach and injecting some new ideas.

Degrees of change

Before you start designing, think about whether you'd like a completely new look, a new feature such as a patio or a pond, or whether you would prefer to keep the same layout but overhaul the planting. If your garden is small or seen as one space (rather than a series of connected spaces), you may want to rethink the entire area; larger plots will take more time and money to redesign from scratch. List the features you consider important and bear in mind that your needs may change in the future, as your children grow, for example.

A COMPLETELY NEW LOOK

Wholesale change can be hard to visualize and often means removing existing structures and mature plants. However, it gives you the chance to do something radically different with a garden and create an innovative space personal to you.

PROS
- An exciting blank canvas upon which to create whatever you want.
- The end result will be more coherent and integrated if you do not have to make compromises around existing elements.

CONS
- Trees and shrubs required for structure.
- New plants take time to fulfil their potential.
- The reality may not match your vision.
- Short-term loss of wildlife habitats – although, depending on your new design, these should return over time.
- Sometimes a completely blank canvas can be more daunting than adapting an existing layout.

COST CONSIDERATIONS
- Potentially expensive – hard landscaping and mature plants, if you don't want to wait for plants to grow, are costly.

DEVELOPING AN EXISTING PLOT

This is the most common approach, and even though you will be working with existing elements, it is still possible to refresh the look. List the features you plan to keep. With multi-level or sloping gardens, a site survey may be needed.

PROS
- This approach is usually less time-consuming and costly than a total makeover.
- You can work in stages and tackle different areas of the garden in sequence.
- You can make use of the existing mature planting, so there is no need to wait for everything in your garden to grow.

CONS
- The end result may lack cohesion. It is important to make sure that the features you add are complementary to existing ones.
- The renovations may not have the dramatic impact you are looking for.

COST CONSIDERATIONS
- Working with the current layout is less expensive than a complete makeover, and makes sense if you want to undertake changes in stages as money becomes available.

ADDING A NEW FEATURE

Making a change to just one part of your garden is the simplest option, but take care to integrate a new feature sympathetically. Pay particular attention to choosing materials and colours that blend in well with the existing design.

PROS
- Adding one new feature should be a straightforward change to manage.
- The rest of your garden will still be usable while this feature is being installed.
- Focusing on just one project means you can concentrate on getting the details right.

CONS
- Making sure that your new feature fits visually with the rest of your garden can be difficult.
- You can't let your imagination run free.
- You may damage other areas of the garden while building the new feature. Lawns and existing plants are particularly vulnerable.

COST CONSIDERATIONS
- This is the least expensive option – unless, of course, you are planning something very glamorous. The budget should be relatively straightforward to manage.

Case study: a new family garden

Every garden overhaul begins with a series of questions, and even when you have made a list of desirables and undesirables, you also need to consider the pros and cons of keeping or removing significant elements. For example, if you are thinking of taking out a mature tree because it casts summer shade, check that this disadvantage is not outweighed by its benefits: it may also provide shelter from wind or privacy and screening from neighbouring buildings. Or, perhaps it adds height to your garden. It is also worth checking if your trees are protected by a tree preservation order (ask your local council).

Making decisions about your garden will be easier if you are very familiar with the plot. If your garden is new to you, be patient and live with it for several seasons to see what appears and what changes before you make any dramatic alterations.

In the case study discussed here, a family garden is the subject of a renovation. The pictures below show some of the options open to the owners, depending on how much change they want.

The original plot
The way you use a typical family garden, and the amount of time you spend in it, will inevitably change as children grow. Design play areas so that they can be adapted.

INTRODUCE

MORE STRUCTURE
New hard-landscaping elements, such as paths, patios, and walls, have immediate impact.

PLAY AREAS
Lay an appropriate surface and add structures that can be changed in the future as needs alter.

OUTDOOR LIVING ROOMS
Extend your living space by creating areas in the garden for eating, entertaining, and relaxing.

ADAPT OR REMOVE

BEDS AND BORDERS
Planting areas can be adapted and new shrubs and perennials added, or they can be totally replanted.

PONDS
Ideal for older children, but fit a grille if you are concerned for the safety of young ones.

UNSIGHTLY PATIOS
It is easy to distract attention from an unattractive terrace with tubs of plants, and garden furniture.

KEEP

OUTBUILDINGS
Sound, useful structures, such as greenhouses, can be integrated into your new design.

MATURE TREES
Try to work around mature, slow-growing trees if possible; they offer valuable structure and height.

PERENNIALS
Keep established plant communities where they are evidently thriving and suit the conditions.

Designing boundaries

Boundaries create a frame for your outdoor space and are among the most important elements in a garden. They may indicate legal ownership, help to create a microclimate, and provide privacy. Most disputes between neighbours concern boundaries, and there are many legal regulations governing them, so before making any changes, first check who owns yours. If your neighbours have ownership, consult with them first and discuss any proposed changes to avoid conflict later.

Evaluating privacy

Before making changes to a boundary, especially if it is to be higher or removed, take time to evaluate the impact of the changes on your own and your neighbours' privacy and light. Check from all doors and windows, in particular upstairs windows, and assess what you can see now and what you will be able to see once the change has been made. Bear in mind that deciduous trees lose their leaves in the winter, which will mean more light but a less secluded garden. Also, raising the ground level on your side – with a deck, for example – may intrude upon your neighbours' privacy.

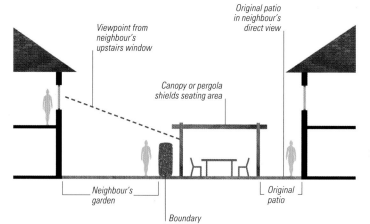

Neighbours' views
Carefully chosen structures can create sheltered areas in your garden, reducing the need for high fences or hedging. A patio or seating area can be screened off from your neighbours' view with a canopy or pergola, allowing you to retain your privacy without loss of light to either side.

Increasing privacy

Increasing the height of boundaries may be illegal, so check with your local planning office first. However, it is possible to increase the privacy within your own garden without altering the boundaries themselves. Strategic positioning of new trees can help, but they will take time to grow. Tall, fast-growing evergreen hedges are now subject to planning control, as well as being high-maintenance, and should be avoided. Consider using trellises, which can support climbing plants and also help to create a sheltered microclimate by allowing air to pass through them (see p.227). Best of all, create spaces in your garden that are not overlooked by your neighbours (see *diagram above*).

Pergola cover
Combined with climbing plants, this is an attractive way to create privacy without blocking light to the rest of your garden.

Sheltered space
Well-placed planting forms a secluded site for seating areas – a garden parasol can give additional privacy and shade.

Temporary screen
A makeshift cover like this one creates shelter and privacy wherever it is needed and can be conveniently packed away.

ASSESSING YOUR GARDEN

Keeping in with neighbours

Although we all want some privacy, it is important to establish good relations with neighbours. You could place tall screens around your patio area and lower fences elsewhere to encourage conversation. When planning your garden, consider anything that could irritate your neighbours, intrude into their space, or block their light.

Communal gardens, on the other hand, are designed to encourage friendship and cooperation. They need careful planning, and you should also consider who will be responsible for the garden's long-term maintenance.

Friendly divide above right
Low fences encourage communication and friendship between neighbours while also allowing more light into both gardens.

Shared space right
Communal gardens encourage community spirit and work well where there is shared responsibility for their care.

BOUNDARY REGULATIONS

Planning permission is needed to build a fence or wall over 1m (3ft) high next to a public highway or footpath, and over 2m (6ft) high on other boundaries, so check with your local planning office first. Fence posts should be on your side to ensure that the fence does not intrude on to your neighbour's property, and plant hedges at least 1m (3ft) away from the boundary on your land. Your title deeds will show you where your garden boundaries lie.

Considering neighbours' light

There are laws governing an individual's right to light. Most light is blocked from gardens by trees, although garden structures and poorly planned building layouts can also create dark zones. Before taking the law into your own hands, seek expert advice. It may be possible to remove part of an offending tree, or to negotiate changes to boundaries to allow your neighbours more light. When planning changes to your own garden, consider the impact they will have on neighbours' light at different times of the day and year, both now and in the future. This particularly applies to trees and hedges, as they will grow in height and width and could potentially cause problems.

Security issues

Boundaries provide security, but it is best to strike a balance between imprisoning yourself and opening your garden to your surroundings. Police recommend that fencing, walls, or hedges at the front of your house are under a metre (3ft) in height, so your doors and windows are visible from the street. Use lights to illuminate your space, but ensure that you do not floodlight your neighbours' property. Spiky evergreen shrubs, such as *Pyracantha*, holly, or blackthorn can be grown to form attractive barriers that will deter most intruders.

Thorny shield below left
Pyracantha is a good choice for a burglar-proof screen, but will take time to grow; combine it with a simple post and wire fence until it matures, then keep it to under 2m (6ft) in height.

Automatic protection below
Electronic gates maximize security for large properties or where burglary rates are high. They can be unattractive, so look for well-designed gates that blend in with your garden.

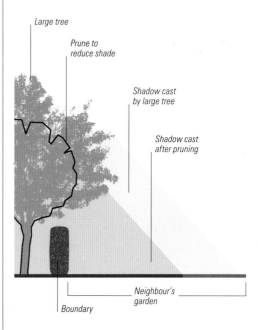

Light idea
Think about how your boundaries, or elements within your garden, will cast shade on to your neighbours' plot. Here, a large tree could be pruned to allow much more light into the adjacent garden.

First principles

Designing your garden is all about finding solutions. It can seem daunting at first, but if you start with a clear idea of your aspirations and practical needs, your basic design will soon begin to take shape.

Start by pulling together all your inspirations, using magazines, photographs, and online sources to create a book or folder of ideas. Your images may include plants and landscapes you love, and perhaps furniture or art you admire. To help clarify your thoughts, you could then draw a simple bubble diagram that identifies areas for different activities, such as eating, seating, or play space for the children.

The routes of paths, shapes of structures, and the spaces between elements all have an impact on the look and feel of a design, and need to be considered before you draw up a finished plan. For example, sinuous paths and organic shapes combine to create relaxed and informal designs, whereas straight paths and symmetrical layouts convey a formal look.

Every site will have its own particular challenges, whether your garden is on a steep slope and needs terracing, or if it is tiny or an awkward shape.

Whatever the problem, an understanding of how to use lines, shapes, height, structure, and perspectives will help. You can also employ a range of techniques to lead or deceive the eye, creating an illusion of space in a small garden, or diverting attention to focus on specific features.

When it comes to creating atmosphere and mood, the colours, patterns, and textures that you choose have a powerful impact. Colour also affects the impression of size and space in the garden – cool blues and whites tend to make an area feel bigger; warm reds and yellows make spaces appear lively and more compact. Pale colours and white reflect light into gloomy plots. Texture can be used to great effect, too, creating exciting contrasts by combining rough with smooth, or shiny with matt.

By ensuring you get the right garden design principles in place, coupled with a clear vision of what you want, your dream garden is not too far away.

Top and above
A strong pattern unifies different materials.
Plans help you to organize design ideas.

Sustainable choices

Finding ways to be more sustainable is an increasingly important part of everyday life, a requirement that also impacts how gardens are made and maintained. Buying new things – be they plants, compost, pots, or construction materials – can be costly and significantly affect our environment, both in the manufacture and transportation of these goods. Through recycling, clever design, and making use of existing materials on site or nearby, it is quite possible to have a first-rate garden with a minimal environmental footprint.

Recycle and reuse

Finding new uses for old materials and objects in the garden rather than buying new ones is a great way to boost sustainability and is good value for money. This is especially true if you already own these items, although charity shops, reclamation yards, and roadside skips (with permission) are great sources of potentially impressive garden features. Recycled objects may need little or no adaption, such as a galvanized bin reused as a plant pot. Some materials will need a bit of work, perhaps weather proofing, for their new role. To prevent a random, piecemeal look, try to keep to a fairly consistent theme: use the same colour wood preservative, for example, or pick similar metal containers. Chosen well, recycled objects can provide the garden with a sense of individuality, while at the same time having less of an impact on the wider environment.

Refined pallet above left
Even a wooden pallet can be transformed; add shelving and a shallow planter, and you have an attractive, versatile feature that is ideal for cheering up a forgotten corner.

Repurposed chic left
Perfect as a hanging basket with its built-in drainage, this upcycled kitchen colander makes a great home for a tumbling strawberry plant.

Pots of flavour top
Planted-up, old galvanized buckets serve as a makeshift herb garden. Just make sure there are drainage holes at the bottom before planting.

Old materials, new garden above
Old wooden floorboards, if treated, can be given a new life outdoors as a decked path. Corrugated iron panels can make effective and attractive fencing.

Harvest rainwater

Perhaps the simplest way to make your garden more sustainable is to add water butts in order to collect and store rainwater (see also p.177). You can then use it during spring and summer when plants need extra water. Using rainwater is better for the plants and does not waste quality drinking water, a valuable and increasingly expensive resource. Butts can be connected to house, greenhouse, or shed downpipes. Linking several butts together ensures you will have a reasonable resource even in drought spells. Water butts can be unsightly, but design solutions allow them to be well integrated into the garden. A cover is important to prevent debris, pets, or wildlife from falling in, and it also prevents mosquitoes using the water as a breeding ground.

Integrated storage
Wood cladding allows this water butt to blend into the wider garden setting.

Use what you have

Finding uses for various materials you already have to hand is a big step forward in terms of sustainability. Coming up with ways to incorporate them may mean you do not need to travel to buy new products, which themselves will probably have already been transported many kilometres, saving both on money and resources. Some raw materials – such as leaves or grass clippings – you will have a steady supply of, and these can be turned into free, sustainable, garden compost. Unused items, such as wood, old bricks, roof slates, and cobbles, may be a more finite resource but can easily find a new role.

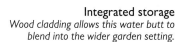

Compost your leaves left
Composted leaves, lawn clippings, and prunings return nutrients and goodness to the garden.

Slate chippings below left
Broken roof slates or unused chippings can be used as a mulch to retain water and keep weeds down.

Pine cone mulch below
If you have pine trees on site, pine cones can make an unusual but effective, long-lasting mulching material that is renewable and sustainable.

Designed-in solutions

The most effective way to ensure a garden is resistant to drought, and thus sustainable in terms of water use, is to consider water usage from the start of the design. Sparse, drought-resistant planting, thick mulches to keep soil moist, harvesting rainwater from the run-off from garden structures, and water rills channelling rainwater run-off into storage reservoirs (via plants that like a little more moisture) are all techniques to help ensure hot, dry summers need not damage your garden.

Plan for rain
Designed effectively, rain-fed gardens can have no need for irrigation with tapwater, even during long, dry periods when gardens require extensive watering.

Consider grey water

Drought is increasingly a problem for gardeners in our changing climate. When reserves of stored rainwater have been depleted, avoid using a hosepipe and put grey water from washing-up, the bath, or shower to good use. Mild detergents in the water will not bother most plants, especially for short periods. You can collect and apply grey water directly to garden plants (except edibles).

Precious resource
Rather than tipping it down the drain, use grey water to water garden plants in times of drought. Use it immediately or store it for no longer than 24 hours.

Future proofing

While gardens are naturally living, evolving places, any successful garden design must be able to accommodate changing conditions and circumstances. These may include the gradual adjustment in requirements of a changing family dynamic; future additions or changes to your buildings; or the less predictable but often more immediate effects of climate change, which increasingly means withstanding extreme summer heat and drought, and/or winter cold and wetness.

Weathering the climate

In our rapidly changing climate, today's gardens have to work harder than ever, providing useable outdoor space through periods of ever-more extreme weather. Summer drought and high temperatures mean that planting must be resilient, able to withstand the varying effects of a less stable climate. Trees, for example, provide cool shade even on the hottest day. Associated landscaping must be capable of withstanding and ideally mitigating rather than exacerbating impacts of torrential summer downpours that may follow.

Winters may be generally less severe, but climate change does not prevent spells of freezing weather. As a result, planting must be hardy enough to tolerate this extreme seasonal variation. Associated problems such as winter waterlogging need to be considered too, with design solutions factoring in this level of unpredictability.

Raised expectations top
Timber sleepers are useful for terracing slopes and forming raised beds. Planted terracing helps slow water flow in times of torrential rain; raised beds improve drainage where waterlogging is problematic.

Natural air conditioning above
Lush planting casts shade and lifts humidity, which has a considerable cooling effect in hot weather and makes a well-planted garden a comfortable place in summer heat.

Hot and cold covered above
This sunken area proves versatile: a fire pit provides warmth on chilly evenings, while cooling planting makes it comfortable for relaxing on hot days.

Preparing for extremes above centre
In this garden, a deck rises above a rain-fed rill, which channels away excess water and prevents flooding. Elsewhere, drought-tolerant planting predominates.

FIRST PRINCIPLES 147

Adapting to change

For most gardens, in particular family gardens, flexibility is an important consideration when a design is created. Over the years, a family – whether adults only or adults and children – goes through many changes, and the function of the garden must be able to adapt to these.

In early years, the space may serve as a nursery, a playground, and then a sports pitch, while providing entertaining space for adults. Later, it may be called on to provide a space for wildlife, for growing fruit and vegetables, or supply an area to rest and relax in, while being easy to maintain in those days when bending and lifting become more of a challenge.

The garden is a precious resource and needs to be somewhere that can be enjoyed at all stages of life. Good design should be able to take such changes into account, allowing adaptation of the garden at key times and ensuring that materials or effort aren't wasted in the long term.

Moving with the times
A simple design with a large lawn that maximizes space, easily maintained planting in sun and shade, and a permeable path to reduce water run-off, makes this garden fit for future changes if needed – as well as the wider use it offers now.

Mid-life flexibility
A family garden will need to adapt over time. Lawns once needed for fun and games can become meadows for wildlife as children become teenagers, with more space for relaxing and perhaps experimenting with new plants.

Staying active
Space to entertain or relax outdoors is important whatever age you are. As you get older, raised beds allow for easier maintenance of the garden, as does a consistent level surface to reduce any mobility problems.

Gathering inspiration

How do we find ideas for our outside spaces? For most of us, inspiration may initially come from other gardens, whether they are our friends' or pictures we have found online or in books, magazines, or newspapers. While this is a good starting point, and probably the best stimulus for anyone who is still developing their confidence in making design decisions, it can ultimately constrain the creative process. Most successful designers look outside their own discipline for other influences to help develop their concepts and push the boundaries, so seek inspiration from a variety of sources or select a theme. You can then create a "mood board" of appealing ideas to help you develop your own unique design.

Finding inspiration

By focusing on aspects of experiences that you like – for example, places you have visited on holiday, natural landscapes that you love, the work of favourite artists or architects, interior designs, ideas you have seen online, such as Instagram, Pinterest, or Houzz, or TV programmes – you can build up a picture of a garden you will enjoy. Also scroll through nurseries' websites for images of plants that you favour, and make a note of these too.

You can collate your images and ideas by printing out pictures and sticking them into a notebook or onto an A3 sheet of paper to create a mood board. Alternatively, source a website that allows you to upload your images to make a mood board online, which you can then easily refer to whenever you need. Whichever method you choose, continue to build up your portfolio of images until you are ready to start the garden design process.

Remember that you do not need to include all of your design influences in your final plan. In fact, professional designers often start with the bare bones of an idea and build on that, rather than cramming in everything on their or their clients' wishlist from the start.

Also narrow down your plant list to about 20 key varieties (you can always introduce more at a later stage), and look through your images for colours that appeal, again keeping to a simple palette – see the information on Introducing colour and the colour wheel on pp.168–69 for guidance.

Using a mood board
Collate photographs, images from websites, and pictures from magazines to create a mood board of creative and planting ideas. You can then use these as the inspiration for a totally new garden design or a starting point for the renovation of an existing plan.

Clockwise from far left
*Beach-themed garden – props?
Bright colours and sculpture – mosaics?
Beach hut style – storage?
Coastal wild plants
Yellow flowers for an accent colour
Mediterranean fishing boat – blues and whites*

Case study: a seaside theme

A coastal theme is a natural choice for anyone who has been inspired by a holiday by the seaside. Study scenes, plants, and other features while you are away, and start compiling a sourcebook of ideas, photographs, and even pressed flowers that capture the essence of the garden you want to create at home.

Also look at colours, shapes, and materials that reflect the location. These may include the turquoise water, local costumes, or landscaping materials used for houses or walls. However, remember that developing a design is not about copying exactly what you have seen elsewhere, nor is it combining all your ideas into one busy area. Good design evolves when a theme is carefully adapted to suit a planned space. So consider all the elements that inspire you and see whether they work together well before you draw up your final plan.

You may also find it useful to sketch a bubble plan (see p.182), marking the different areas and functions you are planning for your new garden. Then file your inspirations under those headings, as shown here.

Main inspiration
An inspiring holiday by the sea will provide a wealth of ideas. Here, the light through the trees adds a romantic ambience.

Seaside planting sources
Recreate coastal shallow soils and drought conditions – for example, with gravel borders – to mimic the environment in which these plants would naturally grow.

Seaside furniture
Furniture that is in keeping with the overall mood, such as these casual deckchairs, helps to create a coherent look, as well as providing a welcome area of relaxation.

Devising play areas

Sand and water continue the seaside theme, and are obvious magnets for children. A micro-environment that includes these elements not only makes a great play area that will provide children with hours of fun, it also looks attractive when not in use. If you have very young children, you may prefer to avoid the potential danger of open water and just install a sand pit. If you are wary of vast quantities of sand ending up in the pool (or in your house), substitute small rounded pebbles to make your "beach".

Sun and sand
A practical play area combined with an organic layout and seaside plants makes a delightful feature.

Swinging idea
If you have room in your garden, allocate a space for a swing. Use recycled, hardwearing rope and driftwood for the seat, and cover the ground beneath with bark chips.

Shapes and spaces

Choosing the basic ground shapes for your plot is a good starting point for a design: one simple shape is best for small gardens, but larger areas can accommodate a variety. How you fill the spaces between the shapes also determines the final look.

How to use shapes

When choosing squares, rectangles, or circles for a design, also consider the size, shape, and location of the surrounding buildings and boundaries. Experiment with different options: try layouts based on existing features, the structure of the house, and the way the garden will be viewed and used. In general, shapes with straight sides are easier and cheaper to build than circles and ovals.

RIGHT-ANGLED SHAPES

A variety of these straight-sided shapes easily divide the garden into separate areas, provide a strong sense of direction, and exploit both long and short views. A long axis running straight down the garden will lengthen it visually; a diagonal layout creates more interest; blocks laid across the plot foreshorten the garden and take the eyes to the sides, making it feel wider.

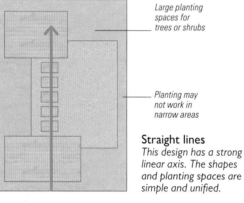

Large planting spaces for trees or shrubs

Planting may not work in narrow areas

Straight lines
This design has a strong linear axis. The shapes and planting spaces are simple and unified.

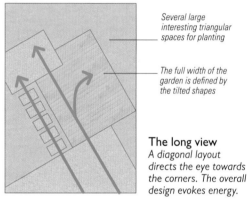

Several large interesting triangular spaces for planting

The full width of the garden is defined by the tilted shapes

The long view
A diagonal layout directs the eye towards the corners. The overall design evokes energy.

CIRCULAR SHAPES

Circles are unifying shapes, and while combinations can create pleasing effects, they do leave awkward pointed junctions that can be difficult to plant or designate. Work with geometric principles: for example, a path should lead you into the centre of the circle; if set to the side, the design will appear unbalanced. Ovals have a long axis, providing direction and orientation.

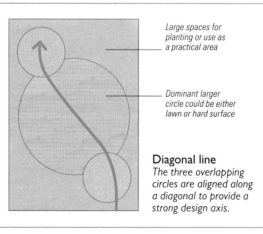

Large spaces for planting or use as a practical area

Dominant larger circle could be either lawn or hard surface

Diagonal line
The three overlapping circles are aligned along a diagonal to provide a strong design axis.

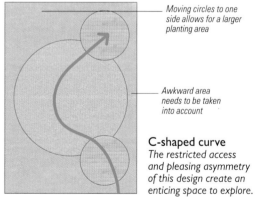

Moving circles to one side allows for a larger planting area

Awkward area needs to be taken into account

C-shaped curve
The restricted access and pleasing asymmetry of this design create an enticing space to explore.

MIXING SHAPES

Combining various shapes creates more interest, but throws up problems when a curve and a rectangle meet, or different materials connect. Generally, keep the layout simple, experimenting with scale and proportion to work out how many opposing shapes can be employed. Planting can be used to "glue" the shapes together, and to blur the joins between awkward junctions.

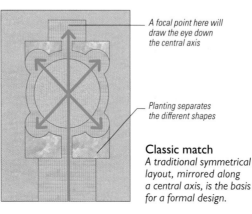

A focal point here will draw the eye down the central axis

Planting separates the different shapes

Classic match
A traditional symmetrical layout, mirrored along a central axis, is the basis for a formal design.

Use planting or a focal point to provide a visual full stop

Planting partly obscures the different areas

Simple approach
Changing the size and orientation of a shape delivers a dramatic and imposing layout.

Using spaces

Densely planted spaces, using height and filling the garden's width, will create an enclosed space, while sparse, airy planting hugging the boundaries gives an open, spacious feel. Spaces can also be used to disguise the size and shape of a garden. For instance, a jungle effect in a small garden can imply the existence of more space by blurring the edges, but exposed boundaries may make it appear smaller. Conversely, in a large country garden, open spaces can blend seamlessly with the surrounding landscape, making the plot appear even bigger. Consider, too, existing planting and structures and work with the spaces they create.

Clean lines
Interlocking, steel-edged rectangular "trays" are the basis for this simple design. The metal cladding on the building creates a focal point and an effective visual full stop.

Mixed moods
This garden is densely planted by the house, allowing close inspection of the flowers and plants, and then opens up on to a spacious lawn, creating two moods.

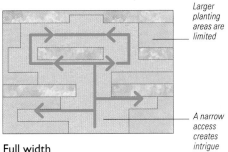

- Larger planting areas are limited
- A narrow access creates intrigue

Full width
A series of parallel divisions, with offset gaps for planting or practical structures, forces movement and views around the garden. The design draws you in.

Open aspect
A narrow space between tall boundaries will be claustrophobic and oppressive. Here, in a design dominated by a lawn or hard landscaping, low vegetation creates an area exposed to more light, longer views, and with a connection to the sky above. It will feel open, but intimate areas may be lost.

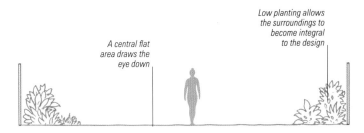

- A central flat area draws the eye down
- Low planting allows the surroundings to become integral to the design

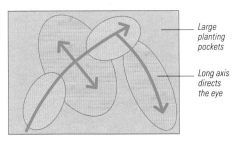

- Large planting pockets
- Long axis directs the eye

Smooth flow
Using ovals instead of circles adds a smoother flow to the layout because the eye is taken along their lengths, rather than in all directions as in a circle.

Enclosed feeling
The same space filled with vegetation of different heights will be darker, much more enclosed, and with no views to the sides. The path will appear as a corridor through the centre and can lead to different parts of the garden, divided by the planting into separately designated areas.

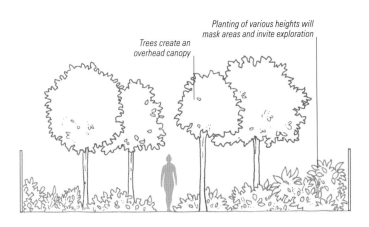

- Trees create an overhead canopy
- Planting of various heights will mask areas and invite exploration

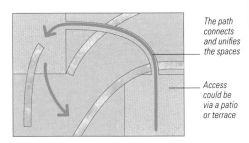

- The path connects and unifies the spaces
- Access could be via a patio or terrace

Secret corners
In this mixture of rectangles and curved hedges, only one part of the garden can be seen at any time. This allows the hidden areas to have different themes.

Balanced approach
The same path now moved to the side also creates a corridor-like effect, but this time views are allowed under the canopy to the right, across a narrower strip of planting into the brighter space beyond. To the left, secret, intimate places can be created with a pergola or arbour among the mixture of high and low planting.

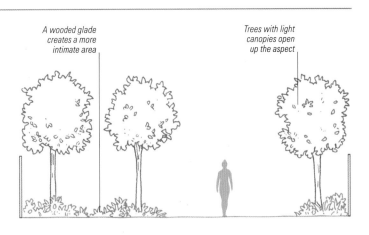

- A wooded glade creates a more intimate area
- Trees with light canopies open up the aspect

Routes and navigation

The location, width, pattern, and choice of materials of your path network will affect the way the garden is used. The routes determine how the area is navigated, as well as revealing views and framing spaces. Not all paths have the same role: some, the primary routes, will dominate the vista and dictate the garden plan. The secondary routes are used occasionally, guiding you off the main thoroughfare to access areas hidden from sight, whether for practical or design purposes.

Primary routes

The main route or pathway through the garden not only links together the different areas, but also determines the basic design. For example, a main path laid straight down the centre suggests formality, while a curved route snaking through the garden creates the template for an informal plan. A wide path offers an open, inviting entrance, welcoming in visitors, and a narrow winding path, flanked by tall planting that obscures the view, adds mystery. To punctuate the end of the route, use a focal point, such as a bench, statue, or container, to create a visual full stop. By its nature, a primary route will be heavily used, so materials need to be durable as well as complementary to the overall garden style. Consider, too, how the shape and appearance of path edges fit into the design.

CENTRAL PATHS

WINDING PATHS

DIAGONAL PATHS

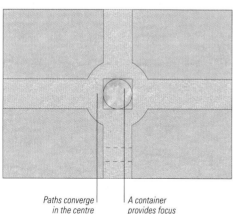

Paths converge in the centre | *A container provides focus*

Classic layout
A formal design is often built around a series of geometric and symmetrical paths. They are used to frame planted areas and meet at a specific focal point. There is usually no opportunity to deviate.

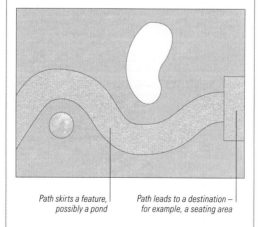

Path skirts a feature, possibly a pond | *Path leads to a destination – for example, a seating area*

Enticing curves
Routes that snake through the plot add a flowing sense of movement and an air of intrigue. They can be used to move around or join up key elements, as well as provide a few unexpected surprises.

The straight path lengthens the plot | *A circular patio adds contrast*

Illusion of size
Setting a path on a diagonal allows the garden to be viewed along its longest axis, thereby creating the illusion of greater space and depth in small spaces, drawing the eye away from the back boundaries.

FIRST PRINCIPLES 153

ROAM FREE

Random paving with planted crevices creates a slightly erratic, informal design. With no defined route, the eye – and body – can move in several directions across the whole area.

CIRCULAR PATHS

The circular path draws you on *A pond, for example, is framed by the path*

Continuous flow
A circular path takes you on a journey around the garden. It can be planned to provide alternative views of key features and different elements, depending on the direction in which you travel.

Secondary routes

While primary routes can determine the style of a garden, secondary routes should be less intrusive and subtly incorporated into the design. They can be both practical and ornamental, providing occasional access to a seating area, shed, or compost heap, or leading you off the main path on an intimate journey to view a concealed corner. They can even cut through large flowerbeds, allowing you to experience colours and scents up close. Access routes need not be as durable as main paths, and can be created from softer, organic materials, or mown through an area of grass or wild flowers.

Access paths
While helpful in offering access to other areas, plan secondary routes carefully and use sparingly to avoid a maze-like confusion of paths that make the design look muddled. They can be obvious (as right), or hidden in some way, either deliberately behind planting (see below left), or concealed within the design (see below right).

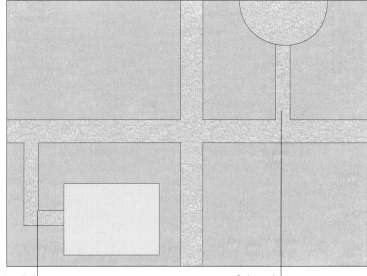

Path to shed *Path to patio*

Practical solution
A path tucked away at the back of this formal design is not obvious, but it provides a practical, hard-surfaced route to the shed and compost bins.

Hidden approach
The gravel to right and left of the path, while part of the design, also provides a direct, hard-wearing pathway to the garden's seating and play areas.

Secret way
Visually, it appears as if the main pathway stops at the lawn, but concealed behind low hedging, a side path takes you off to a secluded area of the garden.

Subtle link
A path laid in the same paving material as the main circular route links the off-set dining area without impinging on the cleanness of the design.

Creating views and vistas

Your garden may look out over countryside or towards a block of flats, but either way, the views within your space can be enhanced with careful planning. A combination of framing and screening, using barriers, archways, and pergolas, can create a memorable experience as you move through your plot, glimpsing the next view as you go.

Planning your route

One ingeniously planned vista is gratifying, but a sequence of changing views is even more rewarding. Different views can be devised by varying the size of open spaces, using screens to mask change of use, and adding focal points. Creating viewing positions by placing a seat or orientating a path along a vista will also direct attention. Remember to consider the view looking back from the end of the plot, as well as the main view from the house. Follow the blue walking route through the plan (*below right*) of this long, thin family garden, by Fran Coulter; the numbered viewpoints correspond to the surrounding images and help demonstrate how these ideas work in practice.

1 View from house
This is the most important view in the garden and dictates the layout. The pergola reinforces and frames the view, and the inclusion of a flower-filled container as a focal point in the middle distance draws the eye forward.

KEY
— route through the garden
➤ direction of viewpoint

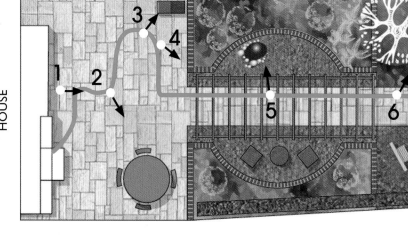

2 Eating outside
The table and chairs are near the house, and are set against a simple green hedge, which creates a comforting sense of seclusion.

3 The tool shed
The slim shed on the patio is both decorative and functional, adding a focal feature to this area of the garden.

4 Looking through planting
From this angle, looking across the planting to the seats beyond, the pergola looks quite different and the garden takes on a more organic, less formal appearance.

5 Water feature
A glance to the side reveals another eye-catching feature. Hostas and grasses frame a discreet, low bubble pool.

FIRST PRINCIPLES

6 Shady corner
Beyond the pergola, the garden is more open and has a different character. This area is hidden from the house and quite shady, providing the owner with an opportunity to use a different range of plants, such as leafy hostas.

7 Relaxing family area
This swing seat is tucked around the corner, just beyond the pergola, and faces towards the brick circle and the shade garden.

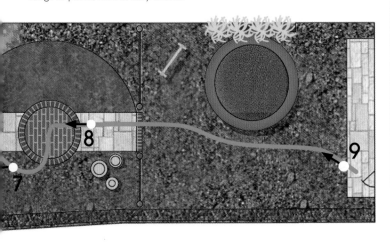

8 Focal point
Circular features break up and soften long, straight lines. The large pot is a focus for this circular space and can be viewed from all sides.

9 Play area
The play area is hidden behind a semi-transparent screen, which separates it, both physically and visually, from the rest of the garden.

Borrowing beautiful views

If you can see the surrounding landscape from your house, try connecting it visually to your own garden. Consider framing a key view, or opening up your garden, using a discreet barrier, such as a low hedge or picket fence, to link it to the wider landscape. Think about the view in different seasons and consider what it will look like in winter when trees and hedges are more open. You may also need to adapt your own garden planting to blend it into the landscape.

Blending in
Here, there is no clear boundary between the garden and the land beyond. One becomes the other, and the garden seems to stretch as far as the horizon.

Framing a view
This "window" to the outside world is focused on a tree-topped hill.

Disguising unattractive views

Not all views are good. Within a garden, especially a small one, there will be areas of utilitarian clutter, such as sheds or household bins, which are not especially attractive and may need screening. Neighbouring houses may overlook the property, spoil the view, and compromise privacy. Tall planting or screens can help to hide eyesores, but if these are not an option, try adding an attractive focal point elsewhere in the garden to distract and lead the eye away.

Covering an old shed
Garden sheds are often unwelcome focal points. This rambling climber is a good summer disguise, less effective in winter.

Screening neighbours
The tall bamboo screen blocks the view to the neighbouring property and provides an attractive backdrop to the planters.

Geometric designs

Small, symmetrical, rectangular-shaped plots, often found in towns and cities, are ideal for geometric layouts, although some large rural gardens are also highly geometric. Most are based on simple combinations of rectangles and squares, with linear elements, such as walls, screens, hedges, and steps used to reinforce the formality of the design.

Descending planes
A progression of levels, low block walls, rectangular beds, strip lighting and matching recliners produces a series of parallel lines, giving this contemporary garden a dynamic feel. The planting is simple, so it does not detract from the strength of the overall design.

Layering shapes

By adding a variety of layers above ground level to offer different views and experiences, gardens can be made more visually exciting and functional. These layers can be set directly above the ground pattern, or angled so that the shapes above eye level have a different, but complementary geometry. Pergolas, clipped-tree canopies, and roof-like structures all offer opportunities to layer your design.

Overlapping layers
The arrangement of elements in this small garden breaks up a dull rectangular plot, and creates different spatial effects.

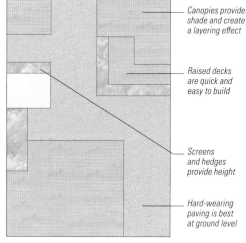

Canopies provide shade and create a layering effect

Raised decks are quick and easy to build

Screens and hedges provide height

Hard-wearing paving is best at ground level

Level changes
To create visual interest, introduce subtle changes of level using a range of different materials, including water.

Circular designs

Layouts based on circles, arcs, and radiating patterns help to create spaces that are full of movement. However, they are difficult to build from hard landscape materials, and getting the geometry wrong will look unattractive. Organic layouts (see pp.160–61) should be considered as an alternative, if this is likely to be a problem.

Formal approach
A central lawn surrounded by a radiating pattern of low beds and clipped hedges combines a sense of order with rhythm and movement.

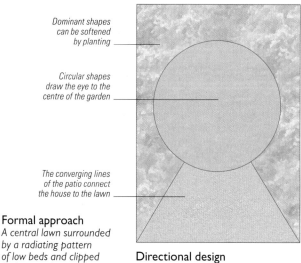

Dominant shapes can be softened by planting

Circular shapes draw the eye to the centre of the garden

The converging lines of the patio connect the house to the lawn

Directional design
This simple design focuses the eye on the centre of the garden. A container or sculpture could be used as a focal point.

FIRST PRINCIPLES

Shapes on a diagonal

A classic design trick for long, linear, and narrow plots is to rotate a rectilinear geometric pattern so that it is orientated along diagonal lines. These layouts on a bias draw your eye down the garden and encourage views to the sides.

Dynamic angles
The diagonal lines of staggered beds, patchwork wooden decking, and a raised pool make a bold statement, and direct visitors through the space.

Twists and turns
A diagonal path with steps traces a zig-zag line through the garden, providing areas to linger and enjoy the wide beds and colourful planting.

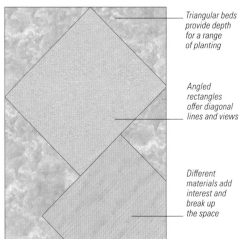

Triangular beds provide depth for a range of planting

Angled rectangles offer diagonal lines and views

Different materials add interest and break up the space

Defining shapes
Here, rectangles of hard landscaping, set side-by-side and edged with planting, make the garden appear wider than it is.

Symmetrical layouts

Throughout the world (except in the Far East), from the middle ages to the early 18th century, gardens were not only geometric, but also symmetrical. Inspired by Islamic and classical designs, they transformed the landscape into a controlled work of art. These formal layouts complemented classical architecture and reinforced the belief that beauty derives from order and simplicity.

Perfect harmony
This sophisticated garden illustrates classical symmetry and demonstrates the importance of proportion and scale.

Contemporary symmetry

Contemporary layouts can adapt classical symmetry to meet the requirements of modern living, such as creating space for outdoor entertaining or for growing herbs and vegetables. Good design also involves an understanding of a wide range of hard landscape materials and the way in which they can be combined to make a simple and elegant framework for the planting.

Cool control
A chequerboard of white paving and emerald grass against a dark hedge offers a modern interpretation of a traditional format.

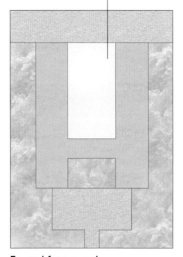

Create a striking central feature to accentuate design symmetry

Formal framework
A combination of rectangles with block planting gives a strong structure that works well in a contemporary setting.

Informal planting

Symmetrical layouts are often less obvious when viewed from eye level, especially when taller plants are used. A variety of forms, textures, and colours will also soften hard lines and sharp edges. The combination of formal design and more relaxed, informal planting is a tried-and-tested formula, but requires skill and discipline if it is to work well. The balancing effect of a restricted colour palette and repeated plants, perhaps mirrored along a path, help to develop and reinforce the symmetrical theme.

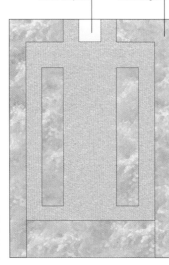

Softened lines
The subtle haze of herbaceous planting spills out from flower beds onto the path and contrasts with the formal garden layout.

Use a focal point to draw the eye to the end of the path

Lush planting can be used to soften edges

Mirror image
In a symmetrical garden, dominant shapes are repeated and guide you through a sequence of harmonious spaces.

Repeated planting
Leading the eye through the garden, this long, airy avenue of grass demonstrates the compositional power of symmetrical planting.

Traditional and formal

Traditionally, it was the symmetrical pattern on the ground, such as a parterre of low hedging laid out around a central axis, that dominated garden layouts. These geometric designs are still popular in vegetable and herb gardens today, where they allow easy access to tend the beds. In the classical gardens of large estates, a sequence of focal points, such as ornamental pools and fountains, dramatic sculptures, or large urns, were added to enhance key points and to make the pattern more interesting from eye level. Nowadays, when many planting styles are used, the geometric approach works best when the overall design can be viewed from a terrace or house above.

Visual journey
Well-positioned focal points, such as this nautilus sculpture, create a strong sense of direction. The domes of box and clipped yew lining the path accentuate this effect.

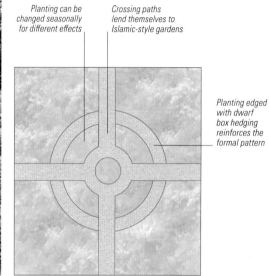

Planting can be changed seasonally for different effects

Crossing paths lend themselves to Islamic-style gardens

Planting edged with dwarf box hedging reinforces the formal pattern

Circles and squares
Reminiscent of a Celtic cross, this layout divides the garden into quadrants with a central focal area, ideal for an ornament.

Permanent patterns
This formal layout of box-edged beds is infilled with spring flowers, which will be replaced as summer approaches.

Organic shapes

Organic shapes and layouts work well in large gardens and are especially suited to rural or semi-rural locations, but they can also work in small spaces. They are characterized by flowing lines, soft curves, sympathetic use of landscaping materials, and relaxed planting schemes. These naturalistic gardens evolve over time as planting matures, blurring the original layout.

Simple curves
Generous curves, wide beds, and the addition of a pinch-point draw the eye around the garden.

Interlocking circles

Developing two areas of the garden, separated by a pinch-point, leads the eye from one space to another, and offers both open and enclosed areas. The organic layout provides a setting where some shrubs and trees can be allowed to grow to their natural size, creating a backdrop for lower plants at the front of the beds. The narrow space between the circular forms can also be used to bring colour and interest into the centre of the design (*right*). This figure-of-eight layout makes the garden appear larger, as all areas are not visible from a single vantage point.

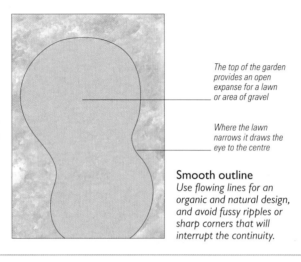

The top of the garden provides an open expanse for a lawn or area of gravel

Where the lawn narrows it draws the eye to the centre

Smooth outline
Use flowing lines for an organic and natural design, and avoid fussy ripples or sharp corners that will interrupt the continuity.

Fluid lines

A simple device to draw the eye along the garden, and to give the illusion of movement and space, is to adopt an S-shaped design. Two circular areas are connected by a single fluid line, which can be developed into a snaking path or a flowing lawn. If used as a path, the spaces at the top and bottom are ideal for planting, a seating area, or an ornamental feature, such as a pool. If these two areas are different in size, the path may be tightly coiled at one point and then more relaxed, providing contrasting experiences.

Serpentine path
A coiling stone path leads through robust planting to a cave-like chamber in this children's play garden.

Curved decking
The sinuous lines of the deck and lawn complement the subtle shades of the surrounding foliage.

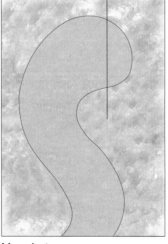

An ideal spot for a pool or feature to be viewed from a winding path

Meandering route
This curvaceous shape provides many different views and vistas as you move through the garden.

Sweeping curves

Curved lines may evolve to avoid an obstacle, such as a tree, pond, or building, or be added to make a path that leads to a particular destination. These are the fluid lines found in the natural world and lend an organic character to shapes and forms. They are frequently used to create calm, relaxing, and unchallenging garden designs.

Bold statement
Curving round a bench, this dynamic feature wall adds colour and momentum to a paved circular terrace.

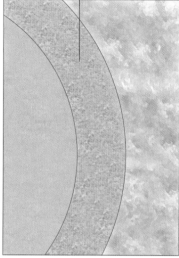

Use gravel or bark for a soft organic look

Gentle arc
Wide curvilinear paths create generous space on either side for deep planting beds or expansive water features.

Continuous journey
This sinuous path guides the visitor between water and soft planting. The view around the curve is partly obscured, which adds a sense of mystery.

Multi-level layouts

Sloping sites provide an opportunity to create beautiful spaces full of movement and drama. Working a plan around the site's natural slope will create a more natural effect, while terraces offer structure and shape for formal and contemporary designs. Drainage is an important consideration, as any changes to slopes will affect the movement of water (see pp.136–37).

Terraced slopes

Terracing makes a dynamic statement and can be used to extend the architecture of buildings into a sloping landscape. Retaining walls and steps are solid, permanent additions and a long-term investment. Measuring and building them are skilled jobs at both the design and construction stages. Wooden decking is a cheaper solution; materials are lighter, but not as long-lasting.

Steep terrace
Tiered wooden sleepers behind a low wall provide perfect conditions for sun-loving plants.

Tree platform
Decked platforms are easier and less costly to build than terraces, which involve major earthworks.

Gentle slopes

Gentle changes of level in a garden offer visual interest and depth to the design. For practical purposes, gardens with only a slight incline can be treated as a flat site. However, if completely level areas are needed, for example, to accommodate a table and chairs, it will be necessary to level the ground and carefully consider the route between changing elevations. A combination of walls, steps, ramps, and terraces can be introduced as required, to suit any design.

Gradual progress
Shallow steps, with space for decorative pots, bridge a small pond and provide an easy route up to the seating area beyond.

FIRST PRINCIPLES

Designing with steps

When building steps, the proportions of the tread (horizontal) and riser (vertical) are both important. Generally, they are more generous outdoors than inside a building, with treads 300–500mm deep (12–20in) and risers 150–200mm high (6–8in). Materials should complement those used elsewhere in the garden, especially adjacent walls.

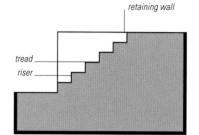

Steep steps
These are a good option if space is limited, or when more drama is required, but they hinder fast movement and can be dangerous, so install a handrail too.

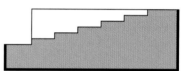

Shallow steps
Although they take up more space, shallow steps allow a relaxed progress through the garden. The depth of the treads also provides space for decorative pots.

Stepped ramp
A stepped ramp is easy to negotiate and, if shallow enough, can accommodate wheeled transport. It can be useful where there is not enough room for a ramp.

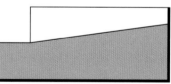

Continuous ramp
Invaluable for wheelchairs, bikes, and so on, ramps also provide a useful route for wheelbarrows. They need seven times more horizontal space than steps.

Natural hillside

The best advice when dealing with a hillside garden is to change a natural slope as little as possible. The soil is likely to be shallow and held together by the existing vegetation. Drainage will be complex and removing the native plant material may result in soil erosion and landslides, as the soil-binding roots are lost. Try to work with the unique contours of the landscape and make small, thoughtful interventions over time rather than significant alterations all at once.

Nature's way
Uneven, weathered stone steps meander romantically up through a secluded and naturalistic woodland setting.

SAFETY ISSUES

Regulations state that any surface higher than 600mm (24in) above surrounding levels must be enclosed by a barrier 900mm (36in) high; railings, walls, or fences are suitable options.

Decorative restraint

Adding a landing
A landing is desirable at the top of a flight of steps, and to provide a resting place every 10 or 11 steps within a long flight. It is also required when there is a change of direction.

Using height and structure

The plants or features that give height and structure to a scheme greatly enhance the way a garden is perceived and used. This is especially true of a straight-sided, horizontal plot, where introducing different heights will create movement and dynamism. There are certain principles to bear in mind, such as the rules of perspective, and it is useful to remember that the closer you are to a structure, the larger it will appear. Use hard landscaping and planting to create the effects you want.

Height levels

It is practical to think about height levels in terms of how they relate to the adult human body, which affects how they are viewed and experienced. Anything below knee height is viewed from above. Waist-high elements are seen at an angle and form a screen, partly blocking views to anything immediately behind them. At shoulder and head height, dense or opaque elements (such as closely planted tall shrubs, hedging, or high screens) will completely block a view. Structures above head height, for example a tree canopy, can create a sense of seclusion as the sky and nearby buildings are obscured. Hard landscaping provides fixed elements but all further interest comes from planting. Indeed, combining plants of different heights is one of the key aspects of a successful garden. Few built elements can compete with a mature tree for interest and drama.

- A see-through trellis distracts the eye from a shed
- The tree lifts the gaze upwards
- A painted, rendered wall forms the boundary
- Low walls double as seating
- The lowest plane is lawn
- Planting is repeated at intervals to provide rhythm
- An outer wall gives a sense of enclosure
- Stones add a change of texture

Varying heights
This multi-level design shows the clever relationship between the fixed height of the parallel low walls, and the natural variations achieved with perennials, grasses, shrubs, and trees.

Height levels explained
This diagram shows the relationship between the human form and height levels within the garden. Planting, hard landscaping, and screens have all been planned to vary viewing angles throughout. The three low walls interrupt the planting but do not obscure the view beyond.

- Planting at waist height is seen at an angle
- Low walling around knee height punctuates the space
- An area laid to lawn creates open space in the scheme
- Paving adds a different texture at ground level
- A see-through screen stands above head height
- The highest element is the rendered wall, creating a backdrop
- Planting breaks up the flat expanse of wall

Above head height
Head height
Waist height
Knee height
Ankle height

FIRST PRINCIPLES

Introducing height

A range of height levels gives variety and interest to a garden, whatever its scale. Elements that create instant height include barriers (walls, fences, screens, or trellis), overhead structures (pergolas, arbours, or canopies), and play equipment, such as a child's swing. Planting options are varied and include trees, many shrubs, bamboos, climbers, hedges, and perennials for seasonal variation. Bear in mind that young trees and shrubs need not be expensive, but take time to gain height. Built structures cost more, but are quickly realized and make permanent features.

Contrasts of height
The stature of these elegant olive trees is given greater emphasis by the low planting below.

Shielding neighbours
A combination of trees and shrubs behind trellis screens provides partial screening and privacy from neighbours. The painted frame adds height and structure to what would otherwise feel like a small space.

Temporary screens

While pergolas and other built structures provide height and solid overhead planes, they need support and can fill small gardens with posts. If uprights would be a problem in your garden, consider suspending temporary canopy screens to create shade and make the garden feel more intimate. Sail-like screens are a good solution and they can be taken down when not required. They need to be attached securely, but can be an excellent way of creating privacy in a small garden.

Nautical screen
A lightweight and elegant sail canopy provides shade, does not clutter the garden with posts, and conveys a feeling of intimacy to small urban gardens.

Using perspective

There are two important principles to consider when using perspective (the way in which objects appear to the eye). The first is that parallel lines in the viewer's sight appear to converge at a point in the distance, known as the "vanishing point". The second is that objects nearer to the viewer appear larger than those further away. A large tree or work of art, for example, may look too dominant placed in the foreground, but in proportion sited farther away. By carefully positioning elements of different heights in the garden, the rules of perspective can be exploited. It is even possible to produce slight optical illusions, for example, by repeating motifs at intervals to make a garden look longer.

Enticing vista
Having a focal point at the end of a line of sight is a great way to create perspective. The glimpse of a sculpture and flowering plants in front tempt the visitor to walk over lush green grass down this tree-entwined tunnel.

Transparent screens

Trellis, glass, and other transparent and semi-transparent screens help to separate garden spaces without diminishing light. They are useful in smaller plots, where they allow visual connections to be made, while breaking up the space into different areas, and adding a change of mood. Transparent screens also make attractive features in their own right.

Versatile trellis
The open latticework of trellis associates well with plants and climbers and may be left open or screened with evergreens.

Glass panels
This patterned glass panel allows light through but slightly obscures the visual connection to the next area of the garden.

Choosing structural elements

Boundaries are the frame within which your garden sits and form the backdrop to the space, especially in a newly planted garden. Screens allow you to divide the garden into smaller areas, and come in a variety of forms and materials, while some garden structures may even be works of art in themselves.

Boundary options

The main boundary choices are walls, fences, or hedges. Walls are an investment, making a permanent addition to the property, and can connect garden and house visually. Fences are cheaper but shorter-lived, so bear in mind that they will need replacing in time. Hedges take time to grow, and need clipping, but form a soft, natural boundary.

Wooden screen
A trellis clad in clematis makes an inexpensive decorative screen.

Mixed materials
Panels of concrete, painted timber, and a planted living wall create striking textural contrasts.

Bright squares
The mix of brightly coloured opaque and transparent screens makes a bold statement.

Green colonnade
An interesting alternative to a traditional continuous hedge, these tall clipped conifers form a strong background feature.

Internal screens

Adding screens and panels within the garden divides it into smaller, more intimate spaces. They are especially useful in predictable rectilinear plots where they can add interest and heighten mystery. Panels below waist height allow views across the garden, taller screens separate different areas, and gaps allow tempting glimpses of the garden beyond. Consider the effect of opaque and transparent screens and introduce colours and textures to add visual contrasts. Supports and other frameworks should form an important part of the design and, if well planned, will help to reinforce the overall composition.

FIRST PRINCIPLES

Using natural forms

Structural elements can be introduced using planting alone. A range of trees and shrubs can be trained to form hedges and screens with great results. Patience is needed while slower-growing plants mature, but this is a rewarding process. Natural forms suit traditional gardens, but are not out of place in a modern design, where clipped shapes, such as "lollipop" trees and sculptural plants like bamboos, add spheres or lines to a design. Accentuate the vertical lines of small trees by placing low-growing plants at the base.

Bamboo screen
This bold planting of tall Phyllostachys sulphurea f. viridis is reflected in the pool in front.

Clipped trees
Here clipped "lollipop" bay trees emerge from box-framed lavender beds, demarcating the dining area. The slate terrace lends textural contrast.

Sculptural structures

Screens and garden dividers of all kinds can be decorative in their own right and, equally, a work of art can play a dual role and have a structural function in a garden. By introducing a strikingly different material, such as glass or metal, into a design filled with plants, you can add exciting accents and heighten the drama. Glass may be frosted or clear, printed with patterns, or moulded in different ways, although even toughened glass may not suit a family garden. Metal adds gleam and reflection to an otherwise matt series of surfaces. Site sculptural structures where they can be fully appreciated.

The path ahead
This unusual elliptical, wire mesh tunnel, a work of art in itself, invites use and functions as both a screen and a walkway.

Frosty looks
The image printed on the transparent and frosted screen acts as additional "planting". Both the screen and the seat appear to float within the garden.

Introducing colour

Colour is a powerful tool in garden design, influencing our senses and the way in which we respond to the environment around us. Colours can also convey an atmosphere, mood, or message: warm, vibrant colours generate a feeling of immediacy, liveliness, and excitement, while cool colours create a calm, spacious, often tranquil atmosphere.

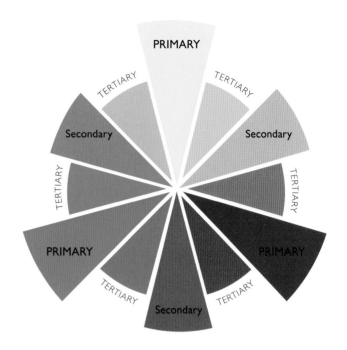

Colour wheel

The language of colour is best understood using a colour wheel – a device employed by many artists and designers to explore the visual relationships between colours and the effects different ones can create when placed together. In particular, it helps us to see why some combinations work better than others, and why one colour can dramatically influence another to produce a startling contrast or confer harmonious continuity.

Primary colours
Red, blue, and yellow, the largest slices of colour on the wheel above, are primary colours, from which all other colours derive. These three hues cannot be mixed or formed by combining other colours.

Secondary colours
Two adjacent primaries will create a secondary colour when mixed together. These secondary hues are green, orange, and purple.

Tertiary colours
These are made by mixing adjacent primary and secondary colours in different quantities, until the wheel becomes a circular rainbow.

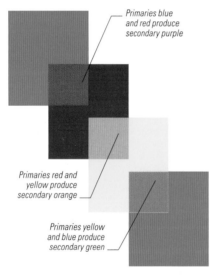

Primaries blue and red produce secondary purple

Primaries red and yellow produce secondary orange

Primaries yellow and blue produce secondary green

Hues, tints, shades, and tones
The true colours or "hues" are in the third ring of this wheel. The two central rings are light "tints", which are mixed with white. The outer rings show how adding black makes darker "shades". If grey were added, it would make a "tone".

Add black to create a shade

Add white to create a tint

True colour or hue

Introducing colour in the garden

Planting combinations
Creating a variety of colour combinations with plants and flowers is exciting. You can alter the palette to produce changing colours for each season.

Hard landscaping
When nothing is in flower, hard landscaping can provide colour and interest. The effect is consistent, although weather conditions may affect the colours.

Paint
Earthy tones, derived from natural pigments, work well in more natural contexts, while bright, bold colours create a feeling of energy, excitement, and optimism.

Combining colours successfully

The opportunity to combine different tints and shades of various colours makes garden design an exciting challenge; using a colour wheel can help our understanding of which combinations create the best effects. The key concept involves working with harmony and contrast to develop a visual experience to engage the viewer. Those colours allocated the most space in your design will become dominant.

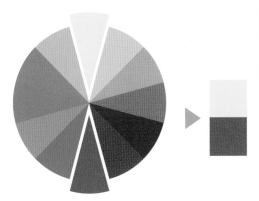

Opposite colours
Two colours from opposite sides of the wheel are considered to be complementary, for example, yellow and purple, and red and green. The high contrast of these colours creates a vibrant look, but they can cause eye strain, too, and should be used sparingly.

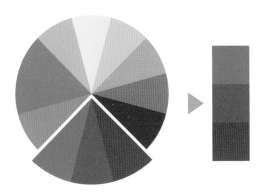

Adjoining colours
Harmonious colours, selected from adjoining hues (also called analogous colours) match well, are pleasing to the eye, and create a sense of order. Choose one colour to dominate, and others to support it. Adjoining colour groups create a "warming" or "cooling" effect.

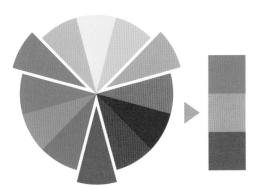

Triadic colours
Selecting three colours that are evenly spaced around the wheel can instil a sense of vibrancy. This works best with flower and foliage colour rather than with hard landscaping materials, where triadic combinations can be overdone and appear chaotic.

Colour effects

In a garden, colour is never perceived in isolation and should always be considered as part of an overall design composition that includes form, line, texture, and scale. Other elements, such as the intensity of sunlight and shadow, can also influence how colours are seen in an outdoor space. It is important to understand how and where to use different colours in your design to achieve the best effects.

Colour influence

You can use colour to attract attention to a particular feature or area; the more an object contrasts with its surroundings, the more visible it becomes. Hues (saturated colours) are dominant and offer the highest level of contrast when placed together. Darker shades or lighter tints contrast less, although small areas of light against dark, or vice versa, can create an accent. Recessive colours, like cool blue or green, give the illusion of distance.

Shorten a view
A dominant colour (red) placed behind a recessive colour (green) will bring the background forward. This is particularly effective if they are complementary.

Lengthen a view
If a dominant colour (purple) is in the foreground with a harmonious recessive backdrop (green), the garden appears longer.

Outline
Without colour, the outline of this tree doesn't stand out from the background.

Red on green
When red is placed on its complementary, green, the colours appear to "vibrate".

Green on red
The intensity is the same, but as red is dominant, the green tree is less clear.

Warm contrasts
This group of yellow flowers is highlighted against the dull red brick wall. The drift of mauve flowers in the distance contrasts with the dark woods behind and the lighter green field.

Bright white
While purple and green are closely related on the colour wheel, adding white creates a stronger composition. As pure white reflects the most light, these pots stand out against the purple wall.

Creating highlights

You can achieve some bold effects in a garden using colour highlights. Try contrasting one hue against another, or combine adjoining hues in close proximity (see p.169). For example, plants with complementary colours (red and green, purple and yellow) will intensify the brightness of each other when placed together, while plants with hues that are close to each other on the colour wheel (see p.169) (purple, red, and pink) blend to form a harmonious effect. The introduction of a single, intensely coloured plant against a recessive background (such as green or blue) will make the bright plant stand out. Combinations of warm and cool colours can also result in eye-catching compositions that highlight the more dominant colour. (Note that white may appear recessive or dominant depending on the light.)

FIRST PRINCIPLES

Colour boosting sunlight
The strong sunlight has a brightening effect on the yellow wall, and on the sizzling intensity of the red flowers in pots and on the hedge in the background.

Nature's neutral colours
Beautiful effects can be achieved by combining a variety of soothing greys, blues, and greens with light-catching whites and yellows, which brighten up a shaded area.

Light and shade

Responding to colour is a sensory reaction, like smell and taste, and the way in which our eyes read a colour is dependent upon the amount, and intensity, of light that is reflected from that colour. Sunny areas make colours appear bolder and more concentrated, while shaded areas reflect more muted hues. This means that flat areas of colour – for example, a painted wall – may look quite different depending upon their aspect and orientation. Similarly, the hues of flowers and leaves will change depending on their location, the degree of shade cast on them, and the time of day.

THE PROPERTIES OF COLOUR

Warm colours (reds, yellows, and oranges) can make spaces appear smaller and intimate. Cool colours (blues, whites) make areas look larger and more open. Green is a neutral colour.

REDS
Reds and oranges suggest excitement, warmth, passion, energy, and vitality. They stand out against neutral greens, and work best in sunny sites but, if over-used, can be oppressive.

YELLOWS
Yellows are sunny and cheerful. Most are warm and associate well with reds and oranges. Greenish-yellows are cooler and suit more delicate combinations.

BLUES
Deep blues can appear very intense, lighter blues more airy. Blues suggest peace, serenity, and coolness. Purples carry some of the characteristics of both reds and blues.

GREENS
The most common colour in the plant kingdom, green comes in many variations, ranging from cool blue-green to warm yellow-green. Greens suggest calm, fertility, and freshness.

WHITES
White is common in nature. It is a combination of all other reflected colours, and suggests purity and harmony. White spaces seem spacious; the downside is they can feel stark.

BLACKS/GREYS
Blacks and greys are the absence of colour, when light rays are absorbed and none are reflected back. Black is glamorous when used sparingly, but depressing when extended over large areas.

Tints, shades, and tones

A general guideline to remember is that pure hues or saturated colours are more intense, while colours that have been mixed together are less vibrant. Black and grey are rare in nature, but they do exist in the form of shadows. A tinted colour, which has been "diluted" with white, will be lightened and appear more airy and farther away. A shaded colour, which has been "diluted" with black, will appear to be nearer. Tones mainly occur when a colour is cast into shade. However, the quality of light in a garden, such as on a bright sunny terrace or in a shady border at twilight, will affect the way that colours are perceived.

Tints
Hue + white = tint. The more white added, the lighter the colour. Tints recede, but pure white may advance.

Shades
Hue + black = shade. Darker shades advance. They are warmer and appear closer than pale tints.

Tones
Hue + grey = tone. Seen mainly in shadows, tones are less intense and appear muted.

Applying colour

We tend to be more adventurous with colour in the garden than we are in our homes, perhaps because the outdoor environment feels brighter and less confined. The neutral greens of foliage and blues and greys of the sky also have a softening effect on more strident or clashing colours.

Vibrant colours

Strong colours can be used to dramatic effect in the garden: as bright pinpoints that energize more subtle plantings, or surprise pockets of colour separated by greenery. In a flower border you can build up from quieter blues and purples to crescendos of fiery reds and oranges. These hot colours will stand out all the more by combining them with a scattering of lime-green, dark bronze, and purple foliage.

Radiant hues
Use glowing flower shades for hot, sunny aspects where the colours will really sizzle in the light.

Hot seats
The colours used in this seating area create an upbeat atmosphere – the ideal setting for stimulating lively conversation.

Relaxing colours

The muted greys, purples, and blue-greens typical of Mediterranean herb gardens create a restrained atmosphere, perfect for a contemplative retreat. Plantings that pick up the heathery colours of distant hills make a space appear larger. However, a calming palette doesn't have to be muted; it can also include fresh greens and pastels, which will work well in most settings.

Refreshment
Fresh white, lemon, and green combine with a brighter pink to create an uplifting but essentially restful planting. Perfect for an intimate seating area tucked somewhere away from the house.

Country calm
The lavender and purple sage add to the serene colour palette of this formal garden with a Lutyens-style seat.

FIRST PRINCIPLES

Neutral colours

Earthy browns and biscuit tones are reminiscent of harvest time and appear warm and nurturing, contributing to a calm, relaxed atmosphere. Weathered wood elements are perfect for gardens with a country look. In urban locations, you can feel closer to nature by utilizing reclaimed timbers, wicker, and bamboo for screens, raised beds, and furniture. For flooring, consider sandstone paving, decking, or a shingle beach effect with pebbles.

Rustic simplicity
Basket-weave stools and a table made from a tree trunk blend seamlessly with a rustic-style garden.

Muted tones
As they die back, perennials and grasses continue to inspire, creating winter interest and a harmonious palette of browns.

Nature room
Blocks of wood provide a muted backdrop for birches and the intermingling greens of the grasses and foliage plants.

Monochrome colours

Hard and soft landscaping in a restrained palette of black, grey, and white, with the addition of green foliage, produce refined, elegant designs. The approach is perfect for smart period gardens with a formal layout. White blooms and silver foliage also work well with metallics in a chic city courtyard. Use cream or white flowers to enliven shade, and combine with variegated and lime-green leaves.

Spring whites
This elegant scheme comprises white forget-me-nots, tulips, daisies, and honesty with hostas and silver astelia foliage.

ARTIFICIAL COLOUR

Colours that are rarely seen in nature tend to be the most attention grabbing. Day-Glo coloured materials and lighting give a space a more futuristic or avant-garde look. You can include these colours with furnishing fabrics, Perspex screens, and LED lights.

Day-Glo colours
Bold, cartoonish colours such as bubblegum pink, lime green, orange, and turquoise are so vivid they seem to glow. Attention grabbing but use sparingly.

Painting with light
LED lighting is available in any colour and can also be programmed to create a sequence of changing hues to produce spectacular effects in the garden.

Black diamonds
Flanked by crisp green woodruff and a low clipped box hedge, this stylish grey and cream gravel pathway, with a black pebble mosaic, makes an eye-catching focus for the small front garden of a town house.

Integrating texture into a design

It is easy to be seduced by colour when selecting plants and materials for the garden, but form and texture are equally important. Whether the design is a success or not depends on how well you combine the various shapes and textures, not only on a large scale but also at a more detailed level. To emphasize the contrasts, try to visualize in monochrome the hard and soft landscaping elements you are considering using. Also pay particular attention to how light affects different forms.

Types of texture

Experiencing different textures in the garden is a crucial part of our sensual enjoyment of the space. You can often tell what something is going to feel like just by looking at it, but there may be more surprises in store as you explore. Certain forms and surfaces invite touch and the visual and physical effect is heightened when there is great textural contrast. There are a number of basic categories describing texture, some of which relate to how something feels and others to how light affects a material's appearance.

Rough
For rough textures choose stone chippings, dry stone walls, wattle hurdles, peeling tree bark, or prickly plants.

Smooth
Choose flat or rounded surfaces like concrete cubes and spheres, plain pots, smooth bark, and water-worn cobbles.

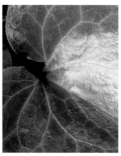

Gloss
Shiny, mirrored surfaces include many evergreens, polished granite, stainless steel, chrome, still water, and glazed ceramic.

Matt
Ideal for combining with glossy elements, matt surfaces include cut timbers, galvanized metal planters, and sandstone.

Soft
Impossible to ignore, soft, felted, furry-leaved plants are irresistible to the touch, as are fluffy seedheads and grass-like stems.

Hard
Non-pliable solid surfaces can be matt or gloss: cast metal, stone and concrete walling, flint, granite setts, and terrazzo pots.

Combining textures

To introduce a variety of textures, combine plain with patterned surfaces, shiny with matt, smooth with rough, and so on, but don't overdo the number of materials or the garden could end up looking too busy. Accentuate the contrast between two elements by making the difference marked. Pair strongly vertical plants with horizontal decking, for example, or a glittering, stainless steel water feature with matt-textured ferns and hostas.

Textural contrasts right
Combinations of textures create the visual excitement in this harmonious design. Horizontal lines on the planter echo the lines of irregular stones bedded in concrete, while the rill provides a glittering contrast.

Rough with smooth
This walled courtyard marries gravel and rough-cut stone with smooth spheres to dramatic effect. The dry stone water feature cuts the sheer rendered wall in half.

Gloss with matt
Shiny glass and metal doors echo the visual qualities of the swimming pool. These elements are separated by the smooth paved terrace and matt rendered wall.

Soft with hard
The wooden walkway, circular terrace, and snaking wall are perfectly opposed by luxuriant "soft" plantings of hostas, irises, grasses, and marginals.

Small garden solutions

Some outdoor gardens may be short on space, but they need not be low on interest, especially if you can be inventive and have an eye for detail. Understand the potential of your space, and maximize opportunities with neat plant selection, clever design, and space-saving techniques. Remember that flexibility is key to keep interest going year-round. Container growing allows plants to be moved, while custom-made storage solutions make use of "dead" space. Put walls and roofs into action and integrate practical needs with aesthetic ones.

Flexible containers

Containers allow displays to be easily changed with the seasons for impressive combinations year-round. Plants with seasonal interest can be placed centre stage, then moved discreetly out of sight once displays have faded. Growing in containers allows a wide range of plants to be kept. You can plant singly or with big pots, or treat them like a pocket border and combine your plants. Remember to use as large a container as is possible – they can make even a small space look bigger and allow for more root zone for your chosen plants.

Creative space above right
Guttering on a wall can be an inventive way of growing salads in a small space.

Moveable feast right
A mix of perennials and a multi-stemmed Rhus tree provide structural impact in these wide pots.

Clipped accents
Pleached hornbeam add height and refined formality in a raised bed, and echo the form of potted standard bay trees in their sleek containers.

Multi-stem marvel
The sinuous trunks of an Amelanchier add sculptural interest, while the plant's canopy casts cool shade for ferns and hostas to thrive.

Room for a tree

Although it may seem counterproductive, using large plants in a small space can be a great way of maximizing interest. Even trees for containers, if carefully selected, can often be included; their canopies tend to lead the eye skywards, providing an impression of space, while dappled shade cast below may allow a host of other plants to thrive. The best ones to pick may be those of limited vigour or those that can be regularly trimmed to keep within bounds. Large specimen shrubs that have had lower branches removed to provide an interesting tree-like form are also great choices. Look out for trees with a long season of interest, perhaps with spring or summer flowers, good autumn leaf tints, and an attractive trunk for winter appeal. Deciduous trees with light canopies often make the best option as they provide summer shade but will not block precious winter light for dwellings nearby. Some trees will even thrive in large containers for several years.

Shedloads of space

Most homes need some outside storage space, but when every square metre is at a premium, a full-sized shed may simply not be an option. Finding a secure place for bicycles is a particular problem in many smaller properties, as is storing wheelie bins out of sight. Purpose-made storage may be the best option. It allows small, unused areas or awkward corners to be put to good use and converted into bike storage, a recycling area, or log store, or somewhere to put garden furniture during winter. Adding a green roof, a green wall, or simply attaching trellis for climbers will help integrate it into the garden and make it an attractive feature, even when it can be seen from the house.

Bike storage solution above right
A purpose-built bicycle shed solves a storage issue; topped with a green roof, it also becomes an attractive garden feature.

Bin today, gone tomorrow right
Wheelie bins are an unsightly modern necessity, but neat storage, also topped with a green roof, hides them away and prevents them blowing over in high winds.

Growing up the wall

With a little ingenuity, walls and sturdy fences can be easily transformed into beautiful and highly productive areas. Containers can be attached and filled with cascading flowers or vegetables. Larger floor-standing pots are best positioned at the base and planted with climbers such as clematis or, in a productive garden, outdoor cucumbers, runner beans, and squash. Trellis or wires will allow these quick-growing plants to soon green up the space. Alternatively, fit outdoor shelves or attach brackets and add hanging baskets. These spaces can make use of recycled material and may be ideal short-term solutions to growing, perhaps in rented properties with little outdoor space.

Ordered production
Recycled plastic water bottles and tin cans can be easily pressed into action as containers, planted with herbs, vegetables, and flowers, and hung on a wall or grown on makeshift shelves.

Water butts in disguise

With increasing concerns about sustainability, and with most new and many older homes on water meters, harvesting and storing rainwater has become an important garden element. You can easily collect rainwater in a single water butt or connected butts, either from water diverted from house guttering or the run-off from greenhouse and shed roofs. If the space that often unsightly water storage takes up is an issue, butts can be hidden by planting or put behind trellis, although increasingly, more attractive ready-made options are available.

Staggered storage top
Far from being an eye-sore, these attractive water butts with integrated planting fill a corner with flowers.

Wood-clad reservoir above
Fitting neatly into a compact space, this water butt could be mistaken for a contemporary raised planter.

Softening hard landscapes

In many garden styles, planting is used to reduce the visual impact of hard landscaping and boundaries, breaking up expanses of wall or paving and bringing a softer, more relaxed feel to a space. Climbers scramble up walls and fences clad with trellis or wires, while more contemporary gardens may feature a green wall.

Planting up your walls

Green walls offer an exciting way to grow a range of plants, even if you have little outside space. In the wild, many plant communities grow vertically on cliffs or even from mossy tree trunks, and these (domestic) plantings emulate those natural effects. Various growing systems are available, the most-simple consisting of fabric planting pockets that contain soil; the most successful are likely to include watering systems, ensuring all plants get a regular soaking. Green walls add a sustainable touch to gardens that fits in with many garden styles, and can be planted with succulents in sunny sites or with shade lovers in sheltered corners. They are more than mere window dressing: in urban areas, they have been proven to have beneficial effects in cooling summer temperatures.

Water and plant cascade top right
A central waterfall splashes down into a verdant clump of Zantedeschia, *while leafy perennials carpet walls.*

Edible artwork above
An expanse of timber cladding is relieved by a green wall with edible planting, including alpine strawberries, nasturtiums, and tumbling rosemary.

Exotic approach right
For a sheltered corner or even under glass, ferns, Fatsia, *begonias,* Melianthus, *and* Schefflera *combine in a dramatic, foliage-led green wall.*

Blurring the edges

Many perennials, low shrubs, herbs, and alpines naturally develop a low, gently spreading habit. These plants are ideal positioned towards the front of beds and borders, where they can mask the boundary between paving and the planted area. This soft, pleasing effect is particularly desirable in certain styles (such as cottage gardens), where the key to achieving the look is to provide informal billowing planting within a formal layout. Natural-style gardens, too, will employ the same trick, helping to make gardens feel more like a natural plant community. Occasionally, plants such as thyme or aubretia may be deliberately planted in planting pockets set within brick or crazy-paved paths for an even more informal effect. There are plants that will spill helpfully over paving in sun or shade; the key is to choose something not too vigorous that will soften the edge of the path and not completely obscure it, and that will stand trimming or an annual prune.

Just in thyme above
Perfect for a sunny, well-drained path or terrace edge, thymes flower well in summer and provide fragrance when stepped on.

Wildflower fringe right
The exuberance of wildflowers, including the white heads of yarrow, is contained yet softened by a brick path edge.

Domed divide below
Repeated along a border edge, evergreen Hebe 'Emerald Gem' gently softens a margin of granite sets.

Social climbers

Climbers and wall shrubs are essential plants to mask and soften boundaries, enhance a feel of privacy, or break up stretches of fencing. Climbers such as wisteria or passion flowers have twining stems or tendrils to scale trellis or wires attached to walls and fencing, while others such as ivy are self-clinging. Wall shrubs like pyracantha or ceanothus will need attaching to their support. Use these plants to hide unsightly features, allow them to cascade down from walls or old trees, or let them scramble through other shrubs for a second hit of flowers and foliage. Climbing roses are often teamed with clematis or honeysuckle. Use quick-growing annual climbers such as *Ipomoea* or sweet peas for summer flowers.

Purple perfection above
The twining stems of Ipomoea *soon scale a fence or trellis, softening without overpowering the structure.*

Traditional favourite below
Climbing roses are among the most reliable flowering climbers, ideal attached to trellis or tied on to wires.

Creating a plan

Drawing up accurate site and planting plans is a crucial stage of any garden design. By bringing all your ideas together on paper, you can see if they are viable within the space available and get a clear visual image of what you want to achieve. Detailed plans also help prevent any costly mistakes before you buy materials and plants or employ contractors.

With a few basic tools, and an assistant to help take measurements, you can draw up a site plan yourself. The process is explained over the next few pages, and includes a few tricks of the trade to make it easier. There is also a variety of computer software packages available for this purpose. However, if you have a difficult site or the prospect of drawing a plan is too daunting, you may prefer to employ a surveyor to help you.

When the site plan is complete, you can start to play around with different design options. Even if you have an idea of the basic shapes you intend to use, it is always interesting to see how redirecting a sightline or introducing a small grove of trees or a collection of containers would change the mood of the garden.

A separate planting plan is also a good idea. Apart from helping you to assess the number of plants needed for your scheme, it will also clarify whether they work well in the overall design and fulfil their intended function. For example, you can use your plan to design a herbaceous bed in a sunny corner, or mark out an area for plants with winter interest that can be seen easily from the house.

Above all, study your plot from all angles and vantage points before you begin. Get to know your soil type and the path of the sun, then relax and enjoy this part of the creative process.

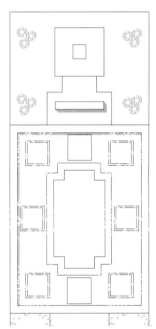

Top and above
A detailed plan, drawn to scale, brings ideas to life.
Plan planting carefully so your schemes work as intended.

182 HOW TO DESIGN

Understanding plans

A plan is a two-dimensional representation of a three-dimensional garden and provides a useful thinking tool. It allows you to develop and share ideas easily with others about how your space can be organized and where various elements should be located. You can produce a simple sketch or a more detailed scale plan to illustrate your design; the plans shown here explain the different types and how to use them.

Working plans

These plans don't need to be accurate or drawn to scale, but they can be used to experiment with ideas, especially the relationship of horizontal surfaces (built and planted) with the locations of walls, screens, trees, and other main features. They can also include connecting elements such as paths and views.

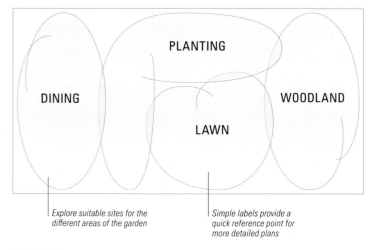

The finished garden
Sara Jane Rothwell, owner of the design practice London Garden Designer, produced both an overhead and a planting plan (opposite, top, and middle) to show clients the new design.

Explore how best to create perspective by siting elements such as trees

Think about whether you want to replace existing elements, like this fence

Consider whether vertical features, such as a wall and steps, will work well

Overlaid photos
Perspective drawings are difficult to master, so cover a photo of your garden with tracing paper and sketch ideas on top to give a three-dimensional view of the changes.

DINING — PLANTING — WOODLAND — LAWN

Explore suitable sites for the different areas of the garden

Simple labels provide a quick reference point for more detailed plans

Bubble diagram
A basic bubble diagram helps you explore relationships between areas within the garden. It is an ideal way to experiment quickly before drawing a more detailed plan.

Garden plan symbols

These common symbols for plans form a visual design language that enables builders and other professionals working in your garden to read the plan quickly and understand what is being proposed. The symbols illustrated here are those that are most often used and most widely understood, and can be reproduced in black and white or colour.

WATER

Still water — Fountain — Water around rocks

PLANTING

Existing tree — New tree — Conifer

Wall shrub

Bulbs

Climber — Perennials

Shrubs — Hedge

LANDSCAPING

Brick – basketweave — Brick – herringbone — Uniform paving — Square-cut stone

Brick – stretcher bond — Decking — Granite setts — Random-cut stone

Cobbles or pebbles — Gravel — Rough grass — Mown grass

Finished plans

Plans that have been drawn to scale and show accurate arrangements, locations, and dimensions of proposed structural elements, planting, and features are known as finished plans (see pp.184–91 for detailed advice on how to draw a plan). These plans are intended mainly for construction purposes and will need to be read and understood by builders or contractors who use them to measure areas and lengths (for costing purposes), and to identify exact locations on the ground. Changing ground levels are shown as separate cross-sections, or by annotating the change of level on the overhead plan.

OVERHEAD PLAN

An overhead plan should show the correct sizes and locations of all proposed elements, such as horizontal surfaces, areas of planting (topsoil), locations and alignments of linear elements (walls, fences, screens, hedges), and singular components (trees, specimen shrubs, pools, stepping stones, steps, lights, drainage points, and so on).

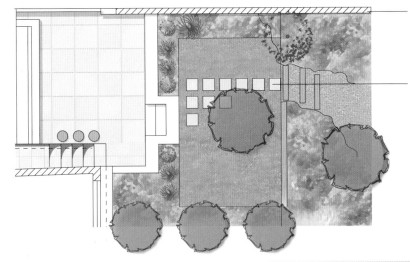

Include the site boundaries and any relevant buildings, doors, and windows on your plan

An overhead plan needs to include the correct materials and measurements of all hard landscaping features

Adding the details
In small-scale overhead plans, the individual materials can be shown; larger scale plans usually illustrate these materials more symbolically (see also p.188).

PLANTING PLAN

A planting plan is important for calculating the correct number of plants in the garden and identifying their exact locations. It also shows the position of larger specimens, as well as groups or drifts of the same species. This plan is most useful, and needs to be most accurate, when planting is being carried out by a contractor without the designer present. If you are doing the planting, a plan can help you accurately calculate the number of plants you'll need and show how to set them out prior to planting (see pp.192–97 for more on creating a planting plan).

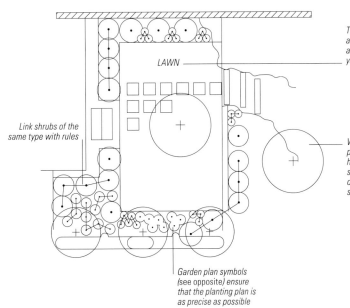

There is no symbol for a lawn, so label the areas on your plan that you want to be turfed

Link shrubs of the same type with rules

When including new plants and trees, check how far they are likely to spread and indicate this on your plan, so you can space them out accurately

Garden plan symbols (see opposite) ensure that the planting plan is as precise as possible

Drawing up a planting plan
Garden plan symbols can be reproduced by hand or by using design software (see also p.191). If you are less experienced in reading planting plans, you may prefer to reproduce these symbols in colour.

CROSS-SECTION

If you have a sloping garden and want to make changes to it, you may need a plan to show the impact of these alterations. For steeply sloping gardens, employ a land surveyor to draw a cross-section, or elevation plan. This will show the significant levels before and after any changes. More complex slopes may need additional plans.

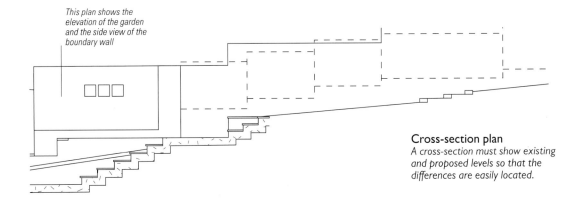

This plan shows the elevation of the garden and the side view of the boundary wall

Cross-section plan
A cross-section must show existing and proposed levels so that the differences are easily located.

Creating a site plan

Now that you have mastered the basic theories of garden design, it is time to put your ideas on paper. There are several different types of plan (see *pp.182–83*), but before creating your final design, you need to draw up a site plan, which shows all the basic measurements in your garden, as well as the position, shape, and size of elements that you intend to keep. You can then use this plan to develop new layouts and planting designs.

Measuring up
Use the right equipment to ensure measurements are accurate. Get it wrong at this stage and your site plan could be rendered useless.

Getting started

The idea of creating a site plan can be a bit daunting if you haven't put one together before, but most plans are easy to produce, especially if you have a small- to medium-size, fairly regularly shaped garden with straightforward topography. However, if you have a large, irregularly shaped or hilly plot, or even one that is very overgrown, it may be wise to employ a land surveyor (see *opposite*).

When drawing up a site plan for your plot, first take a pencil and sketch pad (A4 or A3 are best) out into the garden and study the boundary and position of any elements you plan to keep, such as outbuildings, hard landscaping, and planting. It is also important to take note of the position of your house, including the doors and windows – not only because their location will directly affect your ideas and design, but also because your house is one of the best points from which to measure other features, such as trees, sheds, and so on.

Now, roughly sketch the outline of the garden and the position of the relevant elements within it. Refine your sketch until it is clear enough to mark up with measurements. Then start measuring up (see *below and pp.186–87*). Even if you are only planning minimal changes to your plot, it is worth taking a few basic measurements, such as the length and width of the boundaries, to give you a sense of scale for new features, such as flowerbeds or a water feature. Whatever the size and shape of your garden, you will also find it easier with the help of a family member, friend, or neighbour. Take measurements in centimetres, rather than feet and inches, as the metric system makes it simpler to convert sizes to create a scale plan (see *p.188*).

ESSENTIAL EQUIPMENT

To measure up accurately, you need the right equipment; most items are available from DIY stores and art suppliers. You can use a digital laser measure instead of tapes.

- Spirit level
- Tape measures of varying lengths – e.g. small, medium, and extra-long – or digital laser measure
- Pegs and string
- Sketch pad

Measuring a rectangular-shaped plot

Rectangular and square gardens are the easiest to measure. Ask your assistant to help you measure all four sides of the garden with a long tape measure and add the measurements to the corresponding boundaries on your sketch. Then measure the length of the garden's two diagonals and mark them up on your sketch too. To ascertain the position of features, measure at right angles to the house the distance to the feature/plant you want to keep. Do the same from a boundary, as shown below.

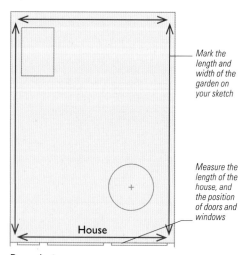

Mark the length and width of the garden on your sketch

Measure the length of the house, and the position of doors and windows

Boundaries
Carefully measure all four sides of your plot. Also measure the house and the distance from the house to the boundary.

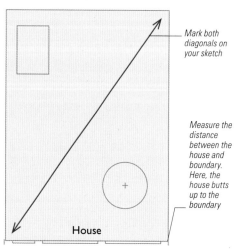

Mark both diagonals on your sketch

Measure the distance between the house and boundary. Here, the house butts up to the boundary

Diagonals
Diagonal measurements help to create an accurate plan of the plot if it is not a perfect square or rectangle.

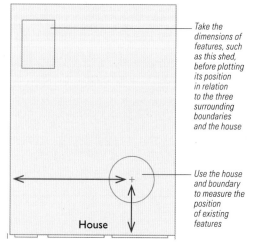

Take the dimensions of features, such as this shed, before plotting its position in relation to the three surrounding boundaries and the house

Use the house and boundary to measure the position of existing features

Features
Plot the position of features that you plan to keep by taking measurements at 90° from the house and boundary.

Site plans for rectangular plots

When you have decided which scale you are going to use, convert your measurements accordingly (see p.188). For large- or medium-sized plots, you may want to create more than one plan for different areas, or use different scales to focus on a planting bed or similar feature that requires more detail. When drawing up your plan, use an A3 pad of graph or squared paper; you can use plain paper and a set square, but it is more difficult and the results may not be as accurate. Then, using a sharp pencil and ruler, plot the measurements on the paper and draw out your scale plan. You can then go over the pencil lines in pen.

You will need
- Metric, A3, squared or graph paper or plain paper
- Set square
- Scale rule and/or clear ruler
- Pencil and pens
- Rubber

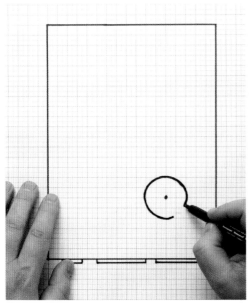

1 Start in the bottom left-hand corner of your page. Draw the wall or walls of your house – including the positions and dimensions of the doors and windows.

2 To draw in the boundaries, mark the length and width on the plan and add the diagonals. Diagonals show if the plot is a perfect square or rectangle, or slightly off.

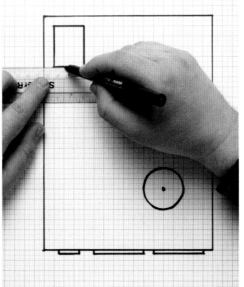

3 Use the measurements you took from the house and the boundaries with a try-square to add trees and major planting – don't forget to include their canopies.

4 Lastly, plot all other features on your site plan. Carefully draw on sheds, greenhouses, patios, pools, paths, and outbuildings, if you are planning to keep them.

Measuring gradients

This method is only suitable for small inclines. It is useful if you want a couple of steps or terraced flowerbed and need to calculate the required heights. For more complex works or difficult sites, employ a land surveyor.

You will need
- 1 length of wood just over 1m (3ft) long
- Spirit level and tape measure
- 2 or 3 wooden pegs

Lay the wood from ground level to the top of the peg
Use a spirit level to ensure that the wood is exactly level
Measure the height of each upright peg from ground level

20cm / 1m (3ft)
50cm / 1m (3ft)
35cm / 1m (3ft)

1 From a specified point on the slope, measure 1m (3ft) down the hill and hammer in a peg. Check it is vertical using a spirit level.

2 Lay the wood from the soil surface at your original point to the top of the peg and use a spirit level to check it is horizontal. Measure the height of the peg.

3 Then, 1m (3ft) further down the slope, hammer in a second peg, as before. Lay the wood from the bottom of the first peg to the top of the second.

4 Measure the height of the second peg. Repeat these steps as necessary until you reach the bottom of the slope. Next, calculate the "fall" or drop.

5 To do this, add up the heights of all the pegs. Here the calculation would be: 35cm + 50cm + 20cm = 105cm over 3m (14in + 20in + 8in = 42in over 9ft).

EMPLOYING A SURVEYOR

You may wish to employ a land surveyor to produce a site plan for you if you have a difficult site. Surveyors in your local area can be found online. Land surveyors come under the jurisdiction of the Royal Institution of Chartered Surveyors (RICS), and it is advisable to check with them that the person you plan to employ is a member.

The cost of employing a land surveyor will depend on the size and complexity of your plot, but expect to pay between £800 and £1,500. This fee will pay for a topographical survey, but a cross-section may cost more. Not all land surveyors are used to surveying gardens, so explain your needs carefully to ensure you employ the right professional for the job.

Measuring an irregularly shaped plot

If your plot is large, has an irregular boundary, is hilly or undulating, or very overgrown, it may be best to pay a surveyor to measure it accurately and draw a site plan. However, the methods shown here are not especially difficult, so try one and see how you fare before calling in the experts.

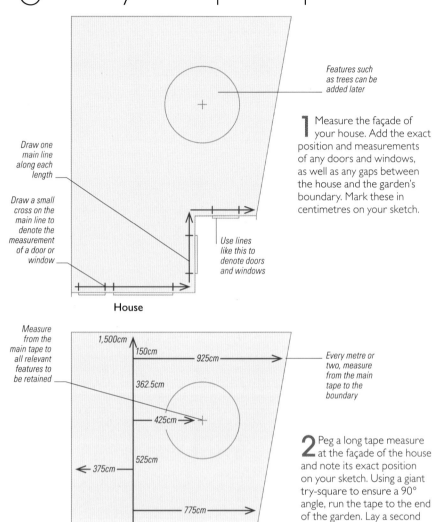

1 Measure the façade of your house. Add the exact position and measurements of any doors and windows, as well as any gaps between the house and the garden's boundary. Mark these in centimetres on your sketch.

2 Peg a long tape measure at the façade of the house and note its exact position on your sketch. Using a giant try-square to ensure a 90° angle, run the tape to the end of the garden. Lay a second tape at 90° to the first to measure points along the boundary and the position of relevant features.

Advanced techniques

Although the measuring techniques shown here are slightly more involved than those used on page 184, they are still relatively straightforward. There are two methods to choose from: "taking offsets" and "triangulation". Start with an outline sketch of your garden on an A4 or A3 sheet of paper (see p.184). Then choose the technique you find easier, but do not use a combination of the two, as this will make the process more complicated, especially when you come to transfer your measurements to a scale plan (see p.188). For both methods, start by taking measurements of the façade of your house, including windows, doors, and gaps between the house and boundary, and mark these on your sketch.

Taking offsets

To take offsets, you need two tape measures – one long and one shorter to measure the length and width of your plot – and a giant try-square, essentially a huge set square. Use the try-square to help you to lay the long tape measure along the full length of the garden on the ground at exactly 90° to the house. Use the second, shorter tape to measure at 90° (again, use the try-square to ensure the accuracy of your right angles) the distances from this main line to points along the boundary and to relevant features you want to keep. Clearly mark these measurements in centimetres on your initial sketch.

Getting some perspective

Whether you want to redesign part or all of your garden, site plans are an indispensable tool. However, unless you have at least some experience in reworking spaces or are naturally adept at imagining change, they may not help you to visualize how your new garden will look in three dimensions.

However, this simple idea will help to convey a sense of scale and proportion. You will need several bamboo canes, each just over 1m (3ft) in length, a tape measure, and a giant try-square. Form a square grid by pushing the canes into the ground at 1m (3ft) intervals, and so that they are 1m (3ft) high (you can clip off the tops with secateurs if necessary). Take a photograph of your garden with the bamboo grid and print it out. Then enlarge it – to A4 or A3 size – on a colour photocopier. Lay a sheet of tracing paper over the photocopy and then use the canes to help you draw your proposed new features in perspective (see p.182). Use the grid to block in areas of planting or to design screens, using the vertical canes to judge the heights.

Mapping your garden
This visualization technique works best in open spaces. Take an initial photograph of the area you want to design from the spot where you will be viewing the garden.

Using triangulation

On paper, this advanced measuring technique looks slightly more complicated than taking offsets, but in practice many garden designers consider triangulation easier and favour it over the offset method.

Triangulation involves marking two spots on the house – usually 1–2m (3–6ft) apart, but they could be further apart on a larger property – and then measuring from each of these spots to one point on the boundary, or a relevant feature, to form a triangle. This triangle and its measurements should then be marked on your sketch. Repeat this process at several points along the boundary – or the edges of a feature, such as a shed or a tree and its canopy. The more measurements you take, the more accurate your site plan will be.

You can then use these measurements to plot points on a scale plan and reproduce the exact dimensions of the garden and position of the boundaries, and any additional structures and key plants (see p.189).

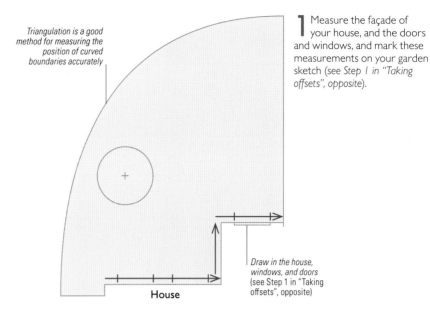

Triangulation is a good method for measuring the position of curved boundaries accurately

1 Measure the façade of your house, and the doors and windows, and mark these measurements on your garden sketch (see Step 1 in "Taking offsets", opposite).

Draw in the house, windows, and doors (see Step 1 in "Taking offsets", opposite)

Measure several points along the boundary. You will need these to get an accurate outline of the garden

2 Measure from one spot on the house to a point on the boundary. Repeat from another spot on the house to the same point on the boundary, and the distance between the two points on the house. Note all three distances on your sketch.

Measure from two points on the house to one point on the boundary to form two sides of a triangle

The façade of the house will form one side of your triangle

Measure to the same point on the feature

3 Measure from one spot on the house to a feature, such as a tree. Repeat from another spot on the house to the same point on the feature, and the distance between the two points on the house. Note all three distances on your sketch.

Measure the canopies of trees, and perimeters of beds and borders too

Measure between two spots on the house

1 Place the bamboo canes 1m (3ft) apart to form a square grid over the whole area – use a tape measure and giant try-square to ensure accuracy.

2 Make sure that the bamboo canes are the same height, 1m (3ft) is a good choice, or the sense of perspective will be lost. Take another photograph of the garden.

3 Print out the photograph and enlarge it on a colour photocopier. Lay tracing paper over the image, then use the canes as a guide to draw your proposed features.

Using scale and drawing more complex plans

Essentially, a scale plan is a proportional visual representation of your garden, and you can draw one easily by converting the measurements you took of your garden (see p.184 and pp.186–87) to one of the scales outlined below. It is also worth investing in a scale rule (a triangular-shaped ruler with scales such as 1:10, 1:20, and 1:50 marked on it) for this job, as it dispenses with the need for calculations. When your site plan is complete, use it as the basis for your design and planting ideas.

Choosing a scale

There are several scales to choose from, including 1:10, 1:20, 1:50, 1:100, and 1:200. Put simply, a 1:1 scale shows an object at its actual size; on a 1:10 scale plan, 1cm on paper represents 10cm measured in your garden; on a 1:20 scale, 1cm on paper represents 20cm on the ground; and on a 1:50 scale, 1cm on paper represents 50cm in your garden. For small domestic gardens, it is best to use scales of 1:20 or 1:50; for a larger plot, you may want to use a 1:100 scale, or even a 1:200 scale for an extensive country garden.

Designers often draw more than one plan and use different scales to show different details. For example, a 1:50 scale can be used for planting plans, and a 1:20 or 1:10 scale is best for structural features, such as a pond.

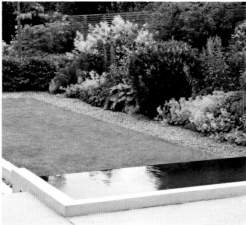

Whole garden 1:100
This is the best scale for an overview of medium-sized to large gardens. If your garden is particularly big, you may have to draw your site plan on an A1 sheet of paper.

Planting plan 1:50
Perfect for most planting plans, this scale is ideal for showing the position of larger architectural or specimen plants and general planting schemes. For more detail, to show exactly how many plants you will need in a 1 x 2m (3 x 6ft) border, for example, 1:20 may be a better option.

Architectural details 1:20
This scale allows you to work out quantities of hard landscaping materials, such as pavers. Use it to calculate the exact numbers you will need if building garden features yourself, or supply building contractors with a 1:20 plan to enable them to make these calculations.

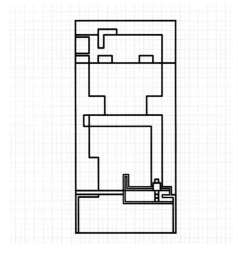
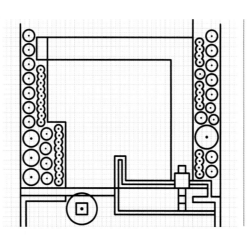
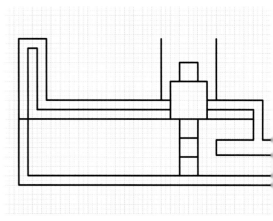

Drawing a plan for an irregularly shaped plot

You will need
- Metric, A3, squared or graph paper, or plain paper and set square
- Large pair of compasses (for triangulation)
- Scale rule and/or clear ruler
- Pencil, pens, and rubber

Regardless of the method – triangulation or offsets – used to measure your irregular plot and its features, start by drawing your house and the doors and windows on your plan.

If you used offsets, draw a line at 90° to the house to represent the tape measure. Using the graph paper's grid and a ruler or a scale rule, plot the boundary measurements at 90° to this line; join the dots to form the boundary. Then add features, also plotting measurements at 90° to the central line.

If you took measurements using triangulation, use the method on the right to draw up your scale site plan.

TOP TIPS
- Use Google Earth to check the shape of your plot. On larger or more open plots, you will see trees, features, and sheds.
- Don't over-complicate your sketch. If necessary, use more than one sheet to record dimensions of the main garden, and a separate sheet for details, such as planting plans.
- If an impenetrable area of vegetation gets in the way, estimate its dimensions from the measurements around it.
- When drawing your site plan, use metric graph paper for a more accurate result.

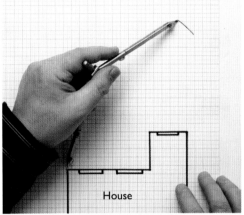

1 Draw the house, doors, and windows, and then set the compasses to the first scaled measurement you took from the house. Place the point where you measured from on the house and draw a small arc.

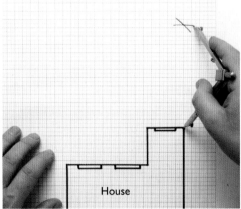

2 Reset the compasses to the second measurement you took from the house to form the triangle. Place the point where you measured from on the house and draw a second arc to cross the first.

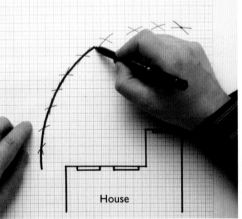

3 Repeat Step 1 and Step 2 for all of your boundary triangulation measurements. With a pencil, join up the centre point of each of the crosses to plot your boundary. You can then go over the line in pen.

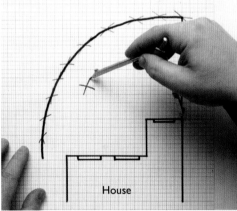

4 Use the same technique outlined in Steps 1 to 3 to plot the position of the garden's features – such as outbuildings, trees, plants, or water features – to create your scale site plan.

The finished site plan

You've taken all the necessary measurements, converted them to your chosen scale, and drawn up your scale site plan (or plans, if you chose to use more than one). This accurate representation of your garden's boundary, and any existing features that you intend to work around, is an important design tool. Take photocopies of your plan, scan it onto a computer, and print out copies or make a few tracings. You can then use these copies or tracings to sketch shapes and ideas that will fit the plot.

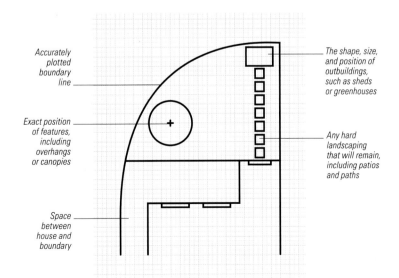

Accurately plotted boundary line

Exact position of features, including overhangs or canopies

Space between house and boundary

The shape, size, and position of outbuildings, such as sheds or greenhouses

Any hard landscaping that will remain, including patios and paths

Using your working plan
As well as creating your own design, you can use a scale site plan to show builders the size and type of surfaces and features you want. Also some design companies offer postal services, particularly for planting schemes, and ask for a site plan to help them produce an accurate plan.

Experimenting with plans

More accurate than a bubble diagram or sketch, a scale drawing enables you to experiment with different layouts in enough detail to ensure that the design fits and works well. Although all proposed elements, such as paths and planting, must be drawn to scale, the drawing does not need to be too technical. Here, designer Richard Sneesby explores four ideas for one simple plot.

One garden: four solutions

This simple plan (see right) shows a rectangular plot, with the rear elevation of the house located along the bottom line. Adjoining the house is a patio, and the garden includes an existing tree and shed. There is also a rear access gate in the top-right corner.

Each of the four plans shows different design options for this site. All feature a lawn, pond, paving/deck area, as well as access to the back gate, and three include a shed. The tree has been removed in two schemes, as it would compromise the suggested layout.

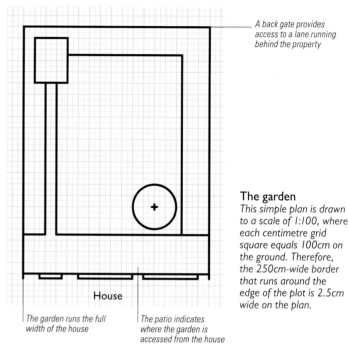

The garden
This simple plan is drawn to a scale of 1:100, where each centimetre grid square equals 100cm on the ground. Therefore, the 250cm-wide border that runs around the edge of the plot is 2.5cm wide on the plan.

- A back gate provides access to a lane running behind the property
- House
- The garden runs the full width of the house
- The patio indicates where the garden is accessed from the house

OPTION ONE

By positioning rectangular areas diagonally, the corner-to-corner orientation of this garden gives it a dramatic appearance. The design provides planting areas that are deep enough for larger specimens, and a triangular pond that can be appreciated from the nearby seating area. This is a garden of two halves, with a hedge dividing (and possibly screening) the two lawn areas, allowing each section to be given a distinct character.

OPTION TWO

The garden here is divided by a series of hedges that create a visual and physical chicane, keeping views short and varied; they also act as a unifying element across the plot. The hedges would be grown to different heights to allow or inhibit views, giving visual variety. Rows of trees reinforce the division created by the hedges but would allow views beneath their canopies. The design also includes rectilinear flowerbeds, a formal pond, and a shed hidden behind tall trees.

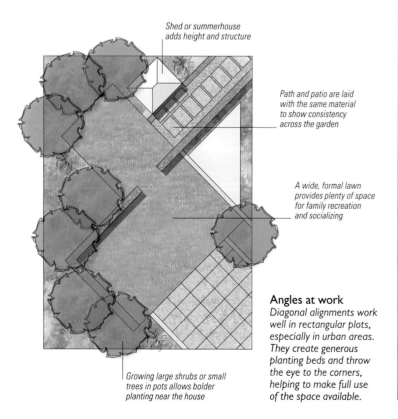

- Shed or summerhouse adds height and structure
- Path and patio are laid with the same material to show consistency across the garden
- A wide, formal lawn provides plenty of space for family recreation and socializing
- Growing large shrubs or small trees in pots allows bolder planting near the house

Angles at work
Diagonal alignments work well in rectangular plots, especially in urban areas. They create generous planting beds and throw the eye to the corners, helping to make full use of the space available.

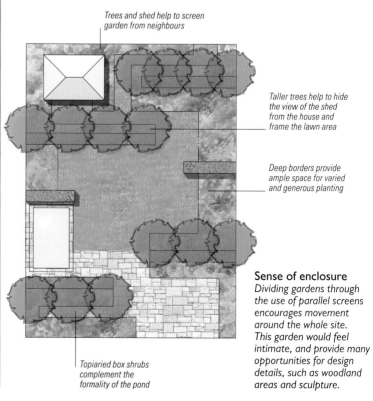

- Trees and shed help to screen garden from neighbours
- Taller trees help to hide the view of the shed from the house and frame the lawn area
- Deep borders provide ample space for varied and generous planting
- Topiaried box shrubs complement the formality of the pond

Sense of enclosure
Dividing gardens through the use of parallel screens encourages movement around the whole site. This garden would feel intimate, and provide many opportunities for design details, such as woodland areas and sculpture.

CREATING A PLAN

OPTION THREE

With its strong diagonal axis, this design works in a similar way to Option one. The oval-shaped lawn provides a central space, further defined by a low, flowering hedge. The trees also help reinforce the geometry and partially enclose the central area. The summerhouse is a focal element here, while a decked area and pool overlap on to the lawn to provide opportunities for attractive detailing. The planting beds are deep and generous.

This hidden area is the perfect place for a compost heap

The pond, crossed via a small bridge, provides a restful setting for the summerhouse

The oval-shaped lawn makes full use of the site, and is kept private by the surrounding trees

Clipped, pot-grown specimens complete the circle of trees closest to the house

Oval approach
Central circular zones can help to unify a space and bring the garden together. Using an oval shape, in particular, gives the garden a sense of direction, and leads the eye across the spacious lawn.

OPTION FOUR

This curvilinear plan would be more complicated to set out on the ground than the other designs, but would accommodate existing features and levels more easily. The lines are sweeping organic curves, the pond much less formal, and there are two distinct seating areas. Planting beds vary in width to allow a wide variety of plants and combinations to be grown. However, as there are no hedges, taller plants would be needed to prevent the garden from looking and feeling too open.

A limited range of materials adds interest without clutter

Decked seating area acts as a focal point and provides space for seasonal containers

Larger trees give shelter and privacy, and help to define the view through the garden from the house

Informal gravelled area offers easy access and long views up the garden

Flexible design
Curved, organic shapes can be used to create a more relaxed feel, and the layout can be adapted to accommodate larger plants as they grow. Such shapes are difficult to build using paving materials.

Using design software

To create a plan on your computer, you can choose from a wide range of garden design software packages available. Look for options appropriate to your level of skill and the amount of detail you want to include. Most are quick to learn and some are free to download, although the price you pay generally determines the quality of the plan you can produce. The quality of packages now is highly sophisticated, some showing different lighting effects, the movement of the sun, different plants, and even animation.

Professional designers use specialist computer-aided design (CAD) software to design accurate 2D layouts for contract drawings and commercial tendering, often combined with SketchUp illustrations to create 3D visuals of their ideas.

Bird's-eye view below
The way computer software replicates level changes, material textures, plant types, and garden furniture can make images look incredibly real, and help the viewer consider what their outdoor space may look like.

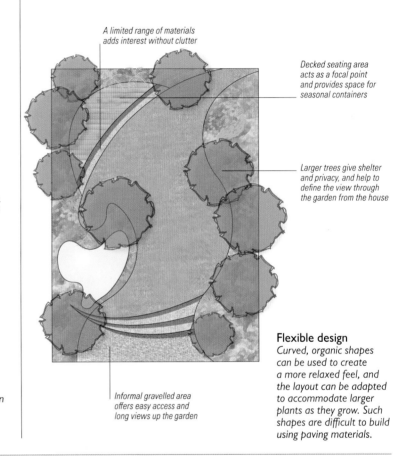

All in the detail above
Getting granular textures and colours helps designers and clients really concentrate on how the end result will look.

Planning your planting

A combination of practicality and artistic flair is required to plan a planting scheme. The practical considerations include soil type, aspect, and the amount of shade and sun the site receives. You may also want to consider using plants to offer shelter, structure, or scent close to a seating area. Your ideas and inspirations inject the all-important artistic input.

Visualization technique
You may find it easier to visualize your planting if you dummy it up by using garden objects of similar sizes, such as bamboo canes, buckets, cardboard boxes, and pots.

First steps

Before planning your planting, draw up a site plan (see pp.184–89). You can then start thinking about the whole design of your garden and how planting fits into the overall look. Sketch in the shapes and sizes of proposed beds and borders, and take photographs of the garden, too – either an aerial shot from a bedroom window or from the area from which your planting will be most often viewed. You can then use these to help judge the scale of planting you need.

Choosing the right plants

You can either start with a list of your favourite plants and work them into your scheme, or decide on the look you want and then find plants to fit the heights and shapes required on your site plan. In reality, though, a planting plan usually ends up being a combination of both.

Whichever approach you take, bear the following points in mind. First, make sure the plants you choose will cope with the site and soil conditions; then when arranging plants on your plan, check their height, texture, and shape in relation to those you will be placing next to them. Flowering period is important if you are looking to highlight a particular season; otherwise focus on foliage attributes first. In a small garden, a planting palette limited to relatively few different types of plants will have the greatest impact. For inspiration, go to the garden centre and group your chosen plants together. Or search online: Pinterest, Instagram, and Houzz offer lots of planting ideas.

Habitat match
In this naturalistic planting, drought-tolerant succulents and alpines, which require free-draining conditions, are planted in a bed of gravel and pebbles.

Balanced forms
Choose a range of marginals with different leaf shapes, such as these irises and astilbes, for a balanced poolside display.

Consider the seasons
Make the most of the available light and moist ground in late winter and spring when planting under deciduous trees.

CREATING A PLAN

Plants with design functions

It is easy to become fixated on flower and, to a lesser extent, leaf colour, but many plants offer other equally attractive attributes that will add an extra dimension to your planting. Perfume is an obvious one and is a must near patios and around doors and windows, while structure — for example, the domed hummocks of *Hebe* and the sword-like leaves of *Phormium* — can be used to give visual emphasis to a planting. Many climbers can be trained over trellis to disguise an ugly view, and tough hedging plants, such as hornbeam or yew, make perfect windbreaks.

Fill the gaps
Bulbs provide seasonal colour and can be squeezed between permanent plantings. Spring bulbs will cheer your border before most perennials appear, and Allium bulbs (left) in early summer are followed by colourful Gladiolus *and* Nerine.

Year-round interest
Flower colour is often a transient feature, but foliage has long-term impact and should be seen as the mainstay of any border throughout different seasons.

Winter colour right
Winter flowers are a treat, so make sure you can see them from a path or the house. Several Hamamelis have the bonus of scent.

Scented plants far right
These are best planted and enjoyed in warm, sheltered areas of the garden where strong winds won't dissipate their perfume.

Drawing up a planting plan

Planting plans don't have to be complicated, but they can be a great aid, helping you to organize your ideas and calculate planting quantities. Just measure your garden fairly accurately and produce a simple scale plan (see *pp.184–89*), then use this to outline areas of plants and, in more detail, the shapes of planting groups and individual specimens.

Grouping plants

The lure of an instant effect often tempts new designers to cram too much into a small space, but overcrowded plants tend to be unhealthy, so always bear in mind their final spreads when drawing up your plan. You can achieve a fuller look by grouping plants together. With perennials, larger groups of three or more of a single species will have a stronger, more substantial effect than single plants dotted around, which can look messy. Grouping plants in sausage shapes (which works well for cottage- and prairie-style plantings), or triangles, is satisfying to the eye and makes it easier to dovetail disparate groups. Also, try placing the occasional plant away from its group to suggest it has self-seeded for a naturalistic look. With shrubs, you can either plant in groups for an instant effect, or singly and wait for them to fill the space. Plant trees at a good distance from your property to prevent subsidence and give them plenty of space to mature.

A formal planting scheme near the house will create a contrast with natural plantings elsewhere. Try a simple parterre formed of squares or rectangles enclosing a cross, and outline your design with box hedging. Avoid making the beds too small, because once planted up, they could look cramped and overly fussy.

Prairie-style drift planting
Interlocking sausage-shaped drifts of plants give a less contrived look. Make the shapes a good size for maximum impact.

Modernist blocks
Strong geometric shapes are emphasized and complemented by bold blocks of planting, such as cubes of hedging.

Parterres
The symmetry and formality of a parterre makes planning fairly simple. Start with the outline hedging, then add the infill plants.

Random planting
To recreate a natural habitat, place plants in random groups. To avoid a chaotic design, use a limited colour palette.

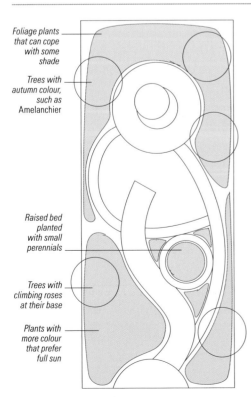

Foliage plants that can cope with some shade
Trees with autumn colour, such as Amelanchier
Raised bed planted with small perennials
Trees with climbing roses at their base
Plants with more colour that prefer full sun

Sketching on a photograph
If you find drawing difficult, doodling over a photograph will help you visualize the design in the context of your garden and get a sense of perspective.

Bubble diagram
This type of simplistic sketch, with rough shapes and annotation, will help you to position larger plants, such as trees, and pinpoint key areas of full sun or shade.

Sketching ideas

One of the simplest ways to visualize a planting plan for a small garden is to sketch the view from an upstairs window. Give full rein to your imagination and don't worry about accuracy at this stage. Next, identify the views from the house at ground level (stand by the back door) and consider whether you want planting to enhance, frame, or block them. Finally, walk around the plot visualizing the overall layout and the shapes and positions of structural plants, such as hedges and shrubs. Mark these on your sketch as simple shapes.

Take photographs as well, so you can refer to them when you come to draw your plan. If you feel confident, you can sketch your ideas directly on to photographs; if not, work on a sheet of tracing paper laid on top. You may find that black and white print-outs are less distracting to work with than colour pictures. Use your rough sketches as the basis for preparing a more organized planting plan.

The final planting plan

If you are preparing a plan for your own use you will not need fancy graphics, but if it is for a client, a professional-looking plan (see *symbols on p.182*) is appropriate.

On your scale plan, first draw the outlines of the areas you want to plant, then add specific plants. To help you position trees or shrubs, draw circles to scale, depicting their likely spread. Mark perennials in as freehand shapes. To help you calculate planting densities, mark out a square metre on the ground and work out plant spacings for different species using their final spreads. Keep a note of them for future reference.

Draw your plan on graph paper or on paper marked with a pencil grid of 1cm squares – you can then erase the latter when you ink in your final design. The scale you choose for your plan depends on the size of the beds or borders you are designing, but for a detailed plan, a scale of 1:50 or 1:20 is appropriate (see p.188 for more on scale).

Use acrylic tracing paper to copy your final sketch and produce a clean, finished drawing. Office suppliers sell tracing paper on rolls or as large sheets. Local printing companies often offer a copying service for large plans. You will need at least two copies: one for best and one that can be taken out into the garden at planting time. Consider laminating plans to make them weatherproof.

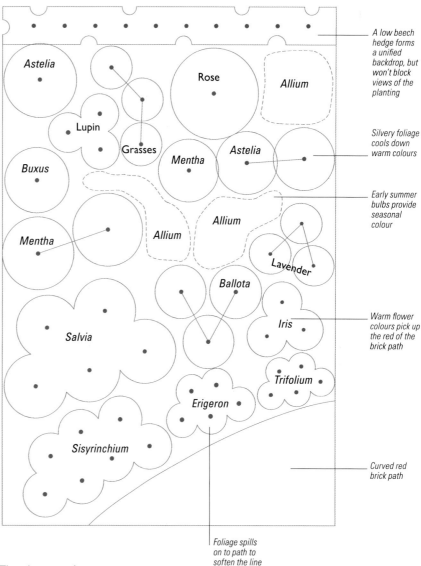

A low beech hedge forms a unified backdrop, but won't block views of the planting

Silvery foliage cools down warm colours

Early summer bulbs provide seasonal colour

Warm flower colours pick up the red of the brick path

Curved red brick path

Foliage spills on to path to soften the line

The finalized plan
This is a planting plan for the border shown below. The shapes indicate the position and number of plants within each group. The plan also shows their final spreads, so you can see how they will fit together.

The planting scheme
Successful plantings will inspire your own creations, helping you to visualize how plants look in situ. Make a note of combinations that work well and use your smartphone or digital camera to take snaps of plantings that catch your eye.

COSTING UP YOUR PLANTS

If you can afford large shrubs and trees, you can produce an instantly mature look; a smaller budget means younger plants and patience while you wait for them to grow. Perennials flower and reach their maximum height in the first couple of years, so don't spend a fortune on big plants.

It is worth asking garden centres and retail nurseries if they give discounts to designers; some also offer a plant-sourcing service. If you can show you are a trade customer, wholesale nurseries allow you to buy plants in bulk.

Examples of planting plans

Irrespective of the style of garden you're designing, whenever you're putting together a planting plan, check first that the plants you choose suit the site, soil, and climate. If working on a design for a client, it is vital that you talk through your planting ideas with them before committing to a final design, not only to help them visualize the finished garden, but also to agree on a scheme that they can easily maintain.

A divided garden

Unless you divide it up in some way, a rectilinear garden holds no surprises. To avoid the "what you see is what you get" effect, designer Fran Coulter created a visual break between a decked terrace along the side and back of the house and the rest of the garden.

Plants used include:
1. *Rosa* 'New Dawn'
2. *Clematis* 'Pink Fantasy'
3. *Trachelospermum jasminoides*
4. *Lonicera nitida* 'Baggesen's Gold'
5. *Buxus sempervirens*
6. *Weigela* NAOMI CAMPBELL ('Bokrashine')
7. *Nepeta nervosa*
8. *Vitis vinifera* 'Purpurea'

Design in focus

When a garden is overlooked by neighbours, especially from an upstairs window, a climber-clad pergola provides privacy for seating or dining areas. However, in this design – the area shown is approximately 3.5 x 2.5m (11 x 8ft) – the pergola is used as a colourful boundary between a decked terrace and the garden beyond. The wood is painted a matt red to match the Scandinavian-style property. In Sweden, the paint is traditionally made with iron and copper ores, and these tones are picked up in the planting: the purple grapevine, wine-red Weigela, and the pink rose and clematis.

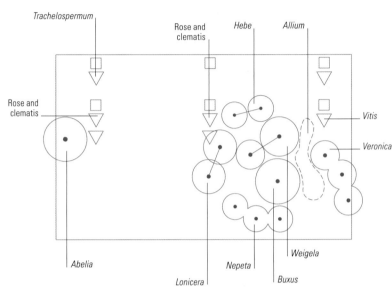

CREATING A PLAN

Shady area

This 3.5m (11ft) square border is backed by a high stone wall and cherry laurel. The owners asked designer Paul Williams for shade-tolerant planting that would mirror the formality of the adjacent garden. The plants here are mostly green with the odd splash of colour.

Plants used
1 *Dryopteris affinis* 'Cristata'
2 *Gazania*
3 *Prunus laurocerasus*
4 *Hosta* 'Krossa Regal'
5 *Taxus baccata*

Design in focus
To emphasize the formality of the garden on the other side of the path, this border (of which this is one section) is broken up with yew "buttresses" every three metres. Each section contains a simple planting and an urn or feature plant. Foliage is important: the plants need to be shapely and shade tolerant. Seasonal plants in the stone urn can contrast with or complement the surrounding plants.

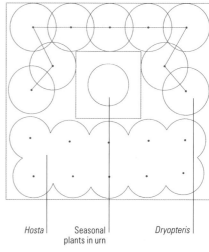

Hosta | Seasonal plants in urn | Dryopteris

City garden

Adam Frost designed this small city garden filled with romantic cottage-style planting. Soft red bricks are the perfect foil for the colour-themed planting, which is a sumptuous mix of crimson, pink, and mauve.

Plants used
1 *Salix elaeagnos* subsp. *angustifolia*
2 *Persicaria bistorta* 'Superba'
3 *Rosa* 'Souvenir du Docteur Jamain'
4 *Heuchera* 'Chocolate Ruffles'
5 *Astrantia major* 'Roma'

Design in focus
At the centre of this border, which measures roughly 1.2 x 2m (4 x 6ft), is a highly fragrant, dark crimson cup-shaped rose, its glossy green leaves forming an open framework for the slim stems of the Persicaria and Astrantia to grow through. These pale pink perennials complement the rich tones of the rose and help reflect light into the scheme, and are fringed at ground level by a wine-coloured Heuchera. The Salix, with its pale green filigree leaves, provides the perfect neutral backdrop to the warm colours.

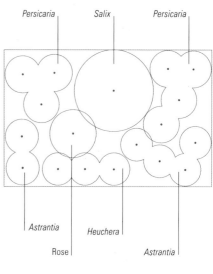

Persicaria | Salix | Persicaria
Astrantia | Heuchera
Rose | Astrantia

Designing with plants

Plants perform at their best when they are grown in the right place. Learn about their needs and the kind of soil they prefer, and this will help you devise the right planting scheme for your plot.

Including a wide range of plant groups should ensure interest all year round. Trees and shrubs give height, depth, and shade, as well as the essential framework. Evergreens retain their leaves, so are useful for all-year interest, and the shimmer of frost-covered deciduous plants is one of the pleasures of a winter morning garden. Scented climbers, grasses, and perennials are all useful, permanent plants that should be considered; annuals, biennials, and spring bulbs bring a burst of seasonal colour, just when fresh growth and energy is most needed in a garden.

Do not underestimate the versatility of plants, which can be selected for different roles in a planting scheme in often unexpected ways. A structural plant can be a single specimen, such as a stunning cardoon taking centre stage in a border, or a group of plants, perhaps an evergreen hedge clipped to enclose a parterre. Focal plants attract and guide the eye. They don't have to be long-lasting: a lovely individual specimen with vivid flowers or leaf tints works as well as an evergreen spiky *Phormium* or sculptural tree.

Midrange plants include shrubs, grasses, and herbaceous perennials, and they can help define the style of your garden. Mix strong leaf shapes and flowers and foliage with different colours and textures for a dynamic display. Ground cover is another potential element; choices range from a neat, evergreen carpet to a blowsy show of flowers or scented drift of herbs.

From the heart-lifting first bulbs of spring, through to summer blooms, and on to autumn foliage and scented winter-flowering shrubs like *Mahonia*, seasonal planting is a constantly evolving delight. You can stick to your chosen style, or throw in the odd surprise for fun. Designing with plants is the exciting – and never-ending – pleasure of gardening.

Top and above
Select plants like dahlias for shape as well as colour.
Use layers of plants to create stunning effects.

Understanding plants

Garden plants come from a great number of different habitats around the world and vary in their needs. Providing them with the same conditions in which they grow in the wild is the best way to ensure that they will thrive in your garden. A plant's appearance – the leaves, in particular – can give you a basic understanding of its requirements, but it is best to read the plant label carefully, too. Remember that plants that share a natural habitat will also look good together in the garden.

Shade- or sun-loving?

Imagine the conditions in which a shade-loving plant grows. Light levels are low, so it probably has dark green leaves full of light-catching chlorophyll. Protected from damaging drying winds and scorching sun, it can also afford to have large leaves. Now imagine a plant that has to cope with sizzling midday sun and buffeting winds. Silver or grey leaves with reflective surfaces and protective hairs are less likely to dry out. Leathery or succulent leaves also indicate good tolerance of heat. Many plants fall between these two extremes, but, in general terms, leaves are a useful guide.

Shade-tolerant plants
Moist and shady, sheltered conditions allow large-leaved plants, such as Rheum, Darmera, *and* Rodgersia, *to thrive. Most shade lovers tolerate some full sun during the day, but leaves may scorch with too much exposure.*

Sun-tolerant plants
*Full sun and dry soil make a testing environment for a plant. Heat- and drought-tolerant plants may have silver, heat-reflective leaves (*Artemisia*), or narrow grey ones (*lavender*), which minimize the exposed surface area.*

Plants for different soils

It is easier to match your plants to your soil than to try to change the character of your land. Heavy clay can be cold and wet, but it is fertile and productive once plants are established. Sandy soils can be worked all year round at almost any time but will dry out fast in summer.

Soil acidity is important if you want to grow ericaceous (acid-loving) plants such as *Pieris, Camellia,* or *Rhododendron*. Be aware that labels don't always state whether plants need acid soil conditions. (*For more information on soil types, see p.134.*)

Clay soil
Plants such as Berberis *that like fertile moist conditions grow well on heavy clay soil.*

Sandy dry soil
If soil is too wet, bulbs, such as alliums, may rot. Free-draining sandy soils suit them best.

Alkaline soil
Soil with a pH value over 7 is considered alkaline – if it is also fairly fertile, roses will love it.

Acid soil
Azaleas are ericaceous plants that require acid soil with a pH value below 6.5.

PLANT GROUPS

ANNUAL
A plant with a life cycle of one year. Usually very floriferous because of the number of seeds it needs to yield in order to reproduce.

BIENNIAL
Plants with a two-year life cycle, producing foliage the first year and flowers the next. Canterbury bells and wallflowers are biennials.

PERENNIAL
Non-woody plants that can live for years. Most die down to the ground in winter and come up again in spring; some are evergreen.

EVERGREEN
A plant that retains its leaves all year round.

DECIDUOUS
A plant that loses its foliage during winter, then produces new leaves in spring.

GRASSES AND SEDGES
A mix of evergreen or deciduous plants with grassy leaves. They can be clump-forming or spreading, and range in height from a few centimetres to 2–3m (6–10ft).

SHRUBS
Evergreen or deciduous plants with a permanent, multi-stemmed woody framework from 30cm–4m (1–12ft) tall.

TREES
Large evergreen and deciduous plants, which usually have a single trunk and are capable of reaching great heights. Trees need careful siting due to their longevity and size.

CLIMBERS
Deciduous and evergreen climbing plants useful for their foliage and flowers. Most need wires or trellis to cling to walls or fences, and can grow to a height of several metres.

AQUATICS
Plants that grow in wet ground or in water fall into three groups: those with leaves held above the water, those that lie on the surface, and those that stay submerged (see p.214).

Growth habits

Understanding a plant's habit helps you to place it in the garden. It also ensures you get the planting density right, so you achieve a balanced border that isn't overwhelmed by plants of unexpected vigour. Height and spread are usually marked on the plant label, but expect some variation due to different growing conditions.

Mat-forming
These plants spread by sending out shoots which then put down roots. Mentha requienii (Corsican mint) will steadily creep over gravel and paving.

Upright
As they often have little sideways spread, upright plants like Verbascum can be planted quite densely. They also provide useful vertical accents in the garden.

Fast-growing
Plants such as Lavatera need space when planted to allow for rapid spread. Plant labels give the size after 10 years, but check with other sources for growth rates.

Clump-forming
Over a few years, plants such as the non-invasive grass Pennisetum alopecuroides form a good-sized clump without threatening to swamp their neighbours.

Climbers
Climbers, including most clematis, take up little horizontal space as they want to grow up rather than out. Train them through shrubs and to clothe vertical structures.

Slow-growing
Many slow-growers will eventually become big, but it can take years. Ilex crenata has a slow growth rate that makes it ideal for low hedging.

Mirroring nature

If you bring together plants from different parts of the world but from a similar habitat, it is possible to create a planting design that is both botanically and aesthetically pleasing. Seeing the plants *in situ* in their natural environment will inspire you — and give you a feel for the conditions they require.

Coastal survivors
A plant's ability to cope with gale-force winds and salty spray will govern your choice for a seaside garden. Luckily, there are some beautiful plants that are perfectly adapted.

Woodland effects
You don't need to be a botanical purist to create a woodland garden. You can combine plants from different countries, so long as they all enjoy cool dry shade in summer.

Alpine inspiration
A rock garden is designed to emulate the free-draining dry conditions of an alpine meadow. This image of the real thing shows the effects you can aim for.

Plants in containers

There is no reason why a container garden can't be as well planted as a border. It is an intimate and very flexible form of gardening that allows an almost continual mixing and matching of your plants. However, growing plants in pots can affect their growth rates and restrict their size, since compost, water, and nutrients are limited.

Big bonus
A wide range of plants will grow successfully in large containers since they can accommodate more roots, water, and nutrients than small, narrow pots.

Tight squeeze
The restricted size and volume of compost in small pots limits your plant choices. You must water and feed plants regularly when grown in these conditions.

Selecting plants

At this stage of the design process, you should be getting a clearer idea of the look you want to create in your garden and thinking about the plants you'll need. Designers often talk about using a "palette" of plants, as if they were paints, and, in many ways, creating a beautiful garden is like painting – except that you are visualizing three dimensions, and your materials, being living, growing things, aren't static. Use the ideas outlined here to help you draw up an inspired planting scheme.

Choosing a planting palette

Focusing your ideas at an early stage in the design process narrows your choices and helps to guide you towards choosing the right plants. It also minimizes expensive mistakes. Sourcing plants is much easier when you have a specific theme, perhaps a favourite colour or style, in mind. A cottage garden, for example, will give you the scope to mix and match a wide range of plants in an informal setting. Something more modern, on the other hand, will demand that you use a limited number of plants in a more organized way. Designing a low-maintenance garden filled with evergreens will, again, focus your choice (see pp.28–129 for garden styles).

Tropical collection
A flamboyant display of annuals with hardy and tender perennials is high-maintenance, but the results are exciting and worth the effort.

Easy-care scheme
The established hardy shrubs and perennials in this formal planting require minimal maintenance. Their structure extends the seasonal appeal right through late autumn and into winter.

Functional planting

Certain garden features design themselves by default. For example, an exposed garden will need a windbreak, while an overlooked plot must have screening for privacy. Other design considerations might include fragrance by the front door, or a tree by the patio to provide shade on a hot sunny day. The design of such schemes is guided by their specific use, and this may limit your choice of suitable plants. The list below details the different design functions plants can fulfil, some of which may be pertinent to your plot.

1 Provide shelter
2 Create a boundary
3 Produce food to eat
4 Offer shade
5 Perfume the garden
6 Screen neighbours
7 Hide an ugly view
8 Provide a wildlife habitat

Sheltered seating area
Hedges do pretty much the same job as a fence or wall, but they have the edge when it comes to absorbing sound and wind. They also create a much softer effect.

Layers of interest

When space is limited, try to select plants that have a long season of interest. As well as those that flower over a long period, there are also many shrubs and perennials with colourful autumn foliage, structural winter stems, and spring buds. Precious few plants will fulfil all your demands, but look for those that tick the most boxes.

Structure and colour
The most useful plants here (peonies) work on several levels, providing structure and colour. In spring, their red shoots are followed by lush green foliage, then flowers.

Foliage and form
A closer look at a peony reveals how its flowers and foliage combine to make it stand out as an individual. Peonies often provide vibrant autumn leaf colour too.

Flower in focus
Close up you can appreciate the folded and crushed petals of this peony's double blooms. With other plants, such as passion flowers, the detail is in the intricate stamens.

Plant types and their design uses

There is, without doubt, a plant for virtually every situation, be it a tree, shrub, perennial, bedding plant, or bulb. When you're working out a planting plan, consider how best to use each plant, and ask yourself if it will create the look you are after, as well as how it will work next to other plants in the border.

Midrange plants
These make up the majority of the plants in a garden and include perennials and small shrubs. The substance of most plantings, they fill the gaps between bigger, more structural elements.

Structural plants
Plants can be structural on two levels. They can define the limits and framework of a garden, or the term can describe the plant itself, for example, if it has large paddle-shaped leaves.

Focal plants
Like ornaments, these are visual treats for the garden. It could be their distinctive colour, leaf shape, or stature that makes them stand out from other plants in the border.

Ground cover
People tend to think of ground-cover plants as being workmanlike. But there's no reason why they can't do a great job of being ornamental while smothering weeds as well.

Seasonal interest
The changing seasons make gardening a real pleasure. Choosing plants that provide an ever-changing display prolongs a garden's interest, changing its character as time passes.

Using structural plants

Structural plants are the backbone of a garden, forming the framework and helping to anchor other plants within a defined space. A beech hedge encircling a garden works in this way, as does a low box hedge around a border. By their sheer physical presence, individual structural plants – such as a *Gunnera* or *Cordyline* – can give focus to a planting scheme. Identifying key plants and deciding where to position them is the first step towards organizing a planting scheme for any garden.

Creating a framework

Hedging is ideal for defining the boundaries of a large- or medium-sized garden. It also provides shelter and increases privacy. Strike a balance between evergreen and deciduous species: evergreens are effective year-round screens, but because of the low winter sun, they can cast a dense gloomy shade, while deciduous hedges allow in some light for most of the year and can offer seasonal colour, too.

Use structural plants within the garden to frame (or block out) views and to lead your eye around the design. Shrubs in a border, perhaps forming a low hedge, provide a setting for midrange plants, and repeating planting helps to create visual reference points. When planting trees, consider their eventual size and the shade they will cast.

Hedges for definition
Hedging plants, both small and large scale (in this instance, beech), can be used to define the internal structure of a garden.

Structure in a border
Here, green and purple maples (Acer) frame a stone statue, while the sculptural Gunnera at the back forms a focal point.

Temporary structure

While the main framework of a garden may be permanent, much of the planting can change throughout the seasons. Some perennials provide vital structure for all but a few weeks in spring, when, as is the case with many handsome grasses, their stems are cut to make way for new growth. Large, shapely foliage plants, such as *Miscanthus*, act as an anchor for smaller species, or contrast with leafy flowering shrubs like *Deutzia*. Airy plantings also benefit from the occasional strong shape as a visual counterbalance to their wispy forms.

Structural accents
Clumps of bold foliage (here cannas) in a busy planting scheme act as a foil for slim-stemmed flowers and provide structural accents in a border.

Reconstructing nature
Using plants in broad interlocking swathes prevents an over-fussy effect, and the resulting planting, although strongly structured, looks natural.

Year-round interest

While evergreens may seem the obvious choice for year-round interest, visually they can be leaden and static. Deciduous trees and shrubs, on the other hand, may perform for several seasons, with new foliage in spring, followed, perhaps, by flowers, and then berries in late summer and vibrant leaf colour in autumn. In addition, trees often have a beautiful winter silhouette. Many species of *Sorbus* offer these benefits and are ideal four-season trees for a small garden.

A winter garden may not offer the obvious charms of summer, but there can still be sufficient interest to draw your eye into the garden – perhaps even enticing you to pull on a coat and venture outside.

Colour and form right
If you mix deciduous and evergreen species, the garden in winter can be both structurally interesting and surprisingly colourful.

Spring offering below
Trees form an important element of the spring landscape, some offering blossom, others vibrant green new growth.

Formal topiary bottom
Formal planting is the ultimate in structural design. This row of clipped evergreen trees is balanced and restful, and the effect can be enjoyed during all four seasons.

Using midrange plants

Midrange plants belong to a broad group that includes bulbs, some small shrubs (often called subshrubs), grasses, and most herbaceous perennials. Their great range of shapes, colours, and textures gives you huge scope for creativity, and you'll find plenty to define your chosen garden style. They are also invaluable as gap fillers between structural specimens, and since many flower and reach their full height in their first season or two, you won't have to wait long to enjoy the full effect.

Shape and texture

Some of the best midrange plants rely on their shape and texture for interest more than their flowers. Those with strong leaf shapes, such as *Acanthus*, *Hosta*, *Ligularia*, and *Rodgersia*, can be grouped together for bold shapely plantings; or they can be used to separate plants with frothy flowers or foliage. Using contrasting shapes and textures throughout a planting design creates visual excitement, with no shortage of interest. Imagine the fine leaves of fennel (*Foeniculum vulgare*) against the large sculptural foliage of the globe artichoke (*Cynara cardunculus* Scolymus Group), or the delicate but busy fizz of gypsophila against bold round *Bergenia* foliage. Grouping plants with similar soft textures creates a different, much gentler, effect: try fennel with *Anemanthele lessoniana*, or *Molinia caerulea* subsp. *arundinacea* 'Windspiel' with *Aruncus dioicus* 'Kneiffii' or *Thalictrum delavayi*.

Multi-layered texture left
This sloping site features layers of beautiful foliage textures and colours, including pompom alliums and feathery fennel.

Spiky foliage below
The structural leaves of crocosmias give season-long interest; the late summer flowers can almost be seen as a bonus.

Shrubby structure

Many small shrubs are useful additions to a herbaceous planting because they add a degree of permanence and a change of character. Plant short shrubby evergreens at the front of a border to act as a foil to the procession of perennials that come and go as the seasons progress. Good front-line plants include *Teucrium chamaedrys*, *Lotus hirsutus*, *Hebe pinguifolia*, and *Iberis sempervirens*.

Staying power left
Once its small trumpet-like flowers fade in late summer, the silvery evergreen subshrub Convolvulus cneorum *remains as a foil for other perennials.*

Good mixers far left
Low subshrubs, such as Helianthemum, *provide useful low level structure and mix well with perennials, but they also make a reliable display on their own.*

Foreground interest below
Block plantings of low evergreen hebes provide a weighty foreground that contrasts well with the lighter, airy grasses planted behind.

Flower and leaf colour

Perhaps the most exciting aspect of gardening is the chance to play with colour. If you include herbaceous perennials, the range of leaves and flowers can provide you with almost any tone or shade for your planting palette. When designing a scheme, consider the effect each plant has on its neighbour and decide if you want to use complementary or contrasting colours (see pp.168–69).

In general terms, a mix of colours generates an exuberant, slightly wild feel to a planting. Single-colour-themed borders look more sophisticated and have a cohesion that is satisfying to the eye. The restricted choice of plants also makes designing that much easier. Don't forget that just a hint of a matching shade in a flower or its foliage can be enough to link two plants.

Within a bigger border, colour combinations using two or three plants are effective. These can be timed for seasonal display, say, yellow wallflowers with the near-black tulip, 'Queen of Night'; or for something less transient, pale yellow Anthemis tinctoria 'E.C. Buxton', fronted by purple-leaved Heuchera 'Plum Pudding', surrounded by the leaves of Hakonechloa macra 'Aureola'.

Early summer border
A jumble of flower colours and textured foliage injects this border with a huge amount of energy. Adding some summer bedding will add to the overall excitement.

Focus on foliage
While still providing a perfect backdrop for other plants in the border, the large ribbed leaves of this luscious blue-green hosta make it a star in its own right.

Using ground cover

Ground-cover plants are used primarily to swamp weeds by creating a densely knitted blanket of leaves, stems, and flowers that exclude light and use up all available moisture. The best examples are also decorative features in their own right, offering a tapestry of colour, texture, and form and providing a foil for other plants. Ground cover does not have to be restricted to very low-growing plants and can include a variety of shapes and sizes, as long as they form a smothering canopy.

Dry sunny sites

Free-draining soils are "hungry"; you can feed them with organic matter but it usually breaks down quickly and its effect is short-lived, so it is best to choose plants suited to the conditions rather than to try to change the soil. Flowering ground-cover plants that thrive on sunny sites include *Helianthemum*, dwarf *Genista*, and low-growing shrubby potentillas, such as *Potentilla fruticosa* 'Dart's Golddigger'. For leafy ground cover, try plants with grey leaves, such as *Hebe pinguifolia*, *Santolina chamaecyparissus*, and sage (*Salvia officinalis*). Several plants suited to hot dry conditions are also aromatic and include lavender and thyme. These conditions are the natural habitat of many bulbs, too. Small irises, such as *Iris reticulata*, and smaller species tulips, such as *Tulipa kaufmanniana* and *T. linifolia* Batalinii Group, can be dotted among the ground cover to add extra colour.

Tough plants for tough sites
This gravel border features mostly Mediterranean-style ground-cover plants, including thyme and catmint (Nepeta).

Sun protection
Perfect for a hot spot, the silvery leaves of Stachys byzantina *reflect the heat of the sun and prevent the plant from drying out.*

Cool shady sites

Ground shaded by a leafy tree canopy is often extremely dry throughout the summer and provides the biggest challenge for both the plants and the designer. Reducing a tree's crown allows more light and moisture through to the plants below, and adding organic matter to the soil also helps to retain moisture. For dense spreading cover, try *Aegopodium podagraria* 'Variegatum' (variegated bishop's weed), *Asperula odorata*, *Cornus canadensis*, *Geranium macrorrhizum*, *Pachysandra terminalis*, or *Hedera* (ivy) species. When shaded by buildings, the soil can be slightly damper, making it easier to establish ground-cover plants. Shade-loving *Bergenia*, *Epimedium*, *Helleborus orientalis*, hostas, and many ferns, especially the dry-tolerant *Dryopteris* species, all produce a lovely effect.

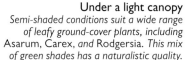

Under a light canopy
Semi-shaded conditions suit a wide range of leafy ground-cover plants, including Asarum, Carex, *and* Rodgersia. *This mix of green shades has a naturalistic quality.*

Dense shade
Many colourful hardy geraniums are tough enough to cope with the difficult conditions under a tree canopy.

Twice the value
Plants with long-lasting foliage make good ground cover; if, like these astilbes, they also offer flowers, their value is doubled.

Easy-care plants

In large gardens, where you can give them the space they need, vigorous spreading plants, such as *Hedera helix*, *Lonicera pileata*, *Trachystemon orientalis*, and *Vinca major*, make ideal low-maintenance ground cover. In smaller gardens, however, giving over large areas to a single species is not always appropriate or practical; it can also be a waste of a good planting opportunity. Where space is limited, it is far better to use a mix of leafy plants, such as *Astilbe*, *Astrantia*, *Bergenia*, and *Geranium endressii*, planted close together. You will achieve the same effect, but it will be more ornamental and can be achieved with very little effort.

Carpet of colour
Low-growing Lysimachia *and* Ajuga reptans *suppress weeds while also providing a colourful foil for larger plants.*

Mat-forming ground cover
Vinca minor *puts down roots from spreading shoots to form a dense mat. Its small leaves contrast well with those of the* Bergenia.

Using focal plants

Focal plants work on several levels: they can entice you into a garden, distract you from ugly views beyond the boundary, or provide an eye-catching feature within a border. Most focal plants are evergreen or have strong shapes or colours and offer a long season of interest, but don't dismiss those that perform for only a few weeks each year. Allow them their brief, glorious time in the limelight and plan the rest of the garden around the show.

Visual trickery

In much the same way as you would use a statue or an attractive container, you can site focal plants to lead the viewer's eye to a particular area of the garden. Positioned strategically, they can also distract attention from unsightly objects or views. Their presence not only makes someone shift their gaze, but can entice them to take a stroll around the garden too. When focal plants are repeated throughout a long border, they act like visual stepping stones, helping to carry the eye along its length. They also hold the planting together, giving it an essential cohesion. Finally, using a clever trick of perspective, when planted in the foreground, focal plants make the garden behind seem like a separate area waiting to be explored.

Handle carefully
Take care that a plant does not overwhelm the garden by grabbing all the attention and becoming an unplanned focal plant.

Scene stealer
Pampas grasses have considerable stature, even when they are not in flower. Their late summer display makes them the natural focus of attention.

Striking shapes

Many plants have naturally architectural or sculptural shapes: *Acer palmatum* var. *dissectum*, *Cornus alternifolia*, *Phormium*, and *Yucca* all make great focal plants. Many more, however, can be enticed over time with pruning and training to take on striking forms. This can be through traditional topiary, using slow-growing evergreens such as box, yew, *Ilex crenata*, or *Ligustrum delavayanum*. (Avoid fast-growing plants such as *Lonicera nitida*, which needs clipping several times over the summer to stop it losing its shape.) In addition, the adventurous gardener may like to experiment with other creative pruning techniques. By trimming off the lower branches of shrubs and trees you can make standards that produce lollipop shapes, or you can manipulate the branches to form tiers or cascading stems. *Carpinus betulus*, *Cotoneaster frigidus*, *Thuja plicata*, and *Viburnum plicatum* f. *tomentosum* 'Mariesii' are just four that respond well to this type of pruning. When trained, the skeletal winter outlines of deciduous plants can be as interesting as their leafy summer profiles.

Worth the wait
A single plant's display (here a Yucca) can be the raison d'être and seasonal climax of a whole section of a garden.

Using colour

Very few plants can offer season-long colour, but you can still achieve some great effects with even just a short burst of activity from foliage or flowers. The following are all good candidates for focal plants: the autumn foliage of Japanese acers, azaleas, *Fothergilla*, and larch; the flowers of *Hamamelis*, *Laburnum*, and *Viburnum plicatum* f. *tomentosum* 'Mariesii'; and the winter stems of many of the birches, dogwoods, and willows.

Plants that provide dramatic colour, however, need careful handling. Remember that bright reds or yellows planted at the furthermost corners of the garden have a foreshortening effect. On the other hand, using paler colours at the end of the garden visually lengthens your plot (see p.170).

Colour care right
Acers are real scene stealers when their foliage fires up in autumn. Position them carefully among more subdued colours so that they can really shine out.

Come closer far right above
The vibrant pink, pea-like flowers of Cercis siliquastrum *appear before the leaves in early spring. The tree's form provides a focus at other times of the year.*

Second innings far right
Hydrangea flowers are great value: colourful when fresh in summer, ethereally beautiful when faded in autumn, and stunning in winter with a dusting of frost.

In the limelight
Large-scale centrepieces, these birch trees are made all the more arresting with dramatic winter sunlight.

Have fun with topiary
Extravagance and humour are two ingredients that turn a feature into a great focal point. Here, yew is being trained through a giant topiary frame.

Seasonal planting

Designing a garden that offers a continuing series of delights throughout the year is both challenging and highly rewarding. Anticipating the emergence of new shoots, flowers, and foliage in spring brings a huge amount of pleasure, which is then matched by the abundance of the summer, followed by warming autumn colours and the stark beauty of winter. With careful planning, you can use plants to decorate your garden 365 days a year with their colour, scent, shape, and form.

Spring awakening

Spring brings welcome colour and energy after the gloom of winter. Nature designed early flowerers for high impact, with brilliant displays from *Amelanchier*, cherries, magnolias, rhododendrons, and *Viburnum*. Bulbs are also keen to impress: flowers of blue (anemone, hyacinth, *Muscari*), yellow (daffodils, tulips), purple (crocus), and red (tulips) all add to the season's vibrant spirit. If you prefer a more subtle effect, choose some of the softer-coloured spring-flowering shrubs and smaller plants, such as *Epimedium*, *Fritillaria*, *Helleborus*, and *Primula*. And nearly all spring bulbs have a white selection to temper a colourful display. However, it is often best to give full head to the season and simply enjoy the exuberance – just remember to plant your bulbs in the autumn or you'll miss the show.

Woodland setting
Plants and bulbs that thrive beneath trees make use of available light and moisture by flowering before the leaves appear.

Natural drifts of bulbs
Yellow daffodils and pink magnolia capture the freshness of spring. For naturalistic drifts, throw handfuls of bulbs across the ground and plant them where they land.

Summer profusion

In summer, the emergence of bees and other pollinating insects coincides with the majority of plants coming into flower. This natural abundance offers a huge choice of colours, heights, and shapes, which makes designing for a specific effect relatively easy. Check flowering times and choose a wide range of plants to prolong the display right through the summer months. Select perennials with beautiful foliage, so that when they have finished flowering they still contribute to the overall luxuriant effect, and set out each type of plant in bold groups of at least three for the greatest impact. Finally, to add to the richness, dot summer-flowering bulbs, such as *Allium*, *Gladiolus*, lilies, and *Triteleia*, throughout the border. Keep the display fresh by removing spent flowers and brown or damaged leaves.

Fiery mix
The variety of plants available in summer makes a colour theme a much easier option – here a "hot border" of sizzling hues creates a unified display.

Autumn colour

In sheltered gardens, many half-hardy and tender plants, such as dahlias and *Canna*, will continue to flower until the first frosts. Hardy perennials, such as asters, *Aconitum*, and *Actaea* (syn. *Cimicifuga*), flower very late, too, and together with forms of *Fuchsia magellanica* make good companions for a range of shrubs with fiery autumn leaves. Several summer-flowering perennials, including some peonies and hostas, provide a brief season of autumn leaf colour, but the main stars are the trees and shrubs, such as *Acer*, *Cornus*, *Prunus*, *Rhus*, and some *Berberis*, *Cotoneaster*, and *Viburnum*.

Seasonal transition
The overlap between fading perennials and the onset of luminescent autumn foliage colours is a delightful twilight period in the gardening year.

Borrowed views
This border has been designed as a stage set for the magnificent beech wood behind, but as the fiery autumn colours of Cotinus, Prunus, *and grasses ignite, all eyes are on the foreground.*

Winter interest

There is no shortage of plants to provide colour and interest during the colder months. Winter-flowering honeysuckles, *Fothergilla*, *Hamamelis*, *Mahonia*, *Sarcococca*, and *Viburnum* offer flowers and scent, and the berries or catkins of *Corylus*, *Cotoneaster*, *Crataegus*, *Garrya*, and *Sorbus* add colour and texture. Evergreens and their variegated forms deliver winter foliage, while the bare bones of dormant perennials, such as *Rudbeckia* and *Sedum*, and the stems of grasses, such as *Miscanthus sinensis*, all add to the beauty of the winter garden. Trees also make stunning contributions to a wintry scene: birches with their stark white trunks; the twisted silhouette of *Corylus avellana* 'Contorta'; and the flowers of *Prunus* × *subhirtella* 'Autumnalis'.

Eyes down
An underplanting of snowdrops brings a glimmer of light to the dark base of shrubs, like this Cornus *(dogwood).*

One garden, four seasons

By underplanting a wide range of shrubs and perennials with naturalized spring bulbs, you can achieve year-round interest without the need for bedding plants. The unsung heroes of winter are deciduous trees – without the distraction of foliage you can better appreciate their attractive bark and shapely forms.

Spring: fresh and vibrant

Summer: lush and leafy

Autumn: fiery colours

Winter: stripped to the bare bones

Planting water features

Water fascinates and captivates like no other garden feature. Its movement, reflections, and sound bring an appealing mix of new sensations to a garden. Water also offers the chance to grow a different range of plants that can attract insects and other wildlife to the garden, whether you are planting up a natural pond or complementing a modern installation.

Siting a pond
Check first that the site does not carry main sewers, drains, or utility pipes. Choose a sunny position with some shade during the day, away from overhanging trees.

- Trees are far enough away to prevent pollution from leaves
- The pond is the focus of the overall design
- Service pipe is a good distance from pond
- The view from the house allows you to enjoy the feature

Positioning your feature

For a natural look, small features like spouting figures and heads or an overflowing urn can be placed among the planting in borders. Ponds do best where there is good light, away from trees and falling leaves, which will rot and pollute the water. Also site them away from service pipes, such as electricity cables. All features should be viewed as an integral part of the design and placed where any filters and pumps can be hidden by plants, rocks, or decking. Child safety is also a prime consideration.

Choosing plants

Plan your waterside plantings exactly as you would your garden border, taking height, colour, and seasonal interest into account. Plants carry a label that show their preferred water depth – the distance from the crown of the plant (or top of their pot) to the surface of the water – and your choice is governed by the size and depth of your pool. Choose a mixture from the four main groups of water plants: oxygenators to keep the water clear; aquatic plants that grow in the water; and marginals and bog plants to soften the edges.

Bog plants
These plants thrive in a moist or wet soil. There is a wide range available, which includes some of the most colourful waterside plants, such as several irises, primulas, Lythrum, and evergreen Lysimachia.

Marginal plants
Growing in a few centimetres of water at the margins, these plants soften the line between water and land. As well as colourful or interesting flowers (Saururus, Orontium), many have dramatic foliage (Sagittaria sagittifolia, Pontederia).

To reduce algal bloom, plant marginals in a low-nutrient compost

Aquatic plants
These deep-water plants root on the bottom of the pond, 50cm (20in) or more beneath the water. There are relatively few plants in this group, but it does include water lilies, which grow in water 50cm (20in) to 1.2m (48in) deep.

Sink aquatic plants in their baskets to the correct depth, as marked on their labels

Oxygenators
An essential element in a pond, oxygenators provide oxygen and absorb the nutrients otherwise used by algae. Some, like Ranunculus aquatilis, flower above the water surface.

Planning ahead
Making planting ledges and boggy ground part of the initial design of a pond allows you to grow plants with different depth requirements.

marginal plant depth

aquatic plant depth

DESIGNING WITH PLANTS

Modern water features

In a contemporary setting, water is often used for its reflective properties and movement, rather than as a place to grow plants. However, several water plants, including species of *Juncus, Carex, Cyperus,* and *Equisetum* complement a modern, architectural style. A clean and unfussy look is important, so limit the variety of plants and use those with strong shapes for the best effect. Evergreens work particularly well in a modern setting.

Dramatic statement
The primitive-looking Equisetum hyemale *(horsetail) is invasive on land, but contained in a pond planter, its stiff, upright shape is very useful to the modern designer.*

Symmetrical planting
The round leaves of water lilies emphasize the squareness of this formal pool, while the dramatic foliage of Zantedeschia *adds some exuberance and links the pool with the surrounding planting.*

Small pools

If space is limited, a small fountain, bubbling millstone, or half-barrel or trough filled with water and aquatic plants can give great pleasure. Place your feature by a seat or close to the house where it will be visible from a window. If you cannot plant into the feature itself, position it among plants (*Hosta, Astilbe, Primula, Myosotis, Filipendula,* and *Iris*) that often surround a pond or pool.

Mini oasis
When planting a miniature pool, take care to avoid vigorous plants and rely on subjects like Nymphaea tetragona, *a small, compact water lily.*

Wildlife ponds

The combination of water and a wide variety of aquatic plants creates an attractive habitat for frogs, dragonflies, and aquatic insects, as well as offering cover for fish. Native plants will attract local insects, but any exotic, non-invasive water plants will be beneficial to frogs, toads, and newts. If there is room, introduce a small waterfall to create the splash and moisture ideal for growing ferns and mosses at the pond edge. Also, provide both deep and shallow water for diverse planting and a more natural look.

Natural habitat
Even a small pond will attract a surprising amount of wildlife, and is a useful way of increasing children's interest in nature and the garden.

OTHER PLANTS TO CONSIDER

FOR MODERN WATER FEATURES
Cyperus alternifolius
Equisetum scirpoides
Isolepis cernua
Juncus patens 'Carman's Gray'
Schoenoplectus lacustris subsp. *tabernaemontani* 'Albescens'

FOR WILDLIFE WATER FEATURES
Butomus umbellatus
Caltha palustris
Iris pseudacorus
Myosotis scorpioides
Ranunculus flammula

FOR SMALL WATER FEATURES
Juncus effusus f. *spiralis*
Orontium aquaticum
Primula vialii

Plants for a changing climate

As our climate continues to change, we need to ensure that the plants we grow are resilient and adaptable to the altering weather patterns. Whether it's too much rain or too little, too much sun, strong winds, or poor soils, we need to design with a range of plants that are beautiful, tough, and adaptable. While climate change is a huge threat, we can still enjoy our gardens with a range of fantastic plants.

Euphorbia characias subsp. **characias** above right
Once established, this handsome perennial tolerates full sun as it originates from rocky Mediterranean slopes; bright yellow-green heads in spring bring interesting colour.

Stachys byzantina 'Big Ears' right
Even with velvety leaves that look soft and delicate, this plant is, in fact, a robust choice for a gravel garden. Its beautiful silver leaves are able to withstand poor soil and sunny conditions.

Ginkgo biloba below
This tough and dependable, irregularly shaped tree is tolerant of a changing climate. It actually comes from a group of trees that predates the dinosaurs. Fan-shaped leaves turn a gorgeous butter-yellow in autumn.

Verbena bonariensis top
Popular for its ability to set seed and pop up in a range of locations, this tough but pretty plant has wildlife-attracting purple flowers that sway in the wind.

Pyrus calleryana 'Chanticleer' above
Tolerating wind, drought, and pollution, this brilliant tree for smaller gardens has scented white flowers in mid- to late spring – it's a superb ornamental pear.

Stipa gigantea above
One of the best grasses to give an airy, informal look, this tough plant takes some drought and plenty of wind in its stride.

Viburnum opulus 'Roseum' left
Tolerant of wet soils and some shade, this deciduous shrub has fragrant white or pink flowers, which are followed in autumn by blue or black berries.

Plants for wildlife

Knowing that the plants you like to grow are loved by wildlife and help to increase biodiversity is a good starting point for all gardeners. Attracting pollinators, offering refuge for mammals, welcoming amphibians, and creating an environment for a balanced ecosystem – where nature and humans exist in harmony – can easily be achieved with a range of plants. Search for "RHS plants for pollinators" online to find a full list of plants that are perfect for pollinating insects.

***Lavandula* (lavender)** right
The epitome of summer gardens, and with that delicious scent and silver foliage, lavenders are plentiful in choice and attract a wide range of bees and butterflies.

Rosa rugosa below left
This medium-sized shrub has fragrant, wildlife-friendly magenta flowers that are single (so easy to access for pollinators) and bright scarlet hips in autumn. It is a tough and hardy plant too.

Fagus sylvatica below right
Perfect for hedging, beech is a popular choice with its lovely crinkled leaves, making it a great wildlife and habitat shelter that attracts beneficial insects and birds.

Origanum vulgare top
Wild marjoram, also known as oregano, is often used in Mediterranean cooking, but it's also loved by bees and butterflies enjoying its nectar-rich flowers.

Ribes sanguineum above
This flowering currant has dark red, tubular early flowers that are filled with nectar, followed by blue-black currants in autumn. It's not too large and offers a good home for wildlife.

Lonicera periclymenum left
This beautiful climber, which can grow in semi-shade, offers plenty of nectar for moths and butterflies, while red berries, later in the season, offer food for birds and small mammals.

Crataegus laevigata right
Often used in hedging but also grown as a tree, hawthorn offers small white flowers for pollinating insects, leaves as a food source in spring, and autumn berries for birds.

Plants for year-round interest

Many plants have a peak season of interest, but some keep on giving through the year – whether it's their foliage, flowers, changing leaf colour, or their habit. At different times of the year, those virtues can be enjoyed by people and wildlife alike. Choose hard-working, year-round interest plants to add colour and vibrancy to any outside space.

Amelanchier lamarckii above right
Snowy mespilus is a brilliant shrub and small tree that gives so much – from white, star-shaped flowers in spring, to black berries in summer, and then superb autumn colour.

Mahonia eurybracteata 'Soft Caress' right
This mahonia is different from others as it has softer, spine-free, architectural foliage and lovely yellow flowers in late summer; it grows in most soils and is reliably hardy. It's a special plant.

Aucuba japonica 'Crotonifolia' below
A hard-working and dependable evergreen shrub, it brings gold-coloured variegation to its leaves, lighting up any corner. Good in sun or shade, it will often grow where most other plants won't.

Camellia species top
With flowers from singles to doubles, there is a huge range of acid-loving camellias to grow, all offering dependable evergreen interest with lovely glossy leaves and a whole range of flowers to delight.

Hebe species above
With a host of leaf shape, foliage colour, and flower colour to choose from, hebes are great plants for evergreen interest. They are available in a range of sizes and attract pollinators too.

Bergenia 'Bressingham White' above
Low maintenance yet incredibly dependable, this plant has glossy evergreen leaves (known as elephant's ears) and beautiful white flowers in spring. It tolerates a range of soils and grows in sun or shade.

Pinus mugo left
Even though it may not have showy flowers or berries, dwarf mountain pine changes subtle colour (green to gold) when the weather turns colder; it's a great architectural plant for evergreen interest in any border.

Plants for urban gardens

Plants growing in towns or cities have a challenging time: the urban heat island increases temperatures, rainfall can be reduced due to the rain shadows of buildings, and soil is often poor (after multiple building projects have degraded the earth). It can be a challenge. But there are a host of plants that are beautiful, durable, and can tolerate all that urban life throws at them, and actively enhance our city-living experience.

Cotoneaster franchetii right
This small-leaved, evergreen shrub grows to about 3m (10ft) and was found by the RHS to be one of the best plants to help capture emissions. Over 10 dry days, a hedge of this plant could capture the emissions from a car travelling more than 800km (500 miles).

Elaeagnus* x *submacrophylla below
This large, silvery-leaved evergreen shrub has small white flowers in autumn. Its dense, evergreen habit is great for absorbing higher levels of pollution.

Ligustrum ovalifolium below right
Well known in many streets in cities and towns, privet is a brilliant plant to help capture rainfall, thereby helping to reduce the risk of flash flooding.

Prunus laurocerasus top
Cherry laurel's glossy, large, rounded leaves are a great way to help reduce the transmission of noise in an urban environment. Cherry-like glossy red berries bring wildlife interest too.

Hedera helix above
Ivy has many virtues. Not only is it tough and will grow in most locations, but it can also provide insulation when grown against a wall, in addition to a wildlife benefit with its late-season-pollinating flowers.

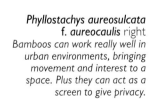

Phyllostachys aureosulcata* f. *aureocaulis right
Bamboos can work really well in urban environments, bringing movement and interest to a space. Plus they can act as a screen to give privacy.

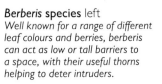

***Berberis* species** left
Well known for a range of different leaf colours and berries, berberis can act as low or tall barriers to a space, with their useful thorns helping to deter intruders.

Seasonal containers

The best plants for containers are all high achievers, promising months of sumptuous displays with minimum effort from gardeners. They need to be tolerant of the odd missed watering or feed, yet still able to provide a never-ending display of flowers or handsome foliage. These hard-working plants are generally seasonal (not long-lasting); those for summer pots need to withstand high temperatures and last well into early autumn; winter choices must be tolerant of wet and cold, and ideally perform into spring.

Echeveria elegans top
This clump-forming, blue-leaved succulent is ideal for a low-maintenance display. It enjoys heat and drought, and has arching spires of pink summer flowers. With protection, it may well survive to the following year.

Begonia 'Glowing Embers' above
A marvellous, compact plant with masses of orange flowers held above silvery-veined, reddish-bronze foliage, its display lasts until frosts. It's great in a mixed planter but best alone in a hanging pot.

Viola top
So easy and so versatile, these miniature pansies are available with flowers in a vast range of colour combinations. They are lovely through autumn and winter into spring if the weather is mild. Deadheading helps to prolong displays.

Nemesia 'Wisley Vanilla' above
With delicious, vanilla-scented white-and-yellow flowers all summer long, this low-growing Nemesia thrives in a planter or hanging basket. Don't let it get too dry and deadhead occasionally.

Felicia amelloides left
One of few plants for summer pots with blue flowers, this daisy is compact and easy to grow in hot, sunny conditions. Deadhead regularly for the best displays. It may survive mild winters.

***Dianthus barbatus* Festival Series** top
Dazzling in late summer and autumn with massed white-marked pink and red starry flowers, this variety of Dianthus is super on its own in a low bowl. It will flower profusely again the following spring.

***Calibrachoa* Million Bells Series** above
For sheer flower power, you can't beat these little petunia-like plants, with flowers in myriad shades that tone in or contrast perfectly in potted or hanging basket displays. They don't like it too dry.

Plants for the senses

It's best to think rather differently when choosing plants that stimulate the senses, as you are looking for friendly plants with a range of different characteristics. Some will stand out in one particular field, perhaps with powerful fragrance, delicious flavour, or soft downy foliage; others will be multi-taskers, good-looking plants that perhaps rustle in the breeze and add scent. Above all, avoid prickly or, indeed, toxic plants – any that you wouldn't want to come into close contact with.

Stipa tenuissima left
This perennial grass is loved for its evergreen clumps of fine foliage that billow beautifully in the breeze. At flowering time, it's silky soft to touch with feathery flower- and seedheads. It likes a sunny open place.

Salvia argentea below left
With the downiest foliage imaginable, this short-lived perennial has large leaves that shimmer silver in the sun; in summer, it also has spires of white flowers. Grow it regularly from seed.

Cosmos atrosanguineus below
This cosmos is a stand-out performer for its handsome, dark purple, late-summer flowers that are famed for their chocolate scent. It likes a sunny, well-drained place and is perennial if you protect the tuberous dahlia-like roots.

Lavandula x intermedia 'Grosso' top
Among the most powerfully scented lavenders, this has purple-blue flowers atop tall stems all summer. It also makes a great hedge and will be buzzing with pollinators.

Melissa officinalis above centre
Easy and aromatic, lemon balm thrives in part-sun or shade and is useful for underplanting. Its citrus-scented leaves can be chewed for a double hit or used to make a calming tea.

Miscanthus sinensis above
These are superb grasses for autumn and winter with rustling foliage and silky, feather-duster-like seedheads. Superb M. s. var. condensatus 'Cosmopolitan' is variegated; others such as 'Purple Fall' have fiery tints.

Lonicera periclymenum 'Scentsation' left
This selection has masses of soft yellow flowers, and the seductive scent of honeysuckle carries on the air in late spring, especially in early evening. Scramble it up a tree or fence.

Choosing materials

It is not just planting that defines a garden. The texture and shape of the hard materials you select, whether for surfaces, boundaries, or structures, are an integral part of the design. Different materials add shape, colour, and movement to lure you in and to determine where the eye is drawn, while materials sympathetic to the house or the local environment produce a more balanced approach.

When making your selection, consider the view from the house. Do you want to soften large areas of hard landscaping by incorporating a mixture of materials – slate with gravel, or wood with crushed shells, perhaps? Paths that are heavily used need to be solid, but a secondary walkway can be constructed from gravel, bark, or stepping stones. Using the same material for a path and a terrace creates continuity; a change further along will suggest a different area of the garden.

Laying materials lengthways or widthways draws the eye onwards or to the side, and obscuring paths invites exploration. Walls and solid screens shut out the vista, while open screens and apertures provide teasing glimpses of what lies beyond.

Furniture should be in keeping with the style of the garden. Ensure any timber pieces carry the Forest Stewardship Council (FSC) logo to show that the wood comes from sustainable forests. Also consider the siting: if you want a large dining table and chairs, you may have to build a terrace big enough to accommodate them.

Most gardens will have a spot for a water feature, as well as a piece of art. If you plan to include lighting, the electricity supply and cables must be installed by a qualified electrician; solar lighting has to be accessible to sunlight. Outdoor heating is becoming popular, too, but consideration should be given to its environmental impact.

Top and above
Tall metal containers form a divide along this path edge.
Permeable materials provide environmentally friendly parking.

Materials for surfaces

Large areas of paving or decking are visually dominant features and have a significant impact on the appearance of a garden. Select materials that reinforce your style, complement the colours and textures used, and mix different types to develop patterns and lead the eye around the garden. (See also pp.350–67 for more on materials.)

Paving and decking

A strong design statement, or simply a block of uniform colour, can be achieved with large paved spaces. Bear in mind that when using slabs, pavers, or bricks, the joints will form a pattern too; the smaller the unit, the more complex the pattern will be. Rectilinear paving can be combined to form larger rectangles or grid layouts, or use fluid materials, such as gravel and poured concrete, for curved edges to make organic shapes. All paving must be constructed on a solid base and should slope to allow drainage (see *opposite*).

Decking with a twist
Decking is easy to cut and a good option for both geometric and organic layouts, and intricate designs such as this, with its inlay of blue tiles.

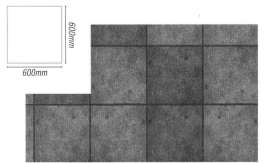

Large pavers may need cutting
When planning an area to be paved, try to avoid cutting by making the overall area an exact multiple of units. If it is not, larger slabs may require more cuts to fit.

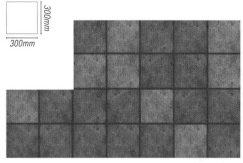

Small pavers fit tighter spaces
Smaller units provide greater flexibility and are more likely to fit exactly the dimension of your patio. They are also easier to cut, when required.

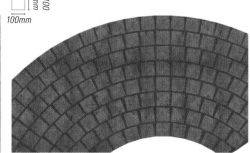

Small units are best for intricate designs
Using small units or even mosaic tiles allows you to create intricate shapes and patterns more easily, but these designs are often very time-consuming to build.

Textured surface
This random paving pattern is framed by a strip made from the same material, giving a clean, sharp edge. Although difficult to construct, the light-coloured textured path works well against the still water of the pond.

Horizontal paving
Bricks are used to frame the edge of this formal path, and stone slabs laid horizontally shift the focus to the planting.

Bricks following direction
The cottage planting is complemented by a traditional brick path that leads the eye to the gate.

Paths and walkways

Paths are the arteries of the garden. Materials should be selected to enhance the journey along the path and to complement the planting on either side. Pavers, and the joints between them, can run lengthways to give a sense of motion, or be laid perpendicular to the direction of travel to slow walking pace and attract attention to the surroundings. Choose paving that matches the garden style: bricks or gravel are good for a cottage-style garden, and more up-to-date materials, such as concrete and composites – or traditional materials used with a contemporary twist – suit a modern space.

Mixing materials

Assorted materials, as well as different textures and levels, can be used to dramatic effect in paving and decking designs. Use different materials to highlight key features, or to define and separate areas, such as a raised wooden deck over a stone-tiled floor. Colours may be complementary or strongly contrasting, but it is best to select pre-sized, coordinating materials to avoid extra work and higher costs; more complex construction techniques may be required when working with materials of varying thicknesses and where a different foundation may be needed.

Wood and slate
This mix of hard and soft materials, with contrasting colours but similar tones, has been combined on four levels to great effect.

Planting opportunities

Plants add colour and texture when squeezed into joints and crevices; take care to choose those that tolerate trampling, are relatively drought-resistant, and ideally produce a scent when crushed. Think carefully about joints when combining paving and plants – a solid foundation, while necessary for most paving, will also contaminate the soil.

Plants between paving
Contrasting colours and textures are combined in this beautifully executed pavement, where mind-your-own-business (Soleirolia soleirolii) frames the paving.

Stones and mosaic
Set on a concrete foundation, these small stone blocks and mosaic tiles create a decorative pattern around the trees and a foil for the gravel.

Complementary textures
Four materials combine here – pebbles, granite, slate, and gravel – to give interest and texture to a threshold between two paths.

Edging ideas

Most paving materials, except *in situ* (poured) concrete, or those set on a concrete slab, will require an edge to contain the material. The edge can be detailed or functional depending on the style of your garden, and also connect or separate different materials or areas of planting. However, you may not need an edge if you intend to allow planting to invade your gravel pathway.

Pebbles
Loose pebbles make an informal edge between the deck boards and the rill.

Slate and setts
This bold design is created by slate paving butting up to stone granite units.

Gravel and paving
Make a design statement with a clear, decorative edging pattern.

Drainage issues

All surfaces should slope to allow water to drain or be collected, and even gravel surfaces may need extra drainage if laid on clay-rich soil. Ensure that rainwater runs away from buildings into collection points, such as gullies; water from small areas of paving can be directed into planting beds.

Slightly sloping patio
Create a slope away from buildings towards a collection point. Patios made from rougher materials will need to slope more steeply than smooth ones.

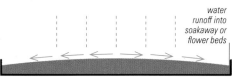

Cambered path
Paths can be profiled to allow water to run off on both sides, where it can be collected in channels or allowed to drain into planting beds.

Materials for screens and boundaries

Walls and boundary features, and the materials they are made from, have a major impact on the look of a garden. Traditionally, boundaries were constructed from local materials, such as stone, brick, timber, or hedging, but today your options are much broader, and modern gardens may make use of smooth rendering, metal screens, or reinforced concrete. If you share a boundary your choice may be limited, but if not, you can make it as subtle or as dominant as you wish, and add a personal touch with your choice of material, colour, shape, and texture.

Stone
Well-constructed stone walls should last for ever but require an expensive initial investment.

Walls and solid screens

Brick, stone, or rendered walls enclose spaces and form a framework around the garden. Solid foundations and specialist construction skills may be required, and these boundaries can demand a large proportion of your building budget.

The colour of stone and brick walls is best left unaltered, so take this into account when making your choice. Consider the size and shape of the units, too, which can range from random rubble to expensive dressed stone blocks. Artificial materials, such as concrete, offer almost endless possibilities in terms of both colour and shape, providing clean lines or fluid structures.

Brick right
Brick has been used for centuries and is durable and useful for creating patterned designs.

Rendered far right
For flexibility and quick and easy construction, consider using rendered concrete walls.

Enhancing walls

Once you've decided on a material, think about any details you could add, whether for aesthetic or practical purposes. You could consider adding colour to all or some of the wall, depending on the material. Masonry walls, especially those made with mortar, render, or clay bricks, benefit from capping or coping to frame the top of the wall and allow water to run off. However, ensure that it is in proportion to the size of the structure. Planting in crevices is another possibility, but select species carefully.

UNUSUAL MATERIALS

As long as walls are stable and shed water, most materials that are suitable for outdoor use can be used. Visit websites, look at books, or visit trade shows, but remember that specialist construction techniques may be required.

Planting pockets
Plants will soon establish in pockets of soil at the top or on the face of a wall. Limited water will be available to them, however, so choose species that can survive and flourish in dry conditions.

Rendered coping
Coping keeps the body of the wall dry and protects it from frost damage. It also forms an important visual element and can make a useful horizontal surface for a decorative effect or for seating.

Textured wall
The walls of this small urban garden have been covered with old billboard vinyl for a dramatically individual, textured look.

Fencing and trellis

Timber and metal fences do not require strong strip foundations or heavy building materials, and so are usually cheap and easy to build. Most are made from strips of material, and you should think about a design based on a combination of these "lines". To unify the design of an existing garden, it may be best to simply repeat or copy the original fencing styles. However, for new designs you can create patterns using different lengths, widths, and shapes of timber. In exposed areas, leave gaps in the fencing to allow some wind to pass through (see *diagrams below*).

Effective windbreaks
Solid screens do not allow any wind to pass through them and create turbulence on the leeward side. Use a perforated screen, such as a trellis, to solve this problem.

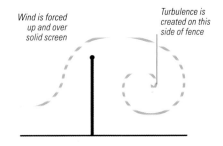

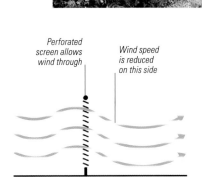

Solid fence
This tall, close-boarded fence creates privacy and has been stained grey to enhance the overall composition.

Perforated fence
The strong pattern of this fence complements the garden, and acts as a decorative windbreak.

Gates and apertures

While screens and boundaries enclose space, they also create barriers that restrict movement and views. Punctuating these with doorways, gates, windows, and other apertures allows access or visual links to other parts of the garden. Importantly, these features provide further opportunities for attractive details and should not be dismissed as utilitarian access points. Choose complementary materials and consider how apertures can frame vistas and views. Also, design doors and gates that look attractive when both open and closed.

Picket fence
When closed, this picket gate blends in with the rest of the fence; the only breaks in continuity are the posts and braces required for structural stability.

Classic doorway
A traditional ledge-and-brace door makes a beautiful contribution to this old brick wall, as well as providing access. When left ajar, it gives an enticing glimpse through to another part of the garden.

Modern aperture
This perforated, reinforced concrete screen would be difficult to construct, but the beautiful results link the contemporary structure to the natural planting beyond.

Materials for slopes and structures

Raised beds, retaining walls, and similar structures that hold soil need to be constructed from water-, frost-, and stain-resistant materials. Natural materials, such as stone and some metals, are obvious choices, but rendered concrete and even sheet metal could be used for a more contemporary look. For garden structures such as pergolas and sheds, choose materials that are lightweight and easy to fit together, and that provide an opportunity to combine colours, textures, and patterns.

Retaining walls

Heavy or strong materials, such as stone, concrete blocks, bricks, timber, sheet metal, or reinforced concrete, are necessary for a retaining wall. Your wall needs to hold water as well as soil, and will require a drain to relieve the build-up of water, unless you have used a permeable material such as dry stone (see *pp.258–59 for more information on building a retaining wall*). You should consult a structural engineer for advice on any impermeable retaining wall above 1m (3ft) in height. Consider coordinating your wall with the house, a water feature, or screen to help unify your garden style.

Wooden walls
Timber walls are reasonably simple to construct: the individual sections will need to be screwed together for added strength and stability.

Dry stone walls
A dry stone wall works well in rural gardens. Place landscape fabric behind the wall to trap soil but allow water to pass through the gaps in the stones.

Raised beds

Essentially low retaining walls, raised beds do not need to be as strong or as heavy as larger structures. They can also be more elegantly designed, rather than serving a purely functional purpose. Line beds with heavy-duty plastic (with drainage holes punched in the bottom) to retain soil moisture and avoid leakage and staining. Also choose materials that complement the plants you plan to use, as well as the composition of your garden.

Contemporary beds
Although susceptible to knocks and dents, metal lends a contemporary note to raised beds. Lighter coloured and galvanized metals do not conduct heat as well as darker metals, and plants are therefore less likely to suffer from scorched roots.

Country charm
For vegetables and native planting, consider woven beds to complement your scheme. They are comparatively short-lived and will need replacing after a few years but add rustic charm to a kitchen or cottage garden.

Elegant containers
Beautifully detailed and finished timber beds can add to the quality of a crisp, modern design. The addition of a gravel margin will keep the timber pristine.

Garden structures

Many suppliers produce prefabricated garden structures, or you may prefer a bespoke design if you have something specific in mind and your budget allows. If you have a small garden, a structure can dominate the space, so plan carefully to ensure that it makes a positive contribution to your design. The materials you choose for the structure can reinforce a particular style. For a sharp, modern look, combine clean-sawn timber with glass and stainless steel, or consider rough-sawn timber for a rustic shed in a woodland-style garden. Hardwood is expensive but durable and does not require treating, but ensure that you use only FSC-certified woods from sustainable forests. A cheaper option is softwood, pressure treated for durability and stained with a coloured preservative, or recycled timber. Metal structures can be light, elegant, and contemporary, and galvanized steel, painted if desired, is a popular choice. Self-oxidizing metals such as Corten steel and copper (ideal for roofs), which develop a green patina as they age, should last indefinitely.

Open structure
This pergola is constructed using powder-coated aluminium combined with a wood trim (see pp.266–67 for more information on constructing a pergola).

Blending in
The choice of dark stain allows this large garden office to recede into the background, while the stainless-steel staircase gives a modern touch.

Step style

To prevent timber and metal steps rotting or rusting, they need to be supported on a solid framework above soil level. Stone slabs can also be constructed in the same way. Alternatively, solid blocks of stone, concrete, or timber can sit directly on the ground on a slope, or smaller units, such as paving slabs, can be used with a retaining edge. Consider the surrounding planting – you can allow it to "intrude" on to or grow through your steps – and the material used for areas around the steps.

Bound chippings
These stylish steps are made from galvanized metal risers and bound crushed CDs (an alternative to gravel).

Metal steps
Strong and durable, these stainless-steel grid steps allow planting to creep between them.

Wooden stairs
Timber steps supported on posts and bearers, like these, can be built to any height.

Materials for water features

When choosing and planning your water feature, make sure that it fits in with the composition of your garden, perhaps using materials that feature elsewhere in the design. Water features can be complex, so consult an expert or research water gardening in detail before planning one. Remember that you will need to ask a qualified electrician to bring an electricity supply into the garden, and some specialist water feature mechanisms and materials may also require expert installation.

Containing water

Waterproof masonry, such as concrete, will seal in the water in your feature, whether it is a raised or sunken pool. Any material with joints, such as bricks, will leak, so add a specialized render to the inside of your pond, which can then be coloured or clad with tiles; alternatively, line it with a waterproof membrane, such as butyl. Take care not to add any decoration that could puncture the waterproof layer or liner, and ensure that any joints where pipes enter the pool are fully watertight.

Wildlife pond
Covering the edge of a butyl liner with flat stones will protect it, but ensure that they are smooth-edged to prevent punctures.

Raised pool
A pond like this can be created with a pre-formed fibreglass liner and enclosed with brick walls that match other garden features or the house.

Edging and lining streams

Natural-looking water features, such as artificial streams or wildlife ponds, are usually irregularly shaped and lined with flexible butyl (see p.270). Ensure that the pond is deep enough in places to allow the required rooting depth for your chosen aquatic plants (see p.214). Streams require a "header pool" or reservoir at the top of the slope, into which water is pumped from the lowest pool. Cover the edges of your pool or stream with planting or flat stones to conceal the waterproof membrane.

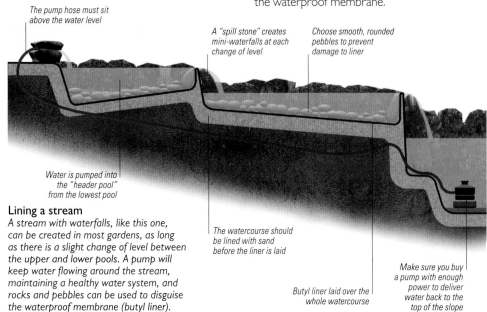

The pump hose must sit above the water level

A "spill stone" creates mini-waterfalls at each change of level

Choose smooth, rounded pebbles to prevent damage to liner

Water is pumped into the "header pool" from the lowest pool

The watercourse should be lined with sand before the liner is laid

Butyl liner laid over the whole watercourse

Make sure you buy a pump with enough power to deliver water back to the top of the slope

Lining a stream
A stream with waterfalls, like this one, can be created in most gardens, as long as there is a slight change of level between the upper and lower pools. A pump will keep water flowing around the stream, maintaining a healthy water system, and rocks and pebbles can be used to disguise the waterproof membrane (butyl liner).

Naturalistic waterfall
This artificial pond is on two levels and has been lined with a membrane covered with flat stones; large stones overhang the edge of each level to protect the liner from damage and to create mini waterfalls.

Design materials checklist

The following table will allow you to quickly compare various materials, and their general suitability for the garden design and features you have in mind. This is intended as a guide, and you should consult other sources (especially product websites) for more comprehensive information when making your choice of materials.

KEY

Durability	Cost
◆ low	£ cheap
◆◆ medium	££ average
◆◆◆ high	£££ expensive

MATERIAL	USE	DURABILITY	IMPACT ON ENVIRONMENT	COST	CONSTRUCTION
POURED CONCRETE	Foundations, walls, pools, surfaces, steps	◆◆◆	High	££	Simple construction easy; can be highly specialist
PRE-CAST CONCRETE	Paving units, blocks, building units, reconstituted stone	◆◆◆	High	££	Easy, but requires skill to achieve high-quality finish
RENDERING	Joints, surface finishes	◆◆	Medium–high	£–££	DIY possible, but skill required to achieve high-quality finish
AGGREGATE	Paving, foundations, drainage, decorative finishes	◆◆◆	Depends on source	£–££	Easy, except wall finishes
BRICK	Paths, surfaces, walls, retaining walls	◆◆◆	Medium	£–£££	DIY possible, but skill required to achieve high-quality finish
EARTH CONSTRUCTION	Walls, retaining walls	◆◆◆	Low	£–£££	DIY possible, but skill required to achieve high-quality finish
LOCAL STONE	Paving, walls, structures	◆◆◆	Medium	£–£££	Variable: irregular stone needs skill for all but basic walling
IMPORTED STONE	Paving, walls, structures	◆◆◆	High	£–£££	Variable: irregular stone needs skill for all but basic walling
CERAMIC TILES	Decorative finishes	Mostly ◆◆◆	High	£–£££	DIY possible, but skill required to achieve high-quality finish
SOFTWOOD TIMBER	Construction timber, fences, gates, decks, paving, structures, furniture	◆–◆◆	Low–medium	£	Easy, but requires skill to achieve high-quality finish
HARDWOOD TIMBER	Decorative details, fences, gates, decks, paving, structures, furniture	◆◆◆	High if from unsustainable source	££	DIY possible, but skill required to achieve high-quality finish
NATURAL WOVEN TIMBER	Fences, hurdles, planters	◆	Low	£	Quite easy, but requires skill to achieve high-quality finish
MILD STEEL	Fences, railings, fixings, structures	◆◆ if not protected	Medium	££	Difficult – requires specialist skills
STAINLESS STEEL	Fences, railings, fixings, structures	◆◆◆	High	£££	Very difficult – requires specialist skills
SPECIAL STEEL ALLOYS	Fences, railings, fixings, structures	Mostly ◆◆◆	Variable	£££	Very difficult – requires specialist skills
ALUMINIUM	Lightweight structures, greenhouses	◆◆◆	Medium	££	DIY possible, but skill required to achieve high-quality finish
COPPER	Pipework, decorative cladding	◆◆◆	High	££	Difficult – requires specialist skills
ZINC	Planters, decorative cladding	◆◆◆	Medium	££	Difficult – requires specialist skills
GLASS	Screens, barriers, windows, surfaces, glasshouses	◆◆	High	£££	Very difficult – requires specialist skills
PLASTICS	Pipes, furniture, fixings, decorative facings	◆◆	High	£	Variable – DIY possible
PERSPEX/ PLEXIGLAS	Screens, structures, windows	◆◆	High	££	Difficult – requires specialist skills

Designing with furniture

A well-placed bench, lounger, or chair is an invitation to spend time relaxing in the garden. Whether permanent or temporary, garden furniture can have a marked effect on the look and feel of an outdoor space. The sculptural qualities of a particularly eye-catching or stylish piece of furniture could even be viewed as garden art. Of course, looks aren't everything, so do ensure that your chairs and tables are comfortable and practical.

Matching your garden style

Furniture has the potential to strengthen a design and create focal points within it. When the style of a plot is distinctive, such as in a Japanese garden, it's best to choose elements that follow the theme faithfully or that have a strong visual relationship to it. For example, cottage garden seating is likely to have a softer, more rustic and homespun feel. You might use wicker or Lloyd Loom chairs or reclaimed farmhouse kitchen furniture. In contrast, seating for contemporary settings works best if it has sleek, minimalist lines and is made from modern materials and fabrics, such as aluminium, plastic, or synthetic rattan. The architecture of the house often influences garden style, and in the grounds of a period property, pieces from the wrong era can stand out like a sore thumb. You don't have to source originals, however: many companies offer quality reproductions.

Secret corner
Simple foldaway furniture, light enough to carry around, is ideal for making use of different areas of the garden. Consider painting it to create highlights.

Integrating furniture into a design

The size and shape of the available space will influence the type of furniture you choose; intimate corners surrounded by planting may, for example, only have room for a couple of foldaway seats. For outdoor dining, carefully calculate the size of table and chairs you can accommodate to ensure a comfortable fit, and select furniture that mirrors the shape of your terrace or patio – a round table on a circular patio not only fits perfectly but also accentuates the curved layout. A decorative seat can make an excellent focal point.

Minimalist lines
Large pieces of furniture, like this modern sunlounger, need space and a simple backdrop to allow their sculptural qualities to be fully appreciated.

Integrated design
Try to match furniture to your design. This quirky, rustic site is enhanced by the bespoke wooden bench seat constructed from reclaimed materials.

Space to lounge
Consider the size of the available space before buying furniture, or design your garden around chosen pieces. This sofa fits perfectly beneath its modern canopy.

Practical considerations

If you want to keep your furniture outside all year, check before you buy that it is resistant to rain and UV damage. Modern synthetic rattan furniture and plastic or resin pieces often come with guarantees, but while sofas and chairs with all-weather cushions will dry quickly after a shower, it is a good idea to cover them when they are not in regular use – an outdoor storage locker could prove useful for this. To retain the original patina on wooden furniture – which weathers and may change colour if left outside all year – clean, oil, or varnish it regularly and, if possible, cover it during the winter.

Outdoor sitting room
Buy plush, upholstered furniture with fade-resistant, shower-proof covers and ensure that the frames are sufficiently durable for outdoor use. Protect heavy pieces in situ.

Dining in style
Hardwood, aluminium, and synthetic woven mesh combine in this stylish yet durable dining table and chairs for a contemporary urban garden.

Environmental factors

Tropical hardwoods like teak have long been used to manufacture garden furniture because of their natural durability. However, this type of timber is not always obtained from a sustainable source, and uncontrolled logging is having a devastating effect on the environment. Always check the source before you buy; temperate hardwoods such as oak or more durable softwoods are likely to be "greener". Also look for furniture that has been manufactured from reclaimed wood, which can add a rustic quality to a design.

Greener options
Check for Forest Stewardship Council (FSC) certification on wooden furniture to ensure that forests have been managed in an environmentally responsible way.

STORAGE IDEAS

In small urban gardens in particular, the lack of space available outdoors to store items such as gardening equipment, furniture cushions, and children's toys can present a real problem. One option is to choose garden seating that also provides storage, such as benches with hinged lids for access. Use a liner inside your storage to create a waterproof area to keep more delicate items safe. Alternatively, buy garden cupboards and boxes specifically designed to store cushions over winter from specialist furniture suppliers.

Storage lockers double as garden seating.

Furniture styles

There are so many ways to buy garden furniture: in garden centres and home stores, of course, but increasingly there are many online stockists with a huge range of furniture styles, designs, and price points. There is definitely something for everyone out there. Once you start looking for furniture, you'll realize that the choice is vast, so persevere to find pieces that fit your garden style perfectly.

Traditional rustic

In more relaxed country- and cottage-style gardens, sleek furniture could well look out of place, though modern pieces with organic forms based on natural shapes may be appropriate. Quirky, reclaimed furniture is worth seeking out, as well as woven and wicker sets. The latter will weather rapidly, so you'll need a convenient storage place, such as a summerhouse or shed. Don't be afraid to mix and match country styles with classic pieces: lightweight, portable foldaway tables and chairs made from wood and metal can work well in period gardens with authentic-looking reproductions, such as Victorian fern seats or Lutyens-style benches.

Simple style
Traditional, hard-working or utilitarian designs add to the relaxed atmosphere of a cottage- or country-style garden.

Willow weave
Though not as durable as wood, wicker furniture, like this circular tree seat, adds romantic charm to an old-fashioned plot.

Chic modernist

A seating area dressed with designer furniture makes a strong statement, particularly in urban courtyards and on roof terraces, where the garden often functions as an extension of the house. Modern, minimalist items made of steel and synthetic mesh fabric or synthetic rattan can add style and comfort to a contemporary design, while all-weather beanbags add colourful highlights. This look is about bringing interior style outdoors, so cushions and matching light fittings and containers play an important linking role.

Sixties model
This up-to-the-minute design echoes the styling of the 1960s. The doughnut form contrasts well with the sparse backdrop.

Comfortable corner
Seating areas are increasingly popular, emulating a style that many people would be pleased to have indoors.

Contemporary looks

It's hard to put your finger on why certain furniture styles have an up-to-date feel. Sometimes a traditional item or seating shape is updated for the 21st century using hi-tech materials; sometimes designs from previous decades experience a revival. In general, clean lines and plain, neutral-coloured fabrics, coupled with human-made elements like steel, glass, and chrome appear modern. Today's designers are developing the architectural role of furniture, as well as working on integrated or site-specific designs.

Spiral appeal
This curving, raised walkway culminating in a seat that "floats" on transparent glass is a piece of sculpture in its own right.

Confident colour
Furniture doesn't have to play second fiddle to a garden's aesthetic – its colour can be a confident focal point in itself.

Furniture as art

There's no doubt that the sculptural qualities of certain furniture items, typically in wood, metal, ceramic, or resin, put them into a different category from everyday functional seating. You can order sculptural furniture online and find artists via their websites, but it is also worth visiting the studios of local craftspeople, as well as gardening shows and galleries, to commission bespoke items. If possible, allow the artist to see the garden and the site for the piece, or provide as many photographs as possible, as this can really affect the success of the design.

Modern abstract *above*
The organic form of snail shells was the inspiration for this original bench with a carved wood seat.

Sleek in steel *below*
These boldly sculptural chairs are constructed from a perforated steel that softens their impact in the overall design.

Integral seating

You can create impromptu seating simply by utilizing steps, sunken areas, and the walls of raised beds: just add a few cushions and you can accommodate a large group of people with ease. Elsewhere, a seat or table could follow the contours of a landscape feature, such as a serpentine wall.

Built-in beauty
Integrated seating can have an intimate feel. A cosy nook for relaxation could be created in a wall alcove, as here, or perhaps carved into a tall hedge.

Temporary seating

As your garden changes through the year, different areas will become more or less attractive or accessible. A portable seat, such as a director's chair, allows you to take advantage of particular settings or to follow the sun around the garden.

Deckchair classic
The wonderful thing about collapsible furniture is that you can easily move it to where it's needed and view the garden from different angles.

Integrating sculpture into a design

When choosing sculpture, you don't need to be limited by what's on offer in your local garden centre. Many objects take on sculptural qualities when placed in a garden, including beautifully shaped ceramic vases, driftwood, rounded boulders, or even pieces of disused machinery, so be as imaginative as possible. Think carefully about the relationship of your sculpture to the rest of your garden, where you will position it for best effect, and how its appearance will change over time.

Choosing sculpture

The appeal of a sculpture depends largely on your emotional response to it. You may prefer abstract shapes for the garden, especially if the style of your plot is sleek and modern, but wildflower gardens or woodland can also provide an exciting setting for a contemporary piece. Equally, classical statuary can add an element of surprise in a modern rectilinear layout and will enhance an urban space. In cottage gardens, try figures of domestic animals, beehives, or rustic farm equipment.

Plant form
This rusting iron sculpture, reminiscent of a flowering plant, works well in the Mediterranean-style setting. As the surface weathers, the patina will subtly change.

Figurative top
With one toe dipping into the water, this figure adds a relaxed and humorous touch to this contemporary landscape.

Topiary above
Clipped greenery, a type of living sculpture, has many forms and includes Japanese cloud pruning.

Abstract left
Bold and strong, some sculpture demands attention in its form and colour. Shadows and light can give an added dimension.

Positioning sculpture

Take time to find the right spot for garden art and to integrate it into your design. Some pieces work best surrounded by reflective water or by plants in a border. Contrast simple, solid shapes with diaphanous grass heads, for example, or view them through a haze of lavender. Intricately detailed sculptures look best with a plain backdrop, such as a rendered wall or clipped yew hedge. Matt surfaces like natural stone or weathered timber create a foil for highly polished metals, and you can use these materials to mount smaller sculptures, too.

Gazing skyward
John O'Connor's bronze child takes your gaze up to decorative fretwork on a pavilion roof above, while the colour blends harmoniously with the timber frame.

Commissioning a piece

You may discover someone whose work you admire by visiting national or regional gardening shows, dropping in at an artist's studio open day, or checking sculpture and land art websites. Help your chosen artist to visualize what you have in mind with rough sketches and photographs and, if possible, organize a site visit for them. Agree at the outset on the design, its dimensions, and the materials to be used, as well as confirming a price and delivery date for the work.

Materials and cost

There are often less expensive alternatives to traditional sculpture materials. Reconstituted stone, terracotta, or ceramic ornaments, for example, are far cheaper than carved stone, and bronze resin costs less than cast bronze, while lead statuary reproductions are relatively inexpensive. You may also find artists working with driftwood or reclaimed wood, rather than expensive hardwoods.

Focal point
This abstract piece appears to hover over the surface of the pool, which also reflects its image, and makes an eye-catching focal point in this small garden.

Hidden torso
Half-hidden by foliage, this weathered terracotta torso appears to grow out of the landscape and would be a fraction of the cost of a bronze piece.

Space to perform
The tall, cartoon-like figure of a girl striding briskly across the garden creates focus but needs a large area to convey her energy and momentum.

Scale and proportion

A small piece of sculpture may be lost in a large, open site, but bring it into an intimate courtyard and you'll find that it's in perfect proportion to its surroundings. Try "anchoring" small ornaments by placing them next to a solid piece like a boulder, a hunk of driftwood, or an oversized vase. Alternatively, mount decorative objects and plaques, fit them into alcoves in walls and hedges, or raise them closer to head height on plinths. To gauge the size of sculpture required for a site – when planning a focal point at the end of a formal path or at the side of a pool, for example – use piles of cardboard boxes or plastic refuse bins to help you visualize how the sculpture will fit into the proposed setting.

THEFT AND PROTECTION

Use common sense when placing your sculptures: try to keep them out of sight of passers-by and consider using alarms or security fixings. For a front garden, choose pieces that are too large and heavy to be carried off easily and keep them close to the house. Ensure that garden sculpture is covered by your home and contents insurance, and let your insurer know about new purchases.

Designing with lights

The beauty of installing creative lighting is that you can design an entirely different look for your garden at night. Soft, subtle lighting, bringing just a few choice elements into focus, is relatively straightforward and makes the most of differing textures and contours. More theatrical styling is possible with the wide range of specialist lighting equipment available. There are important aspects of safety and security to be considered, and you should always discuss your plans with an electrician.

Lighting in the garden

Flooding the garden with light from above creates too harsh an effect, and can cause nuisance to neighbours and add to the problem of light pollution. Avoid strong lights that may shine directly into the eyes of an onlooker. By maintaining areas of shadow, you can accentuate the theatrical effect of any garden illumination and make the night-time experience all the more enchanting. Draw up a plan, taking into account the type of lighting required in each area, such as recessed lighting for a deck, directional spotlighting for a barbecue, or underwater lighting for a fountain. Work out cabling circuits and plug points, and talk through your ideas with a qualified electrician or lighting engineer, preferably before completing any new landscaping work. You can experiment with different lighting effects by simply using a powerful torch, or torches, held at different angles.

Nightlife
Outdoor rooms used for relaxation and entertaining can be lit in a similar way to indoors with low-level lamps and mini spots to highlight decorative elements.

Ways with water
Moving water features such as cascades are easier to light than static pools as the surface disturbance masks the light source, while planting can hide cables.

Coloured glow
In contemporary settings, restrained use of coloured lights can create stylish effects. Programmed, colour-changing LEDs are an option for dynamic shows.

Practical considerations

Unless you plan to use solar-powered lights, you need a convenient power supply. Special waterproof outdoor sockets must be installed by a qualified electrician, and any mains cabling needs armoured ducting to prevent accidents. When using low-voltage lights that run from a transformer, house the transformer in a waterproof casing or locate it inside a building. A transformer reduces the voltage from the mains to a lower level at which many garden lighting products work. The size of transformer you will need depends on the power and number of lights you plan to use. Ask your electrician to install an indoor switch so that you can turn the lights on and off easily. LED (light-emitting-diode) lights are both energy efficient and create no heat, making them particularly safe to use in the garden; you will find a huge selection available. If an area is sufficiently sunny, solar-powered lighting is another good option.

Safe passage
If you plan to use the garden at night, illuminate pathways, steps, and changes in level using low-level lighting and angled recessed lights to avoid glare.

Path lighting
Post lights come in a wide variety of designs, including many solar-powered models, and sets that run from a transformer. Position in the border to light pathways.

Flickering flames
Candles, lanterns, and oil lamps create a magical atmosphere. Never leave them unattended and take care to keep naked flames away from flammable materials.

CHOOSING MATERIALS

Lighting effects

Tiny LED fairy lights running from a transformer are simple to install and create a romantic ambience when woven through climbers on a pergola. Mini spots are great for uplighting an architectural plant or a piece of statuary, or for highlighting textured surfaces. Recessed, low-level lighting in steps, walls, and decks casts gentle light without glare, and coloured lighting can be used to create contemporary effects, floodlight trees or rendered walls, or to light pools. For a contemporary look, try small white or coloured LED spots set into a decked area or a few underwater lights to illuminate a clear, reflective pool.

Mirroring
A single source of illumination bathes this poolside terrace in soft light and produces a perfect reflection in the black, unlit surface.

Uplighting
Matt black mini uplighters are inconspicuous during the day but can be angled to reveal the shape and texture of plants, decorative elements, walls, and screens at night.

Floodlighting
Bright, even lighting is mainly used for security and can be triggered by infrared sensors. LED spotlights can also be used for dramatic up- or downlighting.

Spotlighting
Using a directional spotlight mounted high on a wall and angled in and down towards the subject, you can highlight an area without creating irritating glare.

Backlighting
Low-level backlighting throws the foreground elements into relief and creates dramatic shadow patterns on the wall behind. You can also backlight decorative screens.

Grazing
This term refers to the effect achieved by setting a light close to or along a wall or floor. It can be angled to illuminate an area and reveal texture and form.

Choosing lighting and heating

With such a wealth of creative garden lighting now available, it can be difficult to decide what's right for you: this section looks at the relative merits of each option. Heating systems have increased in their range, style, and popularity over the years, but the impact of some of them on the environment – and cost to run – are important issues to consider.

Types of lighting

With the exception of solar-powered lighting, and candles and oil lamps, all other illumination devices need to be connected to the mains. Lights either work directly from the mains or through a transformer that provides a low-voltage current – ideal for a garden, as water and direct current are a lethal combination.

Garden lighting has been revolutionized by the introduction of efficient LEDs and more reliable and sophisticated solar-powered units. LEDs offer all kinds of "designer" effects, including lights that change colour and systems that can be controlled via a smart phone. While DIY stores carry an increasingly wide range, the largest choice can be found online and via specialist companies.

Always employ a qualified electrician to install mains lighting, to make connections to mains power, and to fit new switches and plugs.

Light show
This courtyard is bathed in light in the evenings, with subtle LEDs grazing the walls and illuminating the modern water feature.

TYPE OF LIGHT

This table shows the pros and cons of the main forms of lighting, but for most types it is also best to discuss your requirements with an electrician or lighting engineer.

WHERE TO SITE

EXTENT OF ILLUMINATION

EXPENSE

INSTALLATION

MAINTENANCE

Heating in the garden

Introducing some kind of environmentally friendly heat source into the garden extends the use of the plot into the cool of the evening or in spring and autumn. Wherever possible, burn logs and prunings cut from your own garden. Never use treated or tanalized timber, and make sure you read the instructions on appliances to check the type of fuel you can burn. Safety gloves are a must as fire grates get very hot, and make sure you allow chimeneas to cool before covering them. Keep a fire extinguisher handy, and use fireguards.

TYPE OF HEATING	PROS	CONS
FIRE PIT	DIY build possible. Some designs portable. Focal point, with potential for 360° seating. Heats and cooks. Burns garden prunings.	Needs space and safety screen. Ash may stain light surrounds. Poses a danger to children and pets – do not leave unattended.
FIREPLACE	Many different models including cast-iron stoves. A range of different materials can be burnt within the garden feature.	Larger designs, including those made from stone, take up space and are permanent fixtures. Cast iron rusts.
CHIMENEA	Fits into a small space. Clay designs often very decorative. Easy to cover and protect from weathering.	Both clay and metal types can crack. Clay may start to crumble after absorbing a lot of moisture. Tricky to clean out ashes.
GAS/ELECTRIC	Convenient and no cleaning up afterwards. Instant heat and/or cooking with flexibility; easily controllable.	Burns fossil fuels. Very inefficient considering amount of energy used and heat produced. Heavy cylinders for gas heaters.

CHOOSING MATERIALS

LED	LIVE FLAME	ELECTRIC	SOLAR-POWERED

Almost anywhere in the garden. Can be used as pool lighting, recessed lighting, fairy lights, spots, or for security.	Candles, oil lamps, and lanterns may be placed on the ground, in wall niches, on tables, hung from hooks, or floated.	Fluorescent and halogen lights are used for security, spotlights, and lamps, although less extensively – LEDs are favoured now.	Edge of pathways/patios; in ponds (floating/rock lights); on walls; by plants. Some types suitable as spotlights.
Very bright for the size of unit. Casings can enhance and focus light output, while diffusers help to soften it.	Low-level, atmospheric lighting. Candelabras and lanterns are suitable for outdoor dining.	Varies according to fixture – halogens can illuminate entire garden. Coloured fluorescents are for special effects.	Units fitted with modern solar-powered LEDs can be quite bright. Strength of illumination depends on battery type.
Initial costs of units vary considerably, but the running costs are very low and the bulbs can last for years.	Candles, gel, and oil lamps are inexpensive compared to electric fittings, but do not offer comparable lighting.	Relatively inexpensive to buy but running costs add up, and the bulbs will need to be replaced more frequently than LEDs.	Costs vary considerably depending on quality. Lights require no mains power; installation and running costs are zero.
The same as conventional bulbs – running off mains power or transformer. Useful for hard-to-reach areas.	Take care to site live flames safely on a non-flammable, level surface in shelter. Never leave a candle or lamp unattended.	Lighting can run off mains power or transformer. Consult a qualified electrician for installation (*see opposite*).	Safe and easy DIY lighting. Needs airy spot to operate well. May not light the garden for as long in winter.
LED bulbs last many times longer than other types, and once installed require very little or no maintenance.	Trim wick to keep candle flame low and efficient. Extinguish with a snuffer. Do not move candles when wax is liquid.	Replace bulbs when they burn out. Keep wall lamps and infrared sensors clean.	Photovoltaic cells need regular cleaning. Good quality rechargeable batteries can last up to 20 years.

Fire pit
An updated version of the campfire, fire pits are a draw at social gatherings and may also be used for cooking.

Chimenea
The chimenea, originally a Mexican device for heating and cooking, comes in several different designs. Ensure that the fire is just below the opening to prevent smoking.

Fireplace
This grand fireplace dominates the garden, creating a dramatic outdoor dining area. Simpler, smaller models for average-sized gardens are widely available.

252 **Building garden structures**
- 254 Laying a path
- 256 Laying a patio
- 258 Building a retaining wall
- 260 Laying decking
- 262 Putting up fence posts
- 264 Laying a gravel border
- 266 Building a pergola
- 268 Making a raised bed
- 270 Making a pond

272 **Planting techniques**
- 274 How to plant trees
- 276 How to plant shrubs
- 278 How to plant climbers
- 280 Laying a lawn
- 282 Meadow-inspired plantings
- 284 Aftercare and maintenance

MAKING A GARDEN

Building preparations

Creating a new garden from scratch, or tackling a major hard landscaping project, is a serious undertaking. If you decide to do the work, but only have weekends free, or do all the ground preparations by hand, it could take months to finish. The upside, however, is the immense satisfaction of having done it yourself and the savings on labour. Detailed preparation is paramount, and it is essential that you calculate the cost of all materials, hire equipment, and any professional fees in your budget.

DIY vs employing professionals

Depending on your experience, you may feel confident about tackling a simple paving project, erecting trellis, or building a deck. In fact, many modern building materials and garden features are specifically designed for ease of construction. When taking on any work yourself, ensure you are equipped with the appropriate safety equipment, such as eye protection when sawing timber and steel-toed boots for any construction work. Jobs involving heavy materials or a high level of skill are often best left to professionals. Natural stone, for instance, often comes in large pieces that require skill to cut and lay. Similarly, in a modern garden, crisp design demands a very high-quality finish to avoid it standing out for all the wrong reasons. Experience and expertise count, especially when it comes to safety. Always ensure that you are confident working with the materials you are using to avoid any problems.

If in any doubt about your ability to take on a project, seek expert advice from garden designers or civil engineers; source them via their professional organizations. Also remember when hiring a contractor that they are responsible for taking out insurance, and ensuring that work complies with all safety standards and building codes.

Laying surfaces DIY style
If you have the necessary strength, skill, and experience (such as in using specialist cutting equipment), you may consider building your own new patio or wooden deck.

Laying paving in difficult places
Building stepping-stones that appear to float on the surface of a pool is not easy, as water shows up the slightest discrepancy in levels. Since the steps are to be walked on, they must also be rock solid to avoid accidents.

Sequencing workflow

The value of an experienced contractor is that they know how long it takes to perform various tasks, such as digging and laying foundations, or constructing brick walls. They should also be able to pull together the necessary skilled workforce, just as the next phase is about to commence.

Any project can be dogged with unforeseen difficulties, such as bad weather or delayed deliveries, which hamper the work. As established contractors often have several projects running simultaneously, delays in these other gardens can also have a knock-on effect on yours. Project managers must maintain good communications with all parties, anticipate problems, and find ways to maintain a free-flowing operation. Sit down with your contractors and go through the details of construction together. Then draw up an agreed schedule and refer to it regularly.

Consider lighting before landscaping
Integrated light effects need to be planned well in advance of construction so that fixtures can be built in and cables suitably camouflaged.

KEEPING TO A BUDGET

If you hire a contractor to run a project from start to finish, and have a contract drawn up detailing completion deadlines, material selection, and costs, you shouldn't run into difficulties over the budget. Problems commonly arise when you make changes to the plan mid-way through the build or alter the specifications of the materials used. Good organization is vital if you run the project yourself, especially when hiring a workforce. Workers standing idle, waiting for materials to be delivered, still have to be paid.

Special effects
Some lighting and water features need expert installation, and many materials also require specialist preparation. Always check that your contractors have the relevant experience.

Pre-construction checklist

Once you have completed a site survey and prepared your design, it is time to work out when the construction and planting should take place and who will do the work. You may decide to do some of the preparations yourself and bring in specialist contractors only for specific jobs. Either way, try to visualize the project from start to finish to make it run as efficiently as possible.

STAGE	PROJECT NAME	DETAILED INFORMATION
1	PERMISSION	Major building work, such as the construction of a conservatory or changing access, may need planning permission. Check if in any doubt, and talk to neighbours to explain plans and settle concerns.
2	HIRING CONTRACTORS	One or more contractors may oversee the project, bringing in specialists as needed. If project-managing the job yourself, you will need to find and hire bricklayers, pavers, joiners, electricians, etc.
3	SELECTING MATERIALS	Ask contractors to provide samples of landscaping materials, or visit stone and builder's merchants and timber yards yourself. Personally select feature items and commission bespoke pieces.
4	MATERIALS ORDER/DELIVERY	Double-check amounts to avoid under- or overbuying. Arrange deliveries to coincide with different construction stages. This avoids materials getting in the way and having to be relocated later.
5	SITE CLEARANCE	Peg out area and hire a skip. Remove unwanted hard landscaping materials and features. If it is to be re-laid, lift current lawn with a turf-cutting machine. Also lift and move existing plants for reuse.
6	TOPSOIL REMOVAL	Save quality topsoil for reuse and do not mix with subsoil. Remove it manually or with a mini digger. Locate topsoil away from the construction site and pile it up on the future planting areas.
7	MACHINERY HIRE/ACCESS	If your plan requires a lot of heavy digging, trenching, and re-levelling, hire a mini digger and operator. Ensure suitable access, clearing pathways and removing fence panels, as required.
8	FOUNDATIONS AND DRAINAGE	Establish different site levels and excavate accordingly. Organize the digging of foundations and drainage channels, then pour foundations and lay drainage pipes. If needed, move existing drains.
9	LIGHTING AND POWER	Bring in an electrician or lighting engineer to install the cabling grid for all garden lighting and powered features. Some of these shouldn't be wired up until the garden has been completed.
10	BUILDING AND SURFACES	Build all hard landscaping features, including all walls, steps, terraces, pathways, water features, and raised beds. Construct timber decks, pergolas, and screens. Prepare new lawn areas.
11	BOUNDARY CONSTRUCTION	Once the contractors, builders, or landscapers no longer require access across the boundary for their machinery, vehicles, and materials, walls and fences can be completed and/or repaired.
12	TOPSOIL AND PLANTING	Some basic planting may have to be done during the dormant season while construction continues. Replace or buy in topsoil to make up levels, then carry out remaining planting.

Preparing an overgrown site

It is tempting, when planning a new garden, to simply remove all trace of what went before, especially when a site has become overgrown and fallen into dereliction. But doing this without first assessing what you actually have may result in the loss of some fine mature specimens and other useful plants that can easily be repurposed and incorporated sustainably into the new garden. Existing walls, fences, and garden buildings may also be repaired or reused, saving both money and resources.

Evaluating the garden

The task of assessing your garden is one that takes time. If you maintain it for a year before making changes, you can observe which areas receive morning, afternoon, or evening sun, or those places that stay moist and cool in summer heat. Look carefully at what you have: existing trees may provide privacy from buildings that overlook the site, which could take years to reinstate. If the garden has an existing layout, assess how well it works, considering perhaps if it could be adapted or refreshed. Existing buildings could be relocated or masked rather than demolished. Check the condition of fences and walls, and note materials such as paving slabs that could be repurposed (see also pp.144–45).

Jungle clearing right
Assessing a completely overgrown site is hard. Some initial clearance of weeds and reduction of overgrown planting will be required from the outset to understand the space and what it potentially has to offer.

Glimpses of a garden above
Sometimes remnants of an old garden layout are evident – hard landscaping or mature shrubs – and it may be possible to renovate, reinvigorate, and add to what you have in situ.

What you see is what you get left
With smaller gardens laid mainly to lawn, there may be limited need to assess. Are these fences sound? Could some of the plants be kept?

Assessing existing plants

Trying to identify existing plants in your new garden can be difficult, but worth the effort when deciding which to keep. Ideally, you will need a full year to observe how existing plants perform – including spring bulbs or winter interest shrubs. That ugly thicket may, in fact, be underplanted with a carpet of snowdrops that could take years to replicate were you to rip the lot out.

Notice areas of perennial weed infestation as these will take special measures to clear. It's also worth taking a look over the fence – what your neighbours grow gives an indication of what is possible. Healthy rhododendrons and camellias indicate acid soil; a thriving *Cordyline* or Canary Island date palm, for instance, shows the area seldom suffers much frost.

Seeking garden treasure above
It can take a calendar year to properly assess which plants to keep in an established garden.

Weeds with attitude right
Perennial weeds such as bindweed take time to eradicate. Dig out roots and smother regrowth if it appears to exclude light.

Site clearance

Once the garden has been appraised, clearance can start. Tree work and jobs involving machinery such as chainsaws and stump grinders should be done safely and swiftly by professionals. Plants to remain *in situ* should be marked clearly with ribbons or spray paint; those to be lifted and moved may be planted temporarily in a purpose-made nursery bed, but ensure they are free of perennial weeds and keep well watered. Store materials to be reused or repurposed, and aim to rehome anything unwanted but of use, such as old bricks or even broken concrete. Social media is a good way to spread the word of what you want to get rid of (or what you need more of).

Daunting task above left
Clearing an overgrown site is the first step to your new garden. Once the area is free of unwanted material, the possibilities are easier to visualize.

A job for professionals above
Unless you are trained to use a chainsaw, it is best to employ professionals to clear overgrown trees quickly and safely.

Rehabilitating soil

The soil in your garden will always impact on your planting. In the early stages of a garden build, it is likely earthworks will take place; these may be as simple as stripping and lifting turf. More complicated projects, such as terracing a sloping garden, will require professional help. You may need additional soil or to dispose of existing materials (such as broken concrete hardcore), so planning is needed for project management and to keep costs under control. Always try to keep as much soil on site as possible to reduce your environmental impact.

From the ground upwards

The messiest and most disruptive part of a garden build is usually in the early stages, when the groundworks are underway. This essential stage quite literally lays the foundations for your new garden. Your project may involve considerable changes to the existing site with adjustments in ground level of certain areas, meaning that soil needs to be removed, moved, or added, terracing put in place, foundations of retaining walls and hardstanding established, and, in an old garden, derelict structures removed. All this work is expensive and time intensive but essential; projects that involve considerable upheaval will need careful management to avoid errors and spiralling costs. Using professionals means the choreography of site work and deliveries should run seamlessly. Experts will also have the necessary equipment, keeping this stage as brief and well executed as possible.

Soil deficit above
The depth of the resulting void when hardstanding is removed can be surprising – most patios are constructed on a hardcore layer that serves as a firm foundation.

Going to ground above centre
Levelling an expanse of new garden for a lawn may be most efficiently carried out by professionals using machinery, especially if it involves spreading large quantities of topsoil.

Removing soil above
Major earthworks are involved in terracing a steeply sloping site. Some soil may need to be removed, but ensure you keep sufficient quality topsoil on site for the new garden.

Test your soil top
Always use a test kit to check soil in your new garden – soil pH will dictate the sort of plants you can grow. If you buy in topsoil, ensure it is compatible with the existing garden.

REHABILITATING SOIL

Smashing job
Take safety precautions seriously when breaking up concrete — wear shoes with steel toe caps, eye protection, and a mask.

Removing concrete

Breaking up expanses of a concrete patio or path is laborious and can be an exhausting task. You can smash up small areas of relatively thin concrete with a sledge hammer, but usually this is a task better — and more safely — carried out by experts with power tools, which will make short work of your old hardstanding. Be aware that you will have more heavy material to dispose of than at first seems apparent. Below the concrete there is likely to be a thick layer of hardcore, which, when removed, will leave a void that needs filling, potentially with bought-in topsoil. Before asking builders to remove rubble, consider carefully your needs: might you be able to recycle the hardcore for any future projects? It is worth checking locally to see if neighbours can make use of material before ordering in a skip.

Expert help
A better option for removing and disposing of hardstanding is to call in the experts, who will swiftly and safely complete the job.

Stripping turf

When making or enlarging new borders, laying a path or patio, or carrying out most other garden earthworks, you must usually first lift existing turf. Although tempting, it would be a mistake to dispose of this. The most fertile part of your garden soil, the topsoil, is found in the top 5–20cm (2–8in). It takes around a century for 2–3cm (¾–1¾in) of topsoil to form naturally, although you can speed this up with careful cultivation.

The topsoil is vitally important to your garden's health. When you strip turf, you also lift a significant portion of the topsoil with it. Buying in topsoil is expensive, carries a hefty carbon footprint, and the product can be variable. For small areas of lawn, you can lift the soil, prepare the new ground, and plant — but that isn't realistic for larger areas. By far the best solution is to carefully remove grass with as little soil attached as possible, and then stack the turfs upside down in the garden and allow them to break down naturally. After a couple of years, the resulting soil will be rich and crumbly and can then be added to the garden soil to improve it. Alternatively, use a hand fork and soil sieve to remove as much soil as possible from each lifted turf before adding grass and roots to compost.

Stripping turf
The top layer of garden soil is valuable. Stack lifted turfs in an out-of-the-way corner for a year or two to rot down, and then add soil back to the borders.

Flowery meadow
An annual meadow mix will flourish on ground with very shallow topsoil, such as found at new-build sites.

Riches of poor soil

In the gardens of new-build homes or even in newly developed borders, the depth of workable topsoil will probably be limited and the soil relatively poor. Although limiting, this can be used to the gardener's advantage — not all plants like deep, rich soil. Many wildflowers, drought-tolerant annuals, and a broad range of plants from Mediterranean regions can thrive in shallow, dry, poor soil. Take advantage of this characteristic by considering annual meadow-style plantings or, if drainage is good, create a gravel garden filled with sun and drought-loving plants. While some initial preparation is required, these plantings will flourish. After a few years of cultivation, the soil gradually improves as organic matter is added and structure slowly develops.

Preparing a new-build site

Creating a dream garden from scratch can be an appealing idea, but the reality of new builds is that they can be difficult projects to get started. The key is in knowing where to begin. Rehabilitating the site after building work must take priority. Take a close look at what you have; investigate the ground, see what lies below the surface – often, it will be necessary to repair and improve the soil to make it workable. It's best to do this before any garden design starts and may be easier with professional help, giving you the best foundations for a lovely garden.

Building the dream

Little is more exciting than moving into a new home. It is likely the garden will not be a top priority initially, but eyes soon turn to the big, empty space outside. In some ways, starting a new garden is an easier proposition than dealing with a previous owner's tastes and choices – you don't need to be constrained by any existing features or planting that you may not like. On the other hand, with a new site, you don't know exactly what you are getting. No one has gardened here before, so there are fewer indicators to site conditions.

Often, new developments are initially rather exposed to the elements, with no existing trees and vegetation to provide any shelter. The legacy of recent construction work will likely have left your garden with a variable mixture of topsoil and subsoil, peppered with waste building materials, such as bricks and concrete. It may be a site full of poorly draining, unstructured soil that is hard to work and difficult to plant, but with remedial work, improving and adding to the soil, as well as design solutions such as raised beds, a beautiful garden is perfectly possible.

A blank canvas above
Part of the joy of a new home is that the garden can be whatever you wish. That sunny corner is asking to be filled with beautiful flowering plants, while a rose trained over the front door will improve kerb appeal no end.

Dividing the spoil right
Gardens of new homes often consist of little more than a thin top dressing of soil and rows of fence panels. You will need to consider site access, and how practical substantial changes will be to make.

Assess the site

First take stock of what you have: the garden's size and shape, its topography, and its aspect – when and where it receives the sun, and where it stays wet and shaded or needs more shelter (see also p.135). Is anything already growing in the garden? New shrubs planted by the developer may not be appropriate for the conditions, or may need repositioning or replanting, depending on their requirements.

New-build gardens may have newly laid turf on a thin layer of topsoil, which usually masks the true situation just a few centimetres below. Try digging a few holes in different spots to see how deep the topsoil is and how much building rubbish might be buried.

Take a look over the fence and see what neighbours are growing. Also look at the broader landscape to give you a clearer idea of the sort of plants that naturally thrive here. There may be an attractive view to the wider landscape – tree canopies just outside the garden you could visually "borrow".

Grassed up right
Often gardens of new homes are simply lawned over before sale, with the turf laid on a thin layer of topsoil spread above compacted subsoil.

Weeds may also tell you something about the site. Some indicate waterlogged ground, such as docks and rushes, or poor, dry soil; others, such as chickweed and groundsel, like fertile conditions. Plants such as burdock and plantain can indicate compacted soil.

All this helps you build a picture about the site and what might be realistically possible. If you are aching to get plants into the garden, try growing a few shrubs in containers; these can be planted semi-mature in a couple of years when you are ready and the site has improved.

Know your weeds left
Weeds soon colonize bare ground around new homes. Most will be annual weeds growing from seeds in the disturbed soil; some may give a clue to site conditions.

Going to ground

The soil of new-build properties is often highly variable. Diggers and other heavy plant machinery can compact soil, damaging its structure. Remove waste building materials and rubbish before you start work to reveal the subsoil underneath. Retain the workable fraction of the topsoil in a mound that is free from compaction and not mixed with other soil. The new garden may well require bought-in topsoil to provide a decent depth to get plantings started. Always buy good-quality topsoil that has been screened (large clods removed), and mix in plenty of well-rotted organic matter. Over time, with ongoing cultivation and the addition of organic matter, soil quality and structure should begin to improve (see also p.134).

What lies beneath
Ideally, topsoil is scraped off during construction and saved elsewhere, to then be spread back over the garden after building work is complete.

Building garden structures

Permanent features and hard surfaces, such as footpaths, patio areas, fences, raised beds, ponds, and pergolas, provide the structural framework for your garden design, underlining and enhancing softer areas of lawn and planting.

Many garden structures are easy to construct, and there are several simple projects that gardeners with few building skills – or none at all – can tackle safely and achieve satisfying results in just a day or two. For example, pergola kits are widely available and quite simple to assemble, and you can buy pressure-treated timber pre-cut to length for features such as raised beds or decking.

When executing your design, start with the hard surfaces, but, before you begin, take time to measure your garden carefully. Check that you have sufficient space for a path that will be easy to negotiate, and that the area for a proposed patio or terrace will accommodate your chosen furniture. It may even be worth selecting furniture before you finalize your design plans; it's surprising how much room you need for a dining table and chairs, allowing for the chairs to be moved back comfortably with space to walk around them. Paths for main routes should be at least 1.2m (4ft) wide, and preferably paved or laid with gravel. These will be easier to navigate than narrow, winding routes or a course of stepping stones. Wide paths also provide space for mature plants to spill over the edges without impeding free movement.

Building patios and some paths can be major DIY projects, and if you intend to pave or deck big areas it may be worth considering professional help, especially if your plans include heavy materials, such as stone or composite slabs. Small setts or bricks laid in intricate designs also require expertise. A gravel surface requires less skill to lay and is ideal for an area around planting or a path.

Informal ponds are beautiful features and quite easy to construct, although for a large site, a digger would be helpful.

Top and above
Small and intimate seating areas are always welcome in a garden.
A range of designs and materials can be incorporated into any scheme.

Laying a path

Small paving units, such as blocks, bricks, and cobbles, offer flexibility when designing a path. For this project we used carpet stones (blocks set on a flexible mat), which are quick and easy to lay. If you use recycled bricks, check they are frostproof and hardwearing; ordinary house bricks are not suitable.

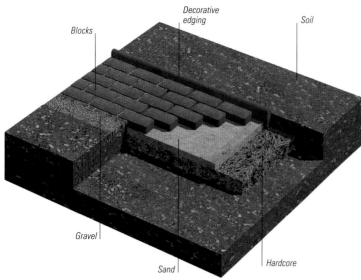

You will need
- Tape measure
- Long pegs and string
- Hammer
- Spade
- Spirit level
- Nails
- Shuttering boards
- Hardcore
- Hand rammer or plate compactor
- Sharp sand
- Carpet stones
- Sledge hammer
- Post-hole concrete
- Edging stones
- Rubber mallet
- Broom
- Sharp knife, trowel
- Compost, herbs
- Gravel

⏱ 1 day

Marking out a path

1 Measure the path and mark with string and long wood pegs, spaced every 1.5m (5ft). Don't forget to allow for shuttering boards (*Step 4*) and decorative edging. Hammer in the pegs gently so they are firm.

2 Dig out the soil between the string to a depth sufficient to accommodate layers of hardcore and sand, as well as the thickness of the blocks. Check levels along the course of the path using a spirit level.

Laying the path

5 Spread an 8–10cm (3–4in) layer of hardcore along the length of the path. You could use excavated soil if the path will only get light use. Use a hand rammer or hired plate compactor to tamp it down.

6 Spread a layer of sharp sand over the hardcore. Level the surface by pulling a length of timber across the path towards you – use the shuttering along the sides as a guide. Fill any hollows with extra sand.

Adding edging and finishing off

9 Carefully knock the shuttering and pegs away and remove the string. Use a spade to create a "vertical face" to the edge – dig down as far as the hardcore base on both sides of the path.

10 Spread a strip of hardcore along each side of the path and tamp firm with a sledge hammer. If you're using heavy edging stones, lay a foundation strip of post-hole concrete mix on top.

11 Position edging stones and bed them in place by tapping them gently with a rubber mallet. Set stones flush with the path, or leave proud to stop soil migrating on to the path's surface. Backfill with soil.

12 Brush sharp sand into the joints (unlike mortar it allows rain to drain away). Remove the occasional block from the edge of the path to form a planting pocket. Carpet stones must be cut from the backing mat.

BUILDING GARDEN STRUCTURES

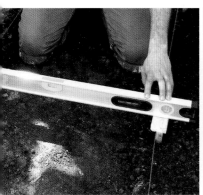

3 To prevent puddles on the surface, the path must slope gently to one side to drain into soil or a soakaway. Angle it away from the house or garden walls to avoid damp problems. Check levels again.

4 Carefully nail the shuttering boards to the pegs to enclose the area of the path. Check the levels once more with the spirit level; any necessary adjustments can be made by easing the pegs up and down.

CUTTING BLOCKS

When you are laying a path, you may need to cut blocks or bricks to fit the pattern or to run around an obstacle, such as a manhole cover or the edge of a wall.

To make a neat cut, place the block on a firm, flat surface. Then, using a bolster chisel, score a line across the block where you want to cut it. Position the bolster on the score line and hit it sharply with a club hammer. Use the chisel to neaten up any rough areas. Remember to wear goggles to protect your eyes while working.

Cutting a block to size.

7 Tamp down the sand (see Step 5), ensuring the surface remains level. Begin laying whole blocks. Carpet stones come prespaced, as do most blocks, but if laying bricks, you will need to use spacers.

8 Once you have finished laying whole blocks, fill any gaps with blocks cut to fit (see top right). Bed the blocks into the sand with a hand rammer on a flat piece of wood or a plate compactor.

13 Use a trowel to remove sand and hardcore from the planting pocket and replace it with loam-based potting compost. Plant a clump-forming aromatic herb, such as thyme. Water well.

14 Brush gravel into the joints between the blocks. If, as here, you have left a strip of soil along one side of the path to act as a soakaway, apply a topping of gravel to keep it looking neat and tidy.

Up the garden path
A well-laid path not only provides safe access through a garden but is also a feature in itself. For period charm, use Victorian-style rope tiles as an edging.

Laying a patio

Pavers and slabs are available in a range of shapes, sizes, and materials (porcelain, natural stone, or concrete) and make a hardwearing surface. Laying large pavers, while heavy work, is quick and easy; preparing the foundations is the hardest part of the job.

Marking out the patio

1 For a rectangular or square patio, mark out the paved area with pegs set at the height of the finished surface and joined with taut string. Use a builder's square to check the corner angles are 90 degrees.

2 Skim off turf using a spade or hire a turf cutter. (Reuse turf elsewhere, or stack rootside up for a year to make compost.) Dig out the soil to a depth of 15cm (6in) plus the thickness of the paving.

You will need
- Pegs and string
- Builder's square
- Spade
- Turf cutter (optional)
- Hand rammer or plate compactor
- Spirit level
- Hardcore, sharp sand
- Rake
- Pavers
- Bricklayer's trowel
- Mortar to suit your material
- Club hammer
- Wood spacers
- Stiff brush
- Pointing tool
- Masking tape

⏲ 2–3 days

Pavers / *Compacted sharp sand* / *Lawn* / *Soil* / *Compacted hardcore*

Laying the paving slabs

5 Top the hardcore with a levelled and compacted 5cm (2in) layer of sharp sand. Lay the first line of pavers along the perimeter string, bedding each slab with mortar as needed, depending on its material.

6 Tamp down each paver with the handle of a club hammer. Maintain even spacing by inserting wood spacers in the joints. Check and keep checking that the pavers are sitting level.

The finishing touches

  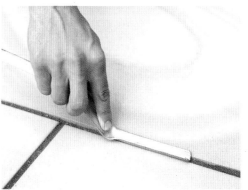

7 Wait about two days before removing the wood spacers. Then, either brush dry ready-mix mortar (or one part cement to three parts builder's sand) into the joints, or, for a neater, more durable finish, you could use a wet mortar mix (see Steps 8 and 9).

8 In dry weather, pre-wet the joints to improve adhesion of the mortar. For wet mortar, add water to the ready-mix mortar and push it into the cracks between the pavers using a bricklayer's trowel.

9 Firm the mortar in place with a pointing tool (*above*). Wet mortar may stain some pavers, but masking tape along the joints will protect them when pointing. Brush off excess mortar before it sets.

BUILDING GARDEN STRUCTURES

3 Use a hand rammer or plate compactor to tamp down the area. Set pegs at the height of the finished surface, allowing for the patio to have a slight slope so rain drains away. Check with a spirit level.

4 Spread a 10cm (4in) deep layer of hardcore over the area, rake level (ensuring you retain the slight slope), then tamp firm with a hand rammer or a plate compactor (*above*).

Cutting a curve into a slab

Although pavers are available in a wide range of shapes, you may have to cut them to size or to accommodate a curve in your design. Pavers, which are usually made from stone or concrete, are surprisingly brittle; to prevent them cracking when they are being cut, lay them flat on a fairly deep, level bed of sand.

1 Protect yourself with goggles, ear defenders, anti-vibration gloves, and a dust mask. Mark the curve on the paver with chalk, then, using an angle grinder fitted with a stone-cutting disc, slowly cut part-way through the paver, going over the line several times.

2 Mark out parallel lines on the waste area with chalk. Cut along the lines part-way through the paver, again going over each one slowly several times. Make sure you don't cross or damage your neatly cut curved line.

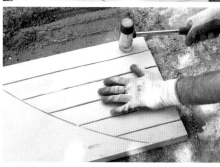

3 Starting on one side of the paver and working across to the other, tap firmly along the length of each cut strip with a rubber-headed mallet. Make sure that the paver is well supported.

4 Grip each strip firmly and snap it sharply along the cut. Remove all the strips in this way. Trim off any roughness along the curved edge with the angle grinder.

Enjoy the space
A well-constructed patio not only provides a practical and level space to sit and walk through, it also makes a garden look professionally finished.

Cutting corners
A few shapely curves can completely transform a rectangular patio. Here, the corners have been opened up to form a planting pocket and to give a gentle sweeping curve to the adjacent area of lawn.

Building a retaining wall

Retaining walls can be made from a range of materials, such as concrete blocks, bricks, or faced stone. They all need to be structurally sound, of course, but if you are unsure, don't build anything higher than 50cm (20in). If in any doubt, consult a builder or engineer. This wall has staggered brickwork with two rows (courses). To cut costs, use facing bricks at the front and concrete blocks at the back, where they won't be visible.

You will need

- Tape measure and spirit level
- String line or line paint
- Builder's square
- Spade
- Lump hammer
- Wooden pegs
- Shovel, mixing tray, bucket, and wheelbarrow (or concrete mixer)
- Concrete (8:1 ballast to cement)
- Block of wood for tamping
- Brick trowel
- Facing bricks and backing bricks
- Mortar mix (4:1 soft sand to cement)
- Weep holes and wall ties
- Circular saw or bolster
- Pointing trowel
- Brick pointer or jointer
- Wire brush

🕑 1–2 days plus time for the footing to dry

Building the wall

1 Mark out exactly where the wall is to go, using a string line or line paint. Dig out a trench to about 30cm (12in) wide and deep. Tap in wooden pegs at the same height to set the base level of your wall.

2 Fill the trench with your mixed concrete. Use ballast (a blend of sand, stone, and gravel) with cement in a ratio of 8:1 ballast to cement. Ensure the concrete is level with the tops of the wooden pegs.

7 It's important to stagger the bricks to create a strong wall. Lay another brick lengthways at the other end to build the corners first. Keep checking the bricks horizontally and vertically with a spirit level.

8 Finish laying the bricks between the two corners (see Step 9). Continue to double-check the bricks with the line and a spirit level. Ensure the mortar is 10mm (½in) deep. between bricks and rows.

Strong wall
A wall built with two courses of bricks will be strong enough to retain any soil behind it.

BUILDING GARDEN STRUCTURES

3 Use a wooden block to tamp the concrete to remove any gaps. Allow a few days for the concrete to dry. Mark out the length and corners of the wall with pins and a tight string line.

4 Mix your mortar – the material between the bricks that holds them in place. A mix of 4:1 soft sand to cement is ideal. Apply mortar to the end of one brick so it is ready to be laid.

5 Set the line at the height you want the first row of bricks. Press each brick with mortar on one end into a mortar bed about 10mm (½in) deep, tapping them into place. Ensure they are level.

6 Once you have finished the first row, lay a brick lengthways at the back of the first row. Then lay a brick on top to span the two bricks below, creating a staggered corner and the beginning of your second row.

9 Moisture will need to be drained from a wall that has soil behind it. Insert weep holes in both rows of bricks at the second layer, at intervals of 1m (3ft) along the length of the wall.

10 Use wall ties to give additional stability between the facing bricks and the backing bricks. These are especially important in taller walls. Position them at regular intervals of about 1m (3ft) along the wall.

11 If the staggered joints get slightly out of alignment, you can use a circular saw or bolster to cut bricks. This isn't usually a problem unless it is visibly obvious.

12 Remove any excess mortar and ensure the wall is clean. Cover it with a tarpaulin or hessian if you aren't able to complete it in one day.

Finishing off

You will need a top layer to finish off your wall. Always consider how a wall will look at the top. Will plants be tumbling over the edge or is there going to be hard landscaping behind it? As well as its appearance, remember that this is the most vulnerable part of the wall as it is exposed to rain, freeze-thaw, weeds, and so on. You can use bricks (as here) for your top layer, or stone or paving slabs, which overhang the wall and help provide a drip line to take water away.

1 Using the same mortar mixture and depth, lay a few bricks at each end of the wall. Run a string line between the ends and continue laying bricks. Check the horizontal and vertical alignments with a spirit level. Tamp bricks into place.

2 Keep checking that the mortar between the bricks is a consistent thickness. It will give a neat appearance and mean that you don't end up running out of space for bricks. If required, use a pointing trowel and some extra mortar to fill in any gaps in the joints.

3 Use a brick pointer or jointer to give the mortar a smooth appearance. Leave the mortar to set for 24 hours (covering it overnight). When the mortar has set, use a wire (or stiff) brush to clean the wall.

Laying decking

Decking is adaptable and blends with most garden styles. It can be built from hard- or softwood, or, more popularly, composite. If using timber, ensure supplies come from responsibly managed sources, and check building regulations and planning requirements for large or above-ground structures.

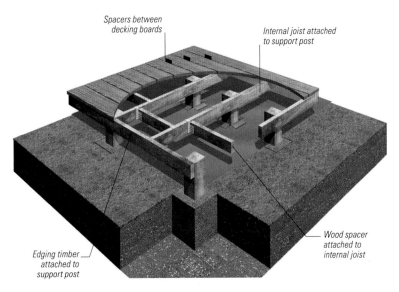

Spacers between decking boards
Internal joist attached to support post
Wood spacer attached to internal joist
Edging timber attached to support post

You will need
- Pegs and string
- Builder's square
- Geotextile membrane
- Tape measure
- Spade
- Hardcore
- Metal pole
- 75 x 75mm (3 x 3in) support posts
- Post-hole concrete
- Spirit level
- 100 x 50mm (4 x 2in) timber lengths
- Drill and router
- Galvanized bolts, washers, screws and nails
- Saw and hammer
- Decking boards
- Chisel, wood spacers

⏲ 2 days

Putting up support posts

1 Mark out a square or rectangular deck with pegs and string. Check the corners are at a 90-degree angle using a builder's square. Mow grass very short, or skim off turf to use elsewhere in the garden.

2 Lay a geotextile membrane over the area, overlapping joins by 45cm (18in). As well as the four corner posts, you will need extra support posts on each side; mark these with pegs about 1.2m (4ft) apart.

Making the deck frame

5 Leave the concrete to set for 24 hours before building the deck frame. Cut edging timbers to length – note that joins should coincide with a post. Predrill bolt holes, countersinking them with a router.

6 Hold the first edging timber in place against the frame (you may need help); mark and drill a bolt hole on the post. Insert a washer and bolt and tighten up, but not too tight; leave a little room for movement.

Building the internal frame and laying the decking

9 Internal joists strengthen the deck. Run them across the shortest span set 40cm (16in) apart. Support joists with extra posts (cut the membrane when you concrete them in) aligned with those on the outer frame.

10 Once all the joists are bolted to the support posts, insert short lengths of wood set 1.2m (4ft) apart to hold them rigid. Nail or screw the joists and spacers in place or use joist hangers (see *top right*).

11 Lay a decking board on the frame (at right angles to the joists) and cut to length, leaving a slight overhang at each end to fit a fascia board (optional). Centre any joins in the board over a joist.

12 Predrill holes in the board, then attach it to each joist using two corrosion-resistant decking nails or countersunk screws. Cut the remaining boards to size and screw them to the joists.

BUILDING GARDEN STRUCTURES

3 Dig out post-holes about 30cm (12in) square and 38cm (15in) deep, and fill the bottom 8cm (3in) with hardcore. Tamp firm with a metal pole, insert the post, and pack upright with more rammed hardcore.

4 Fill the hole with water to dampen the hardcore and allow to drain. Pour in post-hole concrete mixed to a pouring consistency. Use a spirit level to check the post is vertical; adjust as necessary.

JOIST HANGERS

If your deck is at ground level, you can screw or nail the frame together. More robust alternatives are advisable for raised decks or where the joists butt against a wall. Timber-to-timber joist hangers, made from galvanized mild steel, are nailed or bolted on to the joists and then attached to the edging timbers. Stronger steel joist hangers can be mortared into the wall. You may find it easier to bolt a length of timber to the wall first, and then hang the joists from it with timber-to-timber joist hangers.

Timber-to-timber joist hanger.

Joist hanger mortared into a wall.

7 Lift the edging timber into position, use a spirit level to check it's horizontal, and mark and drill the timber where it coincides with a post. Insert a bolt and washer as Step 6.

8 Attach all the edging timbers in the same way, butting the corner joints neatly. Bolt the timbers to all intermediate posts to complete the frame. Cut the tops off the posts flush with the frame.

13 Use a chisel to lever the boards into place, spacing them 5mm (¼in) apart with thin strips of plastic or wood. Spacing allows the decking boards to expand in the heat and lets rainwater drain away.

14 Fascia boards fixed around the edge of your deck make for a neat finish. Overlap them precisely where they meet at the corners. If your decking is built on a slope, fascia boards will hide any ugly gaps.

Wood treatments
Pretreated decking timber can be left natural, or you can choose from a huge variety of coloured stains or treatments. This deck is a contemporary dark brown.

Putting up fence posts

The strength of a fence lies in its supporting posts. Choose 75 x 75mm (3 x 3in) posts made from a rot-resistant timber, such as cedar or pressure-treated softwood, and set them in concrete or metal post supports. Treat the timber with wood preservative every three to four years to prevent it rotting, and replace old posts when you spot signs of deterioration.

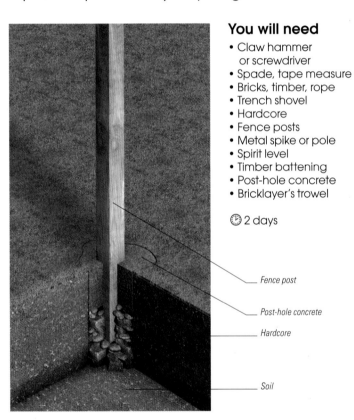

Fence post
Post-hole concrete
Hardcore
Soil

You will need

- Claw hammer or screwdriver
- Spade, tape measure
- Bricks, timber, rope
- Trench shovel
- Hardcore
- Fence posts
- Metal spike or pole
- Spirit level
- Timber battening
- Post-hole concrete
- Bricklayer's trowel

⏲ 2 days

Replacing old fence posts

1 Use a claw hammer or screwdriver to free one end of the panel. Remove metal clips and fixings. Clear soil away from the base of the panel, then free the other end. Leave the top fixing brackets until last for support.

2 Before putting in a new post, first remove the old concrete footing. Once you have removed the fence panels, dig out the soil from round the base of each post to expose the concrete block.

Concreting the posts

7 To test that the post is vertical, hold a spirit level against each of its four sides. Make any adjustments as necessary, and check that the post is the right height for the fence panel.

8 To hold the post upright while you're concreting it in place, tack a temporary wood brace, fixed to a peg driven firmly into the ground, to the post. Don't attach it to the side that you'll be hanging the panels on.

Fixing bolt-down supports

When erecting posts on a solid level surface, such as paving, use bolt-down, galvanized metal plates. These can be fixed in place relatively easily and will help to prolong the life of the timber posts by holding them off the ground.

1 Measure and mark the exact position of the post, as there will be no opportunity to change it later. Position the base plate, marking the position of each of the corner bolt holes with a pencil.

2 Use a percussion or hammer drill fitted with a masonry bit to drill the bolt holes. Keep the drill upright and make sure you penetrate right through the paving into the hardcore underneath.

3 Fill the drilled holes with mortar injection resin and insert Rawl bolts. After the recommended setting time, tighten the bolts using a spanner – the bolts will expand to fill the hole.

BUILDING GARDEN STRUCTURES

3 Using the post as a lever, loosen the block in the hole. Tie a length of timber to the post, balance it on a pile of bricks (*as shown*) and use this simple lever to help minimize any strain as you lift out the post.

4 If a new post is to go in the same spot, refill the hole and compact the soil before digging a new post hole using a trench shovel. Make it about 60cm (2ft) deep and 30cm (12in) across.

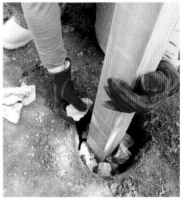

5 Fill the bottom of the hole with a 10cm (4in) layer of hardcore. Stand the post on the base, check it's level with the original fence line, and pack the hardcore around the sides.

6 Use a metal spike or pole to ram the hardcore in place, working the post gently back and forth to help settle the material. Aim to fill the hole to about half its depth with compacted hardcore.

9 Fill the post hole to the top with water and leave it to drain. This will help settle the hardcore and improve adhesion of the concrete. Make up post-hole concrete to a pouring consistency.

10 Pour concrete into the hole, stirring gently to remove air bubbles. Shape it around the post, using a trowel, so rain runs off. Rehang panels after 48 hours. Remove bracing after three weeks.

New posts, new panels
A new fence makes a great backdrop for garden planting: consider painting it a colour or leave a natural stain.

METAL SPIKE SUPPORTS

If you have firm, undisturbed ground, use metal spike supports. Position the spike in place and insert a "dolly", a special post-driver, into the square cup. Hit the dolly with a mallet to drive the spike into the ground. Check the angle of the spike with a spirit level to ensure that it is going in straight – twist the dolly handles to correct any misalignment. When the spike is in the ground, remove the dolly. Clamp the square cup around the post and tighten up.

Laying a gravel border

Gravel isn't just for driveways and paths – when used as a decorative mulch in the border, it sets plants off to great effect. If you spread a thick layer of gravel on top of a reusable weed-control fabric, it will also suppress weeds and help retain moisture in the soil.

You will need
- Scissors or sharp knife
- Reusable weed-control fabric
- Metal pins
- Pea gravel
- Tape measure

⏲ 1 day

Laying the fabric

1 Cut a piece of weed-control fabric to fit your bed or border. For large areas, you may need to join several strips together – in which case, leave a wide overlap along each edge and pin in place.

2 Presoak container-grown plants in a bucket of water for about half an hour. Position plants, still in their pots, on top of the fabric. Check the labels to make sure that each plant has enough room to spread – once the gravel is down, moving them isn't easy.

3 Use scissors or a sharp knife to cut a large cross in the fabric under each plant. Fold back the flaps. Make the opening big enough to allow you to dig a good-sized planting hole.

Planting up the border

4 Remove the plants from their pots and lower each one into its allocated planting hole. Plants should sit in the ground at the same depth as when in the pot. Fill in around the root ball with soil.

5 Firm in the root ball with your hands, then tuck the flaps back around the base of the plant. If necessary, trim the fabric to fit neatly around the plant's stems. Water thoroughly.

6 Cover the fabric with a thick, even layer of gravel. A depth of 5–8cm (2–3in) should prevent any bald patches appearing. Should you need to move plants in the future, pin a piece of fabric over the old planting area to stop weeds popping up through the cut.

AGGREGATE OPTIONS

You can lay most aggregates over a weed-control fabric in the same way as gravel. Other decorative options for a planting area include slate chips (*shown right*), small pebbles, ground recycled glass, crushed shells, and coloured gravels. (*See pp.352–53 for more information on these materials.*)

Permeable paths

The main advantage of using permeable surfacing in a garden is that it allows rainwater to drain through to the soil. But when you discover that the materials are durable, easy to lay, and cost-effective, they're definitely an attractive alternative to paving.

Loose gravel
Look carefully and you'll see that this gravel has been poured into a honeycomb grid. This cleverly designed permeable matting, which you lay like a carpet, prevents gravel migrating all over the garden or driveway.

Self-binding gravel
Gravels are usually washed clean of soil and stones, but self-binding gravels, such as Breedon gravel, are not. When compacted, these fine particles bind the material together to form a strong, weed-free, permeable surface.

Keep it neat
A finished gravel garden gives so much joy – the soft tones and planting blend together, creating a verdant but usable informal space.

Shredded bark
Bark is pleasantly springy underfoot. Lay it over a weed-control fabric or straight on to compacted soil. Whichever you decide to do, the bark will start to break down after a couple of years and should be replaced with fresh bark.

Building a pergola

A pergola is essentially a series of arches linked together to form a covered walkway. The framework provides the perfect support for climbing plants, such as fragrant honeysuckle and roses. Although often made from timbers or metal components, many designers choose to use a wood frame kit, as shown here, the instructions for which are pretty universal.

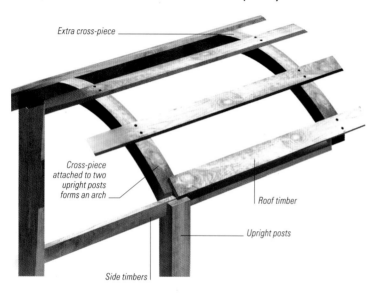

- Extra cross-piece
- Cross-piece attached to two upright posts forms an arch
- Roof timber
- Upright posts
- Side timbers

You will need
- Pergola kit
- Pegs and string
- Builder's square
- Vice
- Drill
- Screwdriver
- Galvanized screws or bolts
- Tape measure
- Hammer
- Wooden battening
- Spray paint
- Spade
- Hardcore
- Metal spike
- Spirit level
- Ready-mix concrete
- 2 days

Making the arches

1 Unpack and identify all the pieces. Mark the layout of the pergola on the ground with pegs and string – use a builder's square to check right angles. If the area is to be paved, lift and reuse the turf elsewhere.

2 Arrange the pieces flat on the ground to check the fit of the joints. Make adjustments as necessary. If the wood isn't predrilled, clamp the timber in a vice and drill holes for the screws and bolts.

Erecting the arches

5 Mark the two upright post positions for the first arch using spray paint. Dig out the holes making them about 60cm (2ft) deep and 30cm (12in) across. Fill with 10cm (4in) of hardcore (see Step 5, p.263).

6 Ram the hardcore firmly in place with a metal spike or pole. Place the upright posts in the holes and test that each one is vertical by holding a spirit level against each of its four sides.

Constructing the roof

9 Dig two holes for the uprights on the second arch (see Steps 5 and 6). Do a final check on the relative position of the two arches by positioning the side timbers on top of their respective uprights.

10 Using a spirit level, check that the side timbers are horizontal and the uprights are vertical before concreting them in position. Repeat Steps 5–10 until all the arches are concreted in place.

11 Leave the concrete to set for 48 hours, then screw or bolt all the side timbers in place, butting the joints tightly together. To avoid splitting the wood, it's best to predrill the holes.

12 Most pergolas have extra cross-pieces to strengthen the roof (these do not sit on uprights so are unsupported). Mark their position midway between the uprights. Predrill screw holes in each piece.

BUILDING GARDEN STRUCTURES

3 To make an arch, attach each end of a cross-piece to the top of an upright post using galvanized screws or bolts. Support the wood on a board to help steady and align the pieces as you work.

4 Measure the distance between the upright posts at the top and bottom of each arch, adjust the posts until the spacing is the same, and then nail wooden battening across to stop them splaying.

7 To hold the upright posts vertical while you're concreting them in, tack a temporary wooden brace to them (see Step 8, p.262). Concrete the posts in place (see Steps 9 and 10, p.263).

8 To position the second arch, lay a side timber on the ground to work out the spacing. Mark the position of the post holes with paint. Allow for a slight overlap where the side timbers will rest on the uprights.

A shady retreat
Walking under a shady, plant-covered pergola is a real treat on a hot summer's day. It would also be the perfect spot for outdoor entertaining.

13 Screw or bolt the cross-pieces in place – you will need someone to hold them steady to stop them twisting when you're drilling. Check that all the fixings on the pergola frame are tight.

14 Position the roof timbers on top of the cross-pieces. Mark and predrill holes, and then screw in place. Leave the bracing on the uprights for three weeks until the concrete has completely set.

WIRING FOR CLIMBERS

A system of wires attached to the upright posts of your pergola will give plants the support they need to start climbing. Fix screw eyes at 30cm (1ft) intervals around the four sides of an upright. Attach galvanized wire to the lowest screw eye, run it through all the eyes on the same side of the upright, and secure it firmly to the top one. Repeat on the other three sides of the upright. Guide shoots of twining plants on to the wires; tie in shoots of stiffer-stemmed climbers.

Set up a system of wires for climbers.

Making a raised bed

Creating a square or rectangular timber-framed raised bed is easy, especially if the pieces are pre-cut to length. Buy pressure-treated wood, which will last for many years, or treat it with preservative before you start. If the bed is to sit next to a lawn, make a brick mowing strip by following the steps opposite.

You will need
- Spade
- Pre-cut wooden sleepers
- Spirit level
- Tape measure
- Rubber mallet
- Drill, screwdriver
- Heavy-duty coach screws
- Rubble and topsoil
- Bark
- 1 day

Pre-sawn timbers for a neat finish
Top timbers rest on the base
Deep layer of topsoil
Mix of soil and rubble for good drainage
Brick mowing strip

Measuring up the base

1 Dig out strips of turf wide enough to accommodate the timbers. Pressure-treated wood is an economic alternative to rot-resistant hardwoods, such as oak. Or consider buying reclaimed hardwood.

2 Lay out the timbers *in situ* and check that they are level with a builder's spirit level (use a plank of wood to support a shorter spirit level). Check levels diagonally, as well as along the length of the timbers.

3 Make sure the base is square by checking that the diagonals are equal in length. For a perfect square or rectangular bed, it is a good idea to have the timbers pre-cut to size at a local timber yard.

Building the bed

4 Using a rubber mallet, gently tap the wood so that it butts up against the adjacent piece; it should stand perfectly level and upright according to the readings on your spirit level. Remove soil as needed.

5 Predrill holes through the end timbers into the adjacent pieces at both the top and bottom to accommodate a couple of long, heavy-duty coach screws. Secure the timbers with the screws.

6 Arrange the next set of timbers on top of the base; make sure they overlap the joints below to give the structure added strength. Check with a spirit level before screwing together (see Step 5).

7 For extra drainage, partially fill the base with rubble. Then add topsoil that is free of perennial weeds. Fill the bed up to about 8cm (3in) from the top with soil, plant up, then mulch with bark or gravel.

RAISED VEGETABLE BED

Raised beds are ideal for growing vegetables, fruit, and herbs. They provide better drainage on heavier soils and a deeper root run for crops like carrots and potatoes. Raised beds also lift up trailing plants, such as strawberries, which helps to prevent rotting. If you buy in fresh topsoil that's guaranteed weed- and disease-free, your crops will have a better chance of doing well.

Raise your profile
As well as providing an eye-catching feature, a raised bed gives you a better view of your plants and, by lifting them up, less strain on your back when tending them.

Laying a mowing strip

Grass doesn't grow well too close to a raised bed, since the soil tends to be dry and any overhanging plants create shade. A strip of bricks, sunk slightly lower than the level of the turf, creates a clean edge to allow for easy mowing.

1 Using a spare brick to measure the appropriate width for your mowing edge, set up a line of string to act as a guide. Dig out a strip of soil deep enough to accommodate the bricks, plus 2.5cm (1in) of mortar.

2 Lay a level mortar mix in the bottom of the trench as a foundation for the bricks. Set them on top, leaving a small gap between each one. (This design is straight, but mowing edges can be set around curves.)

3 With a spirit level, check that the bricks are aligned and slightly below the surface of the lawn (when set in place, you should be able to mow straight over them). Use a rubber mallet to gently tap them into position.

4 Finally, use a dry mix to mortar the joints between the bricks, working the mixture in with a trowel. Clean off the excess with a stiff brush.

A clean cut
The mowing strip makes a decorative feature and allows you to manoeuvre the mower more easily.

Making a pond

Designing a pond with a flexible butyl rubber or PVC liner, rather than a rigid preformed type, allows you to create a feature of almost any size and shape. To work out how much liner you need, add twice the depth of the proposed pond to its maximum length plus the width. Choose somewhere sheltered and sunny for your water feature, avoiding heavy shade under trees.

You will need
- Hosepipe
- Spade
- Pickaxe
- Spirit level/plank
- Sand
- Pond or carpet underlay
- Flexible pond liner
- Waterproof mortar, bucket, trowel
- Sharp knife
- Decorative stone

⏱ 2 days

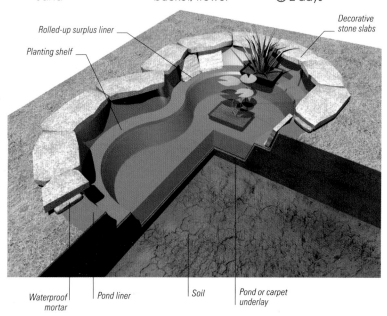

Labels: Rolled-up surplus liner, Planting shelf, Decorative stone slabs, Waterproof mortar, Pond liner, Soil, Pond or carpet underlay

Digging an informal pond

1 Use a hose to mark the outline of the pond. Aim for a curved, natural shape without any sharp corners. To prevent it freezing solid in winter, a section of the pond must be at least 45cm (18in) deep.

2 Before you start digging, skim off any turf for reuse elsewhere. Keep the fertile topsoil (which you can also reuse) separate from the subsoil. Loosen compacted subsoil with a pickaxe.

Lining and edging

 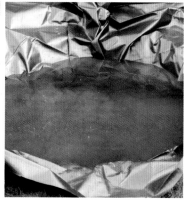

5 To protect the liner, line the sides and base of the pond with pond underlay. If using old carpet underlay, beware stray tacks. On stony soils, spread a 5cm (2in) layer of sand over the base first.

6 Centre the liner over the hole, letting it slide down into the base. Leaving plenty of surplus around the rim, pleat the liner to help fit it to the shape of the pond. Let the pond fill naturally with rainwater.

Making a rill

A rill or channel of water adds light and movement to a garden. Employ a qualified electrician to install a power supply for you.

You will need
- Pegs and string
- Spade
- Sand
- Spirit level
- Reservoir
- Plastic liner
- Sharp knife
- Bricks
- Waterproof mortar
- Submersible pump, flexible pipe, filter
- Gravel, cobbles
- Metal grille
- Reusable fabric

⏱ 1 day

1 Clear and level the site. Mark out the length and width of the rill with pegs and string. Dig out the area to a depth of 15–20cm (6–8in). Cut a shallow shelf all around the rill for the brick edging.

2 Line the rill with sand, compacting it with a piece of wood. Use a spirit level to check the base is flat. Dig a hole at one end and insert the reservoir – check the rim is level with the base of the rill.

3 Line the rill with the plastic liner, smoothing out any creases. Use a sharp knife to trim the liner at the reservoir end so that it drapes over the rim. Leave 20cm (8in) surplus material along the other three sides.

BUILDING GARDEN STRUCTURES

3 Dig out the pond to a depth of 45cm (18in). Make the sides gently sloping. Leave a shelf 30–45cm (12–18in) wide around the edge, then dig out the centre to a further depth of 45cm (18in).

4 Use a spirit level placed on a straight piece of wood to check that the ground around the top of the pond is level. Remove any loose soil and all large or sharp stones from the sides and bottom of the pond.

7 When the pond is full, trim the surplus liner leaving 45cm (18in) around the rim. Pleat the excess liner so it lies flat and bury the edges in the ground. Lay a bed of waterproof mortar for the edging stones.

8 Bed the edging stones into the mortar, overhanging them by 5cm (2in) to hide the liner. When positioning vertical stones, stand them on a piece of rolled-up surplus liner to protect the liner from being torn.

Planting up
Wait a week for the mortar to set before placing water lilies on the bottom of the pond and marginals on the shelf (see pp.214–15 for more on aquatic plants).

4 Edge the rill with bricks on three sides (not the reservoir end). Bed bricks on a 2cm (1in) layer of waterproof mortar, making sure that it doesn't fall into the rill. Mortar between the bricks.

5 Place the pump in the reservoir. Push the pipe on to the pump outlet, run the pipe along the length of the rill, and cut it to fit at the far end. Fit a filter on the free end of the pipe to prevent blockages.

6 Cover over the pipe in the rill with a level bed of gravel. Place a metal grille over the reservoir and top with cobbles. If you sit them on a sheet of reusable fabric, it will stop debris falling into the water.

Finishing touches
Fill the reservoir with water, prime the pump, and adjust the flow, according to manufacturer's instructions. Slate chips make an attractive edging material.

Planting techniques

Having designed a beautiful garden, assessed your soil and aspect, and worked out what plants to buy, it is now time to bring them home, get them into the ground, and put your ideas into practice. Take your time when planting; tackling the task in a measured way will help to ensure your treasures thrive.

Choose a dry, fine spell when the soil is not frozen or too wet. Before starting, gather all necessary tools together – fork, spade, organic fertilizer, and watering can – so you have everything to hand. Also make sure the soil is free of weeds, especially any pernicious perennials, before forking in fertilizer and digging holes. The new plants will need a thorough soaking prior to planting, and the best way to do this is to immerse them in water while they are still in their pots, leave until the bubbles disperse, then remove and allow to drain. Bare-rooted trees, roses, or shrubs should be planted between autumn and early spring; container-grown plants can go in the ground at any time, but hardy plants are best planted in autumn when the soil is still warm and moist. Leave more tender types until spring, as young plants may not survive a cold, wet winter.

Allow space for shrubs and trees to spread – the area needed should be indicated on the plant label. Bare patches can always be filled in with seasonal flowers, or screened by containers or an easily moved ornament, such as a bird bath or light-weight sculpture.

Early spring or early autumn are the best times to establish a lawn, whether you are using turves or sowing seed, and avoid walking on new grass for a few months, if possible. Water it frequently in the early stages and in dry spells.

Giving your new purchases a good start will repay dividends for years to come in the form of strong, healthy plants that continually give a good show, season after season.

Top and above
Planting up a container brings colour to any space.
Aquatic plants can offer a new home for wildlife.

How to plant trees

A well-planted tree will reward you with years of healthy growth. Container-grown trees can be planted at most times of the year, but the best time is in autumn, when the leaves are starting to fall. Bare-root plants are a cheaper option and are available in autumn and winter. Unless it's very frosty or there's been a long dry spell, you should plant them as soon as you get them home.

You will need
- Bucket
- Spade and border fork
- Well-rotted organic matter
- Bamboo cane
- Tree stake
- Mallet and nails
- Tree tie with spacer
- Chipped bark mulch
- ⏲ up to 2 hours

Planting a container-grown tree

1 Soak the tree thoroughly and leave it to drain. Meanwhile, clear the planting area of weeds and debris. Place the tree, still in its pot, in its planting position, making sure that it won't be crowded by other plants.

2 Loosen the soil over a wide area, to the same depth as the tree's root ball. Add organic matter to heavy clay or sandy soils. Dig a large hole no deeper than the tree's pot but ideally three times the root ball's diameter.

Planting and staking

5 With a container-grown tree, you may find that the roots are packed together tightly. If this is the case, gently tease out any encircling roots, as these could prevent it from establishing well.

6 With a helper holding the tree upright, backfill the hole with the excavated soil. Make sure there are no air pockets by working the soil in between the roots and around the root ball with your fingers.

7 Once you are satisfied that there are no gaps or air pockets around the roots, continue to hold the tree upright and firm it in using your foot with your toes pointing towards the trunk.

8 Small trees do not require staking but top-heavy or larger specimens should be staked. Drive into the soil a wooden tree stake at an angle of 45 degrees. Ensure you do not damage the root ball.

Planting a hedge

An informal mixed hedge of native species will provide a rich habitat for wildlife, as well as attractive flowers, fruits, and nuts. The best time to plant a bare-root hedge is autumn, when plants first become available.

You will need
- Spade
- Rake
- Tape measure
- String and canes
- Secateurs

⏲ up to 3 hours

1 A few weeks before planting, remove weeds and dig the area over, working in organic matter (as Step 2 above). At planting time, weed the area again, tread the ground until firm, and rake level.

2 Mark the planting line with pegs and string. If you have space, put in a double row of plants for extra screening. It's also less likely to suffer gaps if plants die. Set the rows 40cm (16in) apart.

3 Set the plants 80cm (32in) apart. Spacing is critical for hedging, so use a tape measure or marked canes rather than guessing. Dig holes large enough to accommodate the roots comfortably.

PLANTING TECHNIQUES

3 Puncture and scuff up the walls and base of the hole to allow for easy root penetration; the result will be a stronger tree. Don't loosen the base too much as the tree may sink after planting.

4 Remove the tree from its pot. Lower it into the hole and check that the first flare of roots will be level with the surface after planting – try scraping off the top layer of compost if you can't see the flare.

9 The stake should be a third of the height of the tree, and the end should face into the prevailing wind. Fit a tree tie with a spacer to the stake and trunk. This can be adjusted as the tree grows.

10 Knock a nail through the tree tie into the stake to prevent it slipping down. Water the tree thoroughly and apply a mulching mat around the trunk, which will keep the area around the tree free of weeds.

Spring blossom in a woodland border
In small- to medium-sized gardens choose compact trees with an attractive overall habit. This hawthorn (*Crataegus*) is ideal, with pretty, pink blossom in spring, followed by ornamental fruits.

4 Plant the bare-root hedging plants at the same depth as they were growing in the field; you'll see a dark soil stain on the stem. Plant roses slightly deeper for stability. Firm plants in with your hands.

5 Stagger the plants on the second row to maximize coverage. Position the first plant 40cm (16in) in from the edge of the front row. Keep bare-root plants wrapped until planted to stop their roots drying out.

6 Check that the soil around the plants is firmed in, and then water each plant thoroughly. Prune back the tips of any tall or leggy shrubs to encourage new, bushy growth from the base of the plant.

Wildlife-friendly screen
A mixed hedge will attract wildlife all year. Don't clip too hard if you want summer flowers and fruit in autumn, and take care not to disturb nesting birds in spring.

How to plant shrubs

Shrubs form the backbone of a planting scheme, providing structure as well as flowers and foliage. Plants grown in containers can be planted all year round if you avoid days when the ground is frozen or excessively wet or dry. Before planting, always check the label for the shrub's preferred site and soil.

You will need
- Spade and fork
- Organic matter
- Bucket
- Mulching material

🕐 1 hour

1 Dig over the soil, removing any weeds and working in plenty of well-rotted manure or garden compost. Make the planting hole twice the diameter of the container and a little deeper.

2 Stand the plant in its container in a bucket of water and leave it to soak. Remove the plant from its pot and tease out any thick, encircling roots. Plant at the same depth that it was in its pot. Backfill with soil.

3 Firm in gently, ensuring the shrub is upright and that it is sitting in a shallow depression to assist watering. Water in well, then spread a mulch of organic matter (*see right*), keeping it away from the stems.

How to plant perennials

Unlike annuals and tender patio plants, herbaceous perennials come up year after year. Many modern varieties need little maintenance other than deadheading and cutting back in spring. Give them a good start by improving the soil at planting time and minimize competition for water and nutrients by controlling weeds.

You will need
- Spade and fork
- Organic matter
- Organic fertilizer
- Bucket
- Mulching material

🕐 up to 1 hour

1 Prepare the planting area, removing perennial weeds and large stones. On dry ground or heavy clay, work in organic matter (*as Step 1, above*). On sandy soil, also apply an organic fertilizer.

2 Dig a hole a bit deeper and wider than the pot. After soaking the plant, remove the pot. Add soil to the hole so that the top of the root ball is level with the soil surface. Backfill and firm in lightly with your hands.

3 Water in well. Apply a thick mulch to conserve moisture, suppress weeds and protect roots from penetrating frosts. Take precautions against slugs and snails, and watch for aphids on shoot tips.

PLANTING TECHNIQUES

Seasonal colour and interest
A mixture of shrubs and perennials provides a rich tapestry of colour, form, and texture that changes in mood as the seasons progress. If space allows, plant the perennials in drifts for greater impact.

Mulch options

Mulches conserve water, which is why they are always applied after planting when the ground is moist. Some improve soil structure and most discourage weeds, which compete with garden plants for water and nutrients. Gravel mulches look attractive, while others, such as leafmould, offer a habitat for beneficial creatures such as ground beetles.

Garden compost
Well-rotted compost and manure lock moisture and nutrients into the soil. As the mulch rots down it releases plant food and improves the soil structure. Apply a layer 10cm (4in) deep in late winter to minimize weed growth.

Leafmould
Although low in nutrients, leafmould is excellent for improving soil and retaining moisture, and it looks good around woodland-style plantings. To make it, fill perforated bin bags with autumn leaves, seal up, and leave for about 18 months.

Chipped bark
A popular mulch, bark comes in various grades, the finest being the most ornamental. It rots down slowly and is a good weed suppressor and moisture conserver but doesn't add many nutrients. Top up worn areas annually.

Gravel mulch
Gravel laid over landscape fabric creates a decorative weed-suppressant foil for alpines and Mediterranean-style plantings. Plant through the fabric by cutting a cross and folding back the flaps before replacing the gravel (see also pp.264–65).

How to plant climbers

Walls, fences, and trellis offer planting space for a wide range of climbers and wall shrubs. Using plants vertically is especially important in courtyard gardens, where space is at a premium. Flowers and foliage soften bare walls and boundary screens, as well as creating potential nesting sites for birds. Avoid over-vigorous climbers that could overwhelm their situation.

You will need
- Vine eyes
- Galvanized wire or trellis
- Border fork and spade
- Bulky organic matter
- Organic fertilizer
- Bamboo canes
- Garden twine
- Trowel or hand fork
- Chipped bark mulch
- 🕐 1 to 2 hours

Preparation for planting

1 Before soil preparation, attach a support of vine eyes and horizontal wires, or a trellis, to the wall or fence. Set the lowest wire about 50cm (20in) above soil level, and space the wires 30–45cm (12–18in) apart.

2 Dig over a large area around the planting site. Work in plenty of bulky organic matter, such as well-rotted manure or garden compost, to combat dryness at the base of the fence.

Planting and aftercare

5 Arrange a fan of bamboo canes behind the planting hole, leaning them back towards the fence. The canes will lead the climber's stems up to the horizontal wires and spread them over a wider area.

6 Plant the climber, backfilling the hole with enriched soil. Untie the stems from their original support and untangle them carefully. Cut off any weak shoots and spread them out ready to attach.

7 Tie the stems to the canes using soft garden twine and a loose figure-of-eight knot. Train the outer stems on to the lower wires and train the central stems upwards to cover the higher wires.

8 Firm the climber in using your fists and then take a trowel or hand fork to fluff up the soil where it has been compacted. Next, create a shallow water reservoir (with a raised rim) around the base of the plant.

Support for climbers

Climbers and wall shrubs scale vertical surfaces in a variety of ways, and the support you provide depends on their vigour and method of climbing. Some, such as jasmine, honeysuckle, and wisteria, are twiners; clematis have coiling leaf stalks; and sweet peas, passion flowers, and vines cling with tendrils.

Horizontal wires
These offer the most adaptable support for climbers, wall-trained shrubs, and fruit trees. Training stems horizontally increases flower and fruit production.

Trees and other host plants
To encourage a rambler rose to clamber up into a fruit tree, plant it 1m (3ft) away from the trunk and give it a rope to climb (peg it to the ground and run it to the lowest branch).

Trellis
Wooden trellis can be used against a wall or as a screen. Climbing roses, honeysuckle, clematis, and passion flower may secure themselves, but tying them in also helps.

PLANTING TECHNIQUES

3 On poor soils, apply a dressing of organic fertilizer (follow manufacturer's instructions). Water the climber well a few hours before planting, or plunge the pot into a bucket of water.

4 Dig a planting hole 45cm (18in) from the fence, and twice the diameter of the root ball. Check the depth is the same as the original compost level, though clematis should be planted 10cm (4in) deeper.

9 Water well, then apply a mulch of chipped bark to help combat weeds, conserve moisture, and keep the roots of plants such as clematis cool. Ensure the mulch doesn't touch the stems.

Fragrant cover
The honeysuckle in this planting sequence will eventually produce a mass of evening-scented flowers, loved by bees and moths. Good ground preparation will ensure that the plant won't run short of water, which can lead to powdery mildew.

Obelisks
These provide ideal support for large-flowered clematis, jasmine, and climbing roses, and annual climbers, such as sweet peas, morning glory, and runner beans.

No support needed
Plants such as Boston ivy have tendrils that adhere to walls without support. Ivy and climbing hydrangea have self-clinging roots on their stems. Some initial support is useful.

PLANTING IN POTS

Large containers, especially glazed ceramic pots or oak half barrels, create the opportunity for covering walls, fences, and screens, even without a bed or border. Some pots and troughs come with integral, freestanding trellis support, but you can also add a trellis fan as shown here. Try small- to medium-sized species and cultivars, such as *Clematis alpina* and *C. macropetala*, as well as annual climbers like *Eccremocarpus scaber* (Chilean glory vine) and morning glory (*Ipomoea*).

Laying a lawn

The best time to lay, or seed, a new lawn is early autumn or spring. Dig the area, adding a margin of 15cm (6in), and improve the drainage of heavy clay and wet soils by working lots of grit into the topsoil. For free-draining soils, dig in an 8–10cm (3–4in) layer of bulky organic matter to conserve moisture and fertility.

You will need
- Spade or fork
- Rake and hoe
- Organic fertilizer
- Sieved topsoil mixed with horticultural sand
- Wooden plank
- Broom
- Hosepipe
- Edging iron

🕒 1 day

Preparing the ground

1 Dig over the lawn area, removing big stones and perennial weeds, and break up the surface into a fine crumb structure. Rake level, then, keeping your weight on your heels, walk over the length of your plot, and then across the width.

2 Rake the ground level to remove any depressions left after walking. Leave for five weeks to allow weed seeds to germinate, then hoe lightly to remove them. Rake level and apply a dressing of organic fertilizer.

Laying the turf

3 Arrange turf delivery a few days after applying fertilizer. Carefully unroll the turf, laying whole pieces and working out from an edge. Stand on a plank to distribute your weight. Tamp down turves with a rake.

4 To ensure that the grass knits together, butt the edges of the turves, lifting them so that they are almost overlapping when pushed down. This helps to combat any shrinkage. Firm again with a rake.

Finishing and shaping

5 Continue to lay the next row of turf, ensuring that the joins are staggered like wall bricks. This produces a much stronger structure. Use an old knife for cutting, and avoid using small pieces at the edges.

6 To help adjacent pieces of turf to grow together and root firmly, brush in a blend of sieved topsoil and horticultural sand. Use a stiff broom to work in the top dressing and raise flattened grass.

7 Water well during dry spells to prevent shrinkage. Shape lawn edges when the turf has rooted (try gently lifting an edge). Lay out curves with a hosepipe and cut using an edging iron or border spade.

PLANTING TECHNIQUES

SPOT WEEDING

During lawn establishment, perennial weeds often take root, especially rosette-forming dandelions and thistles, which can smother the turf. Use an old kitchen knife, forked daisy grubber, or long-handled, lawn-weeding tool to extract them. Try to remove all the taproot.

Seeding a lawn

For large areas of lawn, seeding is the cheapest option and, although it will be about a year before the grass can take heavy use, it should start to green up and look good in under a month. Worn patches in existing lawns can also be repaired by reseeding with an appropriate grass mix.

1 Select a seed mix that suits your conditions and lawn use, such as hard-wearing family or fine, ornamental lawn. Weigh out seed for 1sq m (1sq yd) following pack directions. Pour into a paper cup; mark where the seed reaches.

2 You should have dug, firmed, levelled, and raked the lawn bed at least five weeks previously (see *opposite*). A few days before sowing, remove any weeds and add a top dressing of organic fertilizer. Rake level, removing any stones.

3 Sow in early autumn when the soil is warm and moist, or in spring when plants start to grow actively. Mark out 1sq m (1sq yd) sections using canes, and measure out the grass seed using the marked paper cup.

4 Scatter half the seed in one direction, and then go over at right angles with the remainder, keeping within the template. Move the template along and repeat the process. As a guide, one handful of seed weighs roughly 30g (1oz).

5 Work over the seeded lawn lightly with a rake until the seed is just covered with soil. Protect from birds using netting. Seedlings should appear within 14 days. Once the grass has reached 5cm (2in), cut with the mower blades set high.

A green carpet
The velvet green of a well-maintained lawn is the perfect foil for border flowers. Lawns create a sense of space in the garden and provide colour, even in the depths of winter.

Meadow-inspired plantings

High on the wish list of garden owners are naturally inspired plantings referred to as "meadows" or "prairies". These are plant communities with a natural feel, almost like a border and lawn combined, using a range of different plants and a range of planting densities and techniques. The popularity of these plantings reflects growing understanding of the importance of gardens as wildlife habitats, and an increasing need to provide a visual and physical escape from the outside world.

Achieving the "meadow" look

There are different routes to achieving meadow-inspired effects, all giving rather different results. Simplest of all is to stop mowing an area of lawn and allow it to grow long. Most plants in these "meadows" will be grasses, and reducing grass vigour (see p.283) can allow the introduction of wildflowers such as meadow cranesbill (*Geranium pratense*) and ox-eye daisy (*Leucanthemum vulgare*).

Perhaps the quickest route is to sow a specially formulated seed mix onto prepared ground. Various mixes are available, most using a blend of quick-growing wildflowers and introduced exotics. Seed usually needs resowing each year.

Perennial plantings, as championed by European designers of the New Perennial movement, are drifts of carefully selected perennials to provide a sophisticated effect that lasts year to year.

No-mow meadow
The easiest way to get the meadow look is to let an area of lawn grow long. Mow a path through for contrast.

Immersive appeal
Allowing access through meadow-inspired plantings is important if they are to be fully appreciated – natural timber decking works well.

Perennial perfection
Drifts of Echinacea *and other herbaceous plants that perform year after year are signature plants of the New Perennial movement.*

Appreciating the benefits

Even at its simplest, this form of gardening has many attractions. Reduced lawn mowing means less work; long grass and lawn flowers don't need watering, staking, or feeding. Meadow-like plantings contrast well with other parts of the garden, helping to divide up space; mown paths and margins help reinforce the effect. The plantings provide insects with new habitat, and these visitors will benefit the whole garden: many are pollinators or predators of pests.

Annual flower meadow seed mixes provide spectacular colour at low expense and input: plants need care after germinating, but they are as colourful as a herbaceous border with far less effort. Perennial meadows need more work in establishment, but plants are usually self-supporting and easy to please, providing displays that last into winter with attractive standing stems and seedheads that are also a draw for wildlife.

Five star bug hotel
Adding a bug hotel will help provide overwintering insects with habitat once meadow plants are cut down; it will also fascinate younger gardeners.

Pasture for pollinators
Bee hives are best positioned in out-of-the-way but sunny, sheltered corners of the garden, perfect places for letting grass and wildflowers grow.

Routes to success

Sow seeds of semi-parasitic yellow rattle (*Rhinanthus minor*) in autumn into wildflower meadows to reduce grass vigour and allow young wild flowers to be planted into the mown grass. You can also add bulbs such as *Narcissus* and *Camassia*. Special wildflower turf is available that already contains a mix of flowering plants.

Sow annual flower meadow seed mixes in spring or autumn on soil that is forked over and levelled. Mix seed with dry sand to help achieve an even spread when broadcasting. Firm and water the soil afterwards; you can use netting to prevent animal disturbance.

Perennial meadows are like a herbaceous border, with usually pot-grown perennials set into prepared ground. Position the plants in broad drifts or matrices, with numerous examples of each plant together. There is no front or back, but rather a three-dimensional carpet of planting.

All in the mix
Annual meadow seed mixtures often contain wildflowers mixed with easy-to-grow exotics that help prolong and enrich displays.

Aftercare and maintenance

Making a garden is a process that doesn't end when the construction and planting stages are complete. Even in low-maintenance plots, gardens only thrive when the plants are tended and the soil replenished. Some jobs are regular weekly tasks, but many others are only annual or twice yearly.

When and how to water

No matter where you garden, managing your water supply is essential. As the climate continues to change, gardeners need to use water sparingly and as wisely as possible. Containers, together with some types of vegetable crops and bedding plants, may need regular summer irrigation. Shrubs, trees, and perennials need watering only at planting time and during dry spells in the first year or two, or until they are well established. No matter how brown the grass may turn, established lawns never actually need watering and will eventually recover from drought.

If you need to water, do so in the cool of the morning or evening to minimize evaporation, and water close to the soil rather than overhead, targeting specific plants. Mulches, such as bark and spent mushroom compost, help seal in moisture and reduce competition from weeds. It is better to water heavily, with extended intervals between (allowing moisture to penetrate well into the soil and encourage deep rooting) than to water lightly but more frequently.

Preventing erosion
With shallow-rooting plants like this box, frequent watering can wear away the protective coating of compost. Reduce the problem by directing water on to a large crock or tile so that flow is gently dissipated.

Making watering easy

If you have limited amounts of time or a very large plot, some watering shortcuts are welcome. It makes sense to collect rainwater at sites around the garden and to make use of recycled or grey water, such as from the bath or washing-up (but only if no strong or heavily perfumed products have been used). Automatic irrigation can also be very efficient and, if properly managed, helps to save water.

Leaky hose
A perforated hosepipe (leaky hose) connected to an outdoor tap or water butt will channel water directly to where it is needed, such as snaking through leafy crops or through a newly planted border.

Timed watering
If you are often away from the garden for more than a couple of days or are too busy to water all your patio containers regularly, consider installing an automatic irrigation system with a timer.

Water butts
Raised up high enough so that you can comfortably fit a watering can under the tap, water butts are a convenient way to reduce dependence on the mains supply. Consider fitting extension kits to increase capacity.

Deadheading promotes new flower growth.

BENEFITS OF DEADHEADING

The aim of the plant is to set seed and reproduce: to achieve this it makes flowers and diverts most of its resources to develop a seedhead. To encourage more flowers you need to remove faded blooms before they have a chance to form seed. This is especially important for annuals which can stop flowering altogether and even die if you don't deadhead regularly. But perennials, including so-called patio plants, can also be encouraged to flower for much longer if they are deadheaded. Removing old, blemished heads also improves the appearance of plants and reduces the risk of disease.

The benefits of pruning

It is not essential to prune any plant, but thinning and cutting back to varying degrees or selectively removing whole branches can produce many useful effects. It can rejuvenate an old, congested specimen, giving it a new lease of life; help short-lived shrubs to live longer; increase the supply of flowering or fruiting wood; improve the shape and appearance of a plant; and reduce the incidence of disease.

The right cut
Cut back to just above a strong bud or pair of buds. Cutting halfway between buds causes die-back, which can introduce disease.

Alternate buds
Where buds form alternately along a stem, make a slanting cut, as shown, so that rainwater drains away from the bud.

Removing branches

As a tree matures, it may become too large for its site, or send out branches in inconvenient directions, and require pruning. Damaged or diseased branches and crossing limbs also need to be taken out to maintain the health of the tree. Hire a qualified tree surgeon to tackle very large branches, or those higher than head height. When pruning, take off a branch in sections – if you remove it with one cut close to the trunk, it will be pulled down by its own weight and may tear the bark on the trunk, leaving the tree vulnerable to infection.

1 To cut back branches, make two incisions: one, half way through, from beneath the branch; the second from the top to meet the undercut.

2 Remove the remaining branch stub, starting from the upper surface of the branch, just beyond the crease in the bark where the branch meets the trunk. Angle the cut away from the trunk.

3 This pruning method produces a clean cut, leaving the plant's healing tissue intact. The tree will soon produce bark to cover the exposed area.

Feeding and weeding

Clay loams are naturally fertile, while sandy soils tend to be nutrient poor. Adding bulky organic matter, such as well-rotted manure, improves the quality and structure of both types of soil as well as providing nutrients. During the growing season, wherever you garden intensively, you'll need to add extra organic fertilizer. Control weeds by digging them out or hoeing, or with a natural weedkiller if needed, except on turf, which will require a lawn weedkiller.

Containers
Flowering container plants, in particular, require extra organic fertilizer. Try a convenient, slow-release formula if you can't manage weekly feeds.

Soluble food
Liquid feeds are fast acting and ideal for bedding and patio plants in containers, as well as greenhouse crops such as tomatoes.

Weedkillers
Wherever possible, weed an area by hand to remove unwanted plants. If the weeds are pernicious or over a large area, you could consider a weedkiller. Alternatively, a flame gun can be quite effective for small areas.

Weeding by hand
Among existing plants, remove weed seedlings by hand. Use a hoe on dry days, severing the stems where they meet the roots just beneath the soil, or dig them out with a fork.

288 Plant guide
350 Materials guide

PLANT AND MATERIALS GUIDE

Plant guide

Selecting the right plant for the right place is an essential skill for any garden designer, and this directory will help you to make those critical decisions.

RHS HARDINESS RATINGS
The table (*right*) shows the corresponding lowest temperature range for each of the ratings in the RHS system of hardiness, used in this directory alongside the more general ratings. Please see p.370 for more information.

H1a	warmer than 15°C (59°F)
H1b	10–15°C (50–59°F)
H1c	5–10°C (41–50°F)
H2	1–5°C (34–41°F)
H3	-5–1°C (23–34°F)
H4	-10–-5°C (14–23°F)
H5	-15–-10°C (5–14°F)
H6	-20–-15°C (-4–5°F)
H7	colder than -20°C (-4°F)

Large trees

Acacia dealbata
Mimosa is an evergreen tree with fern-like, silvery grey-green divided leaves. Orange in bud before turning yellow, the fragrant clusters of flowers add colour and scent from winter to spring. Susceptible to frost, so plant in a sheltered site in full sun.
↕15–30m (50–100ft) ↔6–10m (20–30ft) ❄ ❄ H3 ☼ ◊ ♀

Acer campestre
The lobed leaves of the deciduous field maple are red when young, green by late summer, then yellow and red in autumn. The green flowers in spring produce the helicopter fruits that children love to play with. *A. campestre* 'Schwerinii' makes an excellent hedge, or can be grown in a large container.
↕8–25m (25–80ft) ↔4m (12ft) ❄ ❄ ❄ H6 ☼ ☼ ◊ ♀

Acer platanoides 'Crimson King'
The Norway maple is a vigorous, spreading, deciduous tree. 'Crimson King' has large, lobed, dark red-purple leaves that turn orange in autumn. The red-tinged yellow flowers are borne in mid-spring. Fast-growing, it makes a useful screen, but is at its best centre stage as an ornamental specimen.
↕25m (80ft) ↔15m (50ft) ❄ ❄ ❄ H7 ☼ ☼ ◊ ♀

Acer rubrum 'October Glory'
By autumn, the lobed, glossy, dark green foliage of the red maple has turned bright red; erect clusters of tiny red flowers are produced in spring. 'October Glory' is a reliable cultivar, though for best colour, grow it in acid soil. To fully appreciate its beauty give this large deciduous tree plenty of space.
↕20m (70ft) ↔10m (30ft) ❄ ❄ ❄ H6 ☼ ☼ ◊ ♀

Betula nigra
Red-brown when young, becoming blackish or grey-white as it ages, the peeling bark of the black birch is its main attraction. Yellow-brown catkins appear in early spring, and its glossy, diamond-shaped leaves turn buttery yellow in autumn. If space allows, plant in a group for maximum impact.
↕18m (60ft) ↔12m (40ft) ❄ ❄ ❄ H7 ☼ ◊ ◊ ● ♀

Betula utilis var. *jacquemontii*
The smooth, peeling white bark of this Himalayan birch comes into its own in a winter garden. Oval, tapered dark green leaves turn yellow in autumn, and yellow-brown catkins appear in early spring. The reliable cultivar 'Silver Shadow' has an eye-catching pure white trunk.
↕18m (60ft) ↔10m (30ft) ❄ ❄ ❄ H7 ☼ ◊ ♀

PLANT GUIDE

❀❀❀ H7–H5 fully hardy   ❀❀ H4–H3 hardy in mild regions/sheltered sites   ❀ H2 protect from frost over winter   ❀ H1c–H1a no tolerance to frost
☼ full sun   ☼ partial sun   ☀ full shade   ◊ well-drained soil   ◐ moist soil   ● wet soil   ◯◠◡◇△◊◔♤ tree shape

Cedrus atlantica f. glauca
Glaucous blue-green foliage, erect, cylindrical cones in autumn and a silvery-grey bark are the attractions of this coniferous tree. The blue Atlas cedar does well growing on chalk and is striking as a specimen in a sunny lawn, but its eventual size makes it unsuitable for all but the largest of gardens.

↕40m (130ft) ↔10m (30ft) ❀❀❀ H6 ☼ ◊ △–◊

Cercidiphyllum japonicum
The leaves of this fast-growing deciduous tree are bronze when young, turning mid-green, then yellow, orange and red in autumn. Acid soil produces the best colour. Fallen leaves smell of burnt sugar when crushed. The katsura tree is best used as a specimen in a woodland setting.

↕20m (70ft) ↔15m (50ft) ❀❀❀ H5 ☼ ◊ △

Fagus sylvatica (Atropurpurea Group) 'Riversii'
The beauty of this cultivar of the common beech lies in its deep purple leaves, which need full sun for best colour. A spreading, deciduous tree, it can be used for hedging, in a woodland garden, or as a focal point. For dramatic effect, plant next to a golden-leaved tree, such as *Catalpa bignonioides* 'Aurea'.

↕25m (80ft) ↔15m (50ft) ❀❀❀ H6 ☼ ◊ ◯

Liriodendron tulipifera 'Aureomarginatum'
Vigorous, large, deciduous tree forming an upright, rounded crown. The distinctive green three-lobed leaves are margined with bright yellow variegation fading to yellow-green by summer's end; leaves turn yellow before falling. In summer, mature trees bear green-orange tulip-like flowers.

↕20m (70ft) ↔8m (25ft) ❀❀❀ H6 ☼ ◊ ◯

Pinus wallichiana
The Bhutan pine is a graceful, broadly conical, evergreen tree with long, drooping, blue-green leaves and smooth, grey bark, which is grey-green when young but later becomes darker, scaly and fissured. It produces fresh green foliage in spring, and decorative pine cones that ripen to brown in autumn.

↕20–35m (70ft) ↔6–12m (20–40ft) ❀❀❀ H6 ☼ ◊ ◯

Quercus ilex
A majestic, round-headed evergreen tree, the holm oak has glossy, dark green leaves, which are silvery-grey when young. Striking yellow catkins are followed in autumn by small acorns. It makes a good screen and hedge, and thrives on exposed coastal sites. It also does well on shallow chalk.

↕25m (80ft) ↔20m (70ft) ❀❀ H4 ☼ ☼ ◊ ◯

Taxus baccata
A slow-growing evergreen conifer with distinctive dark green, needle-like leaves, the common yew is a familiar sight in churchyards. When closely-clipped, it is excellent for hedging and topiary. The golden-leaved cultivar 'Standishii' is ideal for brightening a shady area. All parts of the plant are poisonous.

↕20m (70ft) ↔10m (30ft) ❀❀❀ H7 ☼ ☼ ☀ ◊ △

Zelkova carpinifolia
Impressive, slow-growing deciduous tree forming a dense oval or vase-shaped crown. The main trunk branches low down, and upright stems rise to form the canopy. Smooth grey bark peels when mature. Leaves are rather rough, toothed, and superficially elm-like. Insignificant green flowers appear in spring.

↕25–30m (80–100ft) ↔20m (70ft) ❀❀❀ H6 ☼ ☼ ◊ ◊

TREES FOR EVERGREEN INTEREST
- Acacia dealbata p.288
- Arbutus unedo p.292
- Cedrus atlantica f. glauca p.289
- Chamaecyparis pisifera 'Filifera Aurea' p.290
- Cupressus macrocarpa 'Goldcrest' p.293
- Laurus nobilis p.294
- Olea europaea p.294
- Picea breweriana p.291
- Picea pungens (Glauca Group) 'Koster' p.291
- Pinus sylvestris (Aurea Group) 'Aurea' p.291
- Pinus wallichiana p.289
- Quercus ilex p.289
- Taxus baccata p.289
- Taxus baccata 'Fastigiata' p.295
- Tsuga canadensis 'Aurea' p.295

Medium-sized trees

Acer negundo 'Variegatum'
There are maples for spring flowers, summer foliage or autumn colour. A fast-growing, deciduous tree, *A. negundo* is known as the ash-leaved maple because of its divided leaves; those of the cultivar 'Variegatum' are splashed white at the margins. It looks good planted near dark-leaved plants.

↕15m (50ft) ↔10m (30ft) ❀ ❀ ❀ H6 ☼ ☼ ◊ ◊ ♃

Carpinus betulus 'Fastigiata'
The dependable, deciduous, spring-flowering common hornbeam has glowing coppery autumn colour and is great for hedging. It is an excellent substitute for beech on drier soils. The narrow, upright cultivar 'Fastigiata' opens up as it matures, making a striking specimen tree.

↕15m (50ft) ↔12m (40ft) ❀ ❀ ❀ H7 ☼ ☼ ◊ ◊

Catalpa bignonioides 'Aurea'
The beautiful, spreading, deciduous Indian bean tree is popular for its large, dramatic heart-shaped leaves, clusters of tubular flowers, and long bean-like seed pods. It makes a striking specimen tree, but can also be grown in a border. The leaves of 'Aurea' are bronze when young, maturing to yellow.

↕12m (40ft) ↔12m (40ft) ❀ ❀ ❀ H6 ☼ ◊ ◊ ♃

Chamaecyparis pisifera 'Filifera Aurea'
This hardy evergreen tree tolerates most soils other than waterlogged sites, and can be grown as a specimen or as hedging. *C. pisifera* 'Filifera' has slender, whip-like shoots and dark green leaves; 'Filifera Aurea' is similar, but has golden yellow leaves and is slower to reach maturity.

↕12m (40ft) ↔5m (15ft) ❀ ❀ ❀ H7 ☼ ◊ ◊ △

Davidia involucrata
The elegant handkerchief tree is so known because of the conspicuous white bracts that surround the small flowerheads in spring. It is deciduous, with sharp-pointed, red-stalked leaves and smooth grey bark. Ridged fruits hang from long stalks in autumn. A fine specimen tree.

↕15m (50ft) ↔10m (30ft) ❀ ❀ ❀ H5 ☼ ☼ ◊ ◊ △

Gleditsia triacanthos f. inermis 'Sunburst'
Also known as honey locust, this striking deciduous tree has delicate, fern-like foliage, spines on the trunk and branches, and long, curved seed pods in autumn. The cultivar 'Sunburst' is fast-growing and thornless, with golden yellow foliage in spring and autumn. Best as a specimen tree.

↕12m (40ft) ↔10m (30ft) ❀ ❀ ❀ H6 ☼ ◊ ◊ △

Liquidambar styraciflua 'Worplesdon'
Grown for its spectacular red, orange, and yellow autumn leaf colour, this handsome deciduous tree forms a broad crown bearing in summer glossy green, hand-shaped leaves with five slender lobes. Stems and young branches have corky bark; this selection often produces spiny fruit.

↕15m (50ft) ↔15m (50ft) ❀ ❀ ❀ H6 ☼ ☼ ◊ △

Morus nigra
The black mulberry forms a rounded, deciduous tree with heart-shaped leaves that have rough upper surfaces and toothed margins. The fruit is green, turning red and then purple-black, becoming edible only when fully ripe. Beware of planting next to pale paving as the fruit will stain it when it falls.

↕12m (40ft) ↔15m (50ft) ❀ ❀ ❀ H6 ☼ ◊ ◊ ♃

Nyssa sinensis
Grown for its pretty foliage and brilliant autumn colour, the Chinese tupelo forms a broadly conical, deciduous tree. The slender, tapered leaves turn bright shades of orange, red, and yellow in autumn, making it a valuable ornamental. Grow as a specimen tree; it looks very effective alongside water.

↕12m (40ft) ↔10m (40ft) ❀ ❀ ❀ H5 ☼ ☼ ◊ ◊ △

❀ ❀ ❀ H7–H5 fully hardy   ❀ ❀ H4–H3 hardy in mild regions/sheltered sites   ❀ H2 protect from frost over winter   ❀ H1c–H1a no tolerance to frost
☼ full sun   ☼ partial sun   ☼ full shade   ◊ well-drained soil   ◊ moist soil   ◆ wet soil   ♀♀♉♊♋♌♍♎ tree shape

Paulownia tomentosa
This fast-growing, deciduous tree is grown for its graceful habit, attractive large leaves, and showy, foxglove-like flowers. The fragrant, pinkish-lilac flowers, marked yellow and purple inside, open in late spring before the leaves appear. The tree can be pollarded, which will result in very large leaves.

↕12m (40ft) ↔10m (30ft) ❀ ❀ ❀ H5 ☼ ◊ ♀

Picea breweriana
The popular Brewer's weeping spruce is a hardy, slow-growing, blue-green conifer with horizontal branches and long, slim, pendent branchlets that give it a distinctive appearance. Purple cones decorate the branches in autumn. It can be grown as an effective windbreak or as a specimen tree.

↕15m (50ft) ↔4m (12ft) ❀ ❀ ❀ H6 ☼ ◊ ◆ ♀

Picea pungens (Glauca Group) 'Koster'
A hardy evergreen tree with scaly, grey bark and sharp, stout, bluish-green leaves. Cultivars of the Colorado spruce make wonderful ornamentals where space permits; 'Koster' has needle-like, silvery-blue leaves that fade to green with age and cylindrical light brown cones with papery scales.

↕15m (50ft) ↔5m (15ft) ❀ ❀ ❀ H7 ☼ ◊ ◊ ◊–♀

Pinus sylvestris (Aurea Group) 'Aurea'
The golden Scots pine is widely grown for its timber, but its cultivars make excellent garden trees, either planted singly or in groups. Upright conifers, they have whorled branches when young, and develop a rounded crown with age. 'Aurea' has striking golden yellow leaves in winter.

↕15m (50ft) ↔9m (28ft) ❀ ❀ ❀ H7 ☼ ◊ ◆ ♀

Prunus padus 'Watereri'
A deciduous, spreading tree, the bird cherry produces slender, pendent spikes of fragrant, star-shaped white flowers in mid-spring, followed by small black fruits. The leaves turn red or yellow in autumn. The conspicuous long flower spikes of the cultivar 'Watereri' create a spectacular spring display.

↕15m (50ft) ↔10m (30ft) ❀ ❀ ❀ H6 ☼ ◊ ◆ ♀

Salix alba var. *sericea*
The silver willow is a fast-growing, deciduous, spreading tree, conical in shape when young. The leaves are long, narrow and an intense silver-grey, and emerge at the same time as the yellow catkins in early spring. The foliage sparkles in the breeze, and it makes an elegant specimen tree.

↕15m (50ft) ↔8m (25ft) ❀ ❀ ❀ H6 ☼ ◊ ◆ ♀

Salix x *sepulcralis* var. *chrysocoma*
A wide-spreading, deciduous tree with supple yellow stems that reach the ground, the golden weeping willow is grown for its beautiful cascading habit. Slender yellow or green catkins are borne with the narrow yellow-green leaves in spring. It looks particularly striking when planted by water.

↕15m (50ft) ↔15m (50ft) ❀ ❀ ❀ H5 ☼ ◊ ◆ ♀

Toona sinensis 'Flamingo'
This spectacular tree bears astonishing pink new foliage in spring, which fades to cream then green after a couple of weeks. The large leaves are feathery. Trees may be single or multi-stemmed with strongly ascending branches forming a narrow canopy. It bears panicles of white summer flowers.

↕12m (40ft) ↔8m (25ft) ❀ ❀ H4 ☼ ◊ ♀

TREES AS FOCAL POINTS

- *Acer griseum* p.292
- *Albizia julibrissin* p.292
- *Betula nigra* p.288
- *Betula utilis* var. *jacquemontii* p.288
- *Carpinus betulus* p.210
- *Carpinus betulus* 'Fastigiata' p.290
- *Cornus controversa* 'Variegata' p.293
- *Cornus kousa* var. *chinensis* 'China Girl' p.293
- *Davidia involucrata* p.290
- *Dicksonia antarctica* p.293
- *Gleditsia triacanthos* 'Sunburst' p.290
- *Laburnum* x *watereri* 'Vossii' p.294
- *Nyssa sinensis* p.290
- *Paulownia tomentosa* p.291
- *Prunus serrula* p.294
- *Pyrus salicifolia* 'Pendula' p.295
- *Toona sinensis* 'Flamingo' p.291

Small trees

Acer griseum

The chief attraction of this deciduous maple is its unusual bark, which is orange to mahogany-red and peels laterally in papery rolls. The dark green leaves turn bright crimson and scarlet in autumn, and the ornamental bark gives this spectacular tree a valued winter role in small gardens.

↕10m (30ft) ↔10m (30ft) ❋ ❋ ❋ H5 ☀ ☼ ◊ ◊ ♤

Acer japonicum 'Vitifolium'

A pretty, deciduous tree with broad, fan-shaped leaves that turn scarlet, gold, and purple in autumn. The leaves are similar to those of a grapevine, hence the cultivar name. In mid-spring it bears clusters of small, delicate, reddish-purple flowers. Can be grown as a bushy tree or large shrub.

↕10m (30ft) ↔10m (30ft) ❋ ❋ ❋ H6 ☀ ☼ ◊ ◊ ♤

Acer palmatum 'Bloodgood'

Japanese maples make lovely ornamental trees. 'Bloodgood' forms a deciduous, bushy-headed shrub or small tree and is grown for its deeply cut, dark reddish-purple leaves, which turn bright red in autumn. Small purple flowers are borne in mid-spring, followed by attractive red-winged fruits.

↕5m (15ft) ↔5m (15ft) ❋ ❋ ❋ H6 ☀ ☼ ◊ ◊ ♤

Acer palmatum 'Ōsakazuki'

A stunning Japanese maple for autumn colour. The mid-green leaves are larger than average and turn a brilliant scarlet before falling. Dainty red-winged fruits appear in late summer. It can be grown in a large container but must not be allowed to dry out, and needs shelter from cold winds.

↕6m (20ft) ↔6m (20ft) ❋ ❋ ❋ H6 ☀ ☼ ◊ ♤

Acer palmatum 'Sango-kaku'

For colour interest all year, this delicate Japanese maple is a perfect choice. The divided leaves are orange-yellow in spring, maturing to green, then turning yellow in autumn before they fall. In winter, the new shoots, borne on ascending branches, turn coral-pink, deepening in colour as winter advances.

↕6m (20ft) ↔5m (15ft) ❋ ❋ ❋ H6 ☀ ☼ ◊ ♤

Albizia julibrissin

Delightful summer-flowering deciduous tree forming a broad, umbrella-shaped crown atop a slender trunk. The leaves are large (to 30cm/12in or so long) and finely divided, almost fern-like. In summer, fluffy flowerheads appear with showy clusters of pink stamens. Needs shelter and warmth.

↕8m (25ft) ↔8–10m (25–30ft) ❋ ❋ H4 ☀ ◊ ♤

Amelanchier lamarckii

With abundant white flowers in spring and brilliant red leaf colour in autumn, this deciduous hardy shrub or small tree provides plenty of seasonal interest. The young oval leaves unfold bronze before the star-shaped flowers emerge, and the small red fruits that follow are attractive to birds.

↕10m (30ft) ↔12m (40ft) ❋ ❋ ❋ H7 ☀ ☼ ◊ ◊ ♤

Arbutus unedo

This handsome evergreen with flaky, red-brown bark and attractive, glossy green leaves forms a large shrub or small tree in sheltered gardens. Lily-of-the-valley-like blooms appear in early winter and the rounded fruits, ripening to red in autumn, give rise to the common name, strawberry tree.

↕8m (25ft) ↔8m (25ft) ❋ ❋ ❋ H5 ☀ ◊ ♤

Cercis canadensis 'Forest Pansy'

A pretty, multi-stemmed tree or shrub with vivid, reddish-purple, heart-shaped leaves that are velvety to the touch. Magenta buds open to pale pink, pea-like flowers in mid-spring before the characteristic leaves appear. Impressive as a single specimen but also useful for the back of the border.

↕10m (30ft) ↔10m (30ft) ❋ ❋ ❋ H5 ☀ ☼ ◊ ◊ ♤

PLANT GUIDE

✿ ✿ ✿ H7–H5 fully hardy   ✿ ✿ H4–H3 hardy in mild regions/sheltered sites   ✿ H2 protect from frost over winter   ✾ H1c–H1a no tolerance to frost
☼ full sun   ☼ partial sun   ☼ full shade   ◊ well-drained soil   ◊ moist soil   ● wet soil   ♤♧◊△◊◊♀ tree shape

Cercis siliquastrum
The Judas tree is an eye-catching, spreading, bushy tree, with bright purple-rose spring flowers and long, purple-tinted pods that appear in late summer. Its heart-shaped leaves are bronze when young, turning yellow in autumn. Although hardy it originates from the Mediterranean, so avoid very cold sites.
↕10m (30ft) ↔10m (30ft) ✿ ✿ ✿ H5 ☼ ☼ ◊ ● ♤

Cornus controversa 'Variegata'
This elegant deciduous tree with horizontally-tiered branches creates a distinctive architectural profile. Flat heads of star-shaped white flowers appear in summer, followed by blue-black fruit. 'Variegata' has bright green leaves with creamy white margins, and makes a beautiful focal point.
↕8m (25ft) ↔8m (25ft) ✿ ✿ ✿ H5 ☼ ◊ ♤

Cornus kousa var. *chinensis* 'China Girl'
A broadly conical deciduous tree, this dogwood has tiny green flowerheads in summer surrounded by decorative petal-like white bracts. Fleshy red fruits develop later, followed by rich, purple-red autumn leaves. 'China Girl', free-flowering even when young, has large creamy-white bracts that age to pink.
↕7m (22ft) ↔5m (15ft) ✿ ✿ ✿ H6 ☼ ☼ ◊ △

Corylus avellana 'Contorta'
The corkscrew hazel is a slow-growing, small deciduous tree or shrub with unusual twisted shoots, which are seen at their best in winter when the long yellow catkins appear. Ideal as a focal point in a winter garden, the stems can also be cut for striking indoor displays.
↕5m (15ft) ↔5m (15ft) ✿ ✿ ✿ H6 ☼ ☼ ◊ ● ♤

Crataegus orientalis
Hawthorns are widely used for hedges and as ornamentals. Many are thorny but *C. orientalis* is almost thornless. It is an attractive, compact, deciduous tree with deeply cut, dark green leaves. White flowers appear in profusion in late spring, followed by yellow-tinged red fruit.
↕6m (20ft) ↔6m (20ft) ✿ ✿ ✿ H6 ☼ ☼ ◊ ● ♤

Crataegus persimilis 'Prunifolia'
An excellent small deciduous tree, with rich brown bark and long, dramatic thorns. It is grown mainly for its polished, deep green leaves that turn brilliant orange and red in autumn. Dense heads of white flowers are produced in early summer followed by clusters of long-lasting, bright red berries.
↕8m (25ft) ↔10m (30ft) ✿ ✿ ✿ H7 ☼ ☼ ◊ ● ♤

Cupressus macrocarpa 'Goldcrest'
The Monterey cypress is a coastal tree in the wild and will tolerate dry growing conditions, which makes it useful as a hedge or windbreak in exposed sites. 'Goldcrest' is a handsome, narrowly conical tree with lemon-scented golden foliage. It looks stunning grown against a dark background.
↕5m (16ft) ↔2.5m (8ft) ✿ ✿ H4 ☼ ◊ ♧

Dicksonia antarctica
A spectacular and hardy tree fern, *D. antarctica* brings drama into the garden. In spring its arching pale green fronds unfurl from the top of a mass of fibrous roots that form the trunk. It is evergreen in mild climates, but in cold winters protect the crown by covering it with straw.
↕6m (20ft) ↔4m (12ft) ✿ ✿ H3 ☼ ☼ ◊ ● ♀

TREES FOR SPRING INTEREST

- *Acacia dealbata* p.288
- *Acer palmatum* 'Sango-kaku' p.292
- *Amelanchier lamarckii* p.292
- *Betula utilis* var. *jacquemontii* p.288
- *Cercis siliquastrum* p.293
- *Crataegus orientalis* p.293
- *Crataegus persimilis* 'Prunifolia' p.293
- *Davidia involucrata* p.290
- *Laburnum* x *watereri* 'Vossii' p.294
- *Malus* 'Evereste' p.294
- *Malus* 'Royalty' p.294
- *Paulownia tomentosa* p.291
- *Prunus* 'Mount Fuji' p.294
- *Prunus padus* 'Watereri' p.291
- *Prunus* 'Spire' p.294
- *Prunus* x *subhirtella* 'Autumnalis Rosea' p.295
- *Pyrus salicifolia* 'Pendula' p.295
- *Salix alba* var. *sericea* p.291
- *Toona sinensis* 'Flamingo' p.291

Small trees

Ficus carica 'Brown Turkey'
A popular variety of fig that thrives in cool climates, 'Brown Turkey' has large lobed leaves and pear-shaped edible fruits, green at first, maturing to purple-brown. Grow as a fan against a sunny wall or as a freestanding tree; in cold areas keep in a pot and move under cover in winter.

↕3m (10ft) ↔4m (12ft) ❀ ❀ H4 ☼ ◐ ◊ ♤

Laburnum x watereri 'Vossii'
This elegant, spreading, deciduous tree has glossy green leaves, cut into oval leaflets, and bears magnificent long golden chains of pea-like flowers in late spring. It makes an impressive specimen tree in a small garden, but can also be trained over a pergola. The leaves and seeds are poisonous.

↕8m (25ft) ↔8m (25ft) ❀ ❀ ❀ H5 ☼ ◊ ♤

Laurus nobilis
Bay laurel is a conical evergreen tree grown for its aromatic, leathery, dark green leaves, which are used as flavouring in cooking. Clusters of small, greenish-yellow flowers appear in spring, followed by black berries in autumn. It can be grown in a pot, and looks attractive when trimmed into formal shapes.

↕to 10m (30ft) ↔to 8m (25ft) ❀ ❀ H4 ☼ ◐ ◊ ♤

Malus 'Evereste'
This crab apple is an excellent choice for a small garden as it forms a neat, conical shape. A profusion of white, shallow, cup-shaped flowers open from pink buds in late spring, followed by small, red-flushed, orange-yellow fruit. The green leaves turn yellow and orange in autumn before falling.

↕7m (22ft) ↔6m (20ft) ❀ ❀ ❀ H6 ☼ ◐ ◊ ♤

Malus 'Royalty'
This pretty crab apple is smothered in deep pink to bright purple flowers, which open from dark red buds in spring. The glossy leaves are dark red-purple and maintain their colour well through the season, turning red in autumn. Inedible small purple fruits follow the flowers. A fine specimen tree.

↕8m (25ft) ↔8m (25ft) ❀ ❀ ❀ H6 ☼ ◐ ◊ ♤

Olea europaea
An elegant, slow-growing evergreen, the olive tree has grey-green leaves and tiny, fragrant, creamy-white flowers in summer. The green olives only ripen to black in hot, dry conditions. It makes a stunning feature in a sunny, sheltered spot, or grow in a large pot and move under cover in winter.

↕10m (30ft) ↔10m (30ft) ❀ ❀ H4 ☼ ◊ ♤

Prunus 'Mount Fuji'
Ornamental cherries make very attractive specimen trees for small gardens. This beautiful deciduous tree has pale green young leaves, darkening to deep green, then turning orange and red in autumn before they fall. Clusters of fragrant, white, cup-shaped flowers are borne in mid-spring.

↕6m (20ft) ↔8m (25ft) ❀ ❀ ❀ H6 ☼ ◊ ♤

Prunus serrula
A dramatic choice for winter interest, this deciduous tree is prized for its glossy mahogany bark with pale horizontal lines. Small white flowers are produced at the same time as the new leaves in late spring, followed by small inedible cherries on long stalks. The leaves turn yellow in autumn.

↕10m (30ft) ↔10m (30ft) ❀ ❀ ❀ H6 ☼ ◊ ♤

Prunus 'Spire'
Attractive over a long season, the leaves of this upright, deciduous cherry are bronze when young, green in summer, then orange and red in autumn. In spring, bowl-shaped, soft pink flowers emerge in clusters against the new leaves. Makes a beautiful feature in a small garden.

↕10m (30ft) ↔6m (20ft) ❀ ❀ ❀ H6 ☼ ◊ ♤

PLANT GUIDE

❋❋❋ H7–H5 fully hardy   ❋❋ H4–H3 hardy in mild regions/sheltered sites   ❋ H2 protect from frost over winter   ❋ H1c–H1a no tolerance to frost
☼ full sun   ☼ partial sun   ☼ full shade   ◊ well-drained soil   ◐ moist soil   ● wet soil   ◯◯◯◯◯◯◯♦ tree shape

Prunus × subhirtella 'Autumnalis Rosea'

A popular tree for its early-flowering nature, this delicate spreading cherry is perfect for a small garden. Clusters of tiny, double, pale pink flowers appear in winter during mild spells. The green leaves are narrow and bronze when young, turning golden-yellow in autumn.

↕8m (25ft) ↔8m (25ft) ❋❋❋ H6 ☼ ◊ ◐ ◯

Pyrus salicifolia 'Pendula'

This delightful ornamental pear tree has an elegant weeping habit and silvery-grey, willow-like leaves. An abundant show of creamy-white flowers in spring is followed by small, hard, inedible pears in late summer. Grow as specimen tree on a lawn, where its graceful habit can be seen to advantage.

↕5m (15ft) ↔4m (12ft) ❋❋❋ H6 ☼ ◊ ◯

Rhus typhina

Known as the stag's horn sumach because of its red velvety shoots, this distinctive deciduous tree is particularly fine in autumn when its deeply divided leaves turn shades of orange and red. The fruits are formed in dense, hairy, crimson-red clusters on female plants. Plant singly or in a shrub border.

↕5m (15ft) ↔6m (20ft) ❋❋❋ H6 ☼ ◊ ◐ ◯

Sorbus aria 'Lutescens'

A pretty deciduous tree, this eye-catching whitebeam has striking silvery-grey young foliage that gradually turns grey-green. White flowers in late spring are followed by orange berries in autumn. A freestanding tree of great beauty, it can also be used for mass planting or screening.

↕10m (30ft) ↔8m (25ft) ❋❋❋ H6 ☼ ☼ ◊ ◯

Sorbus commixta

Sorbus are excellent ornamental trees for city gardens as they tolerate atmospheric pollution. *S. commixta* bears large white flowerheads in spring and has elegant foliage, which turns shades of yellow, red, and purple in autumn. 'Embley' has bright red leaves in late autumn, and plenty of crimson fruit.

↕10m (30ft) ↔7m (22ft) ❋❋❋ H6 ☼ ☼ ◊ ◐ ◯

Stewartia sinensis

A good choice for autumn foliage colour, this small deciduous tree is also prized for its unusual peeling red-brown bark and showy, white fragrant flowers that appear in midsummer. Autumn brings an impressive display of red, orange, and yellow leaves. It prefers acid soil.

↕6m (20ft) ↔3m (10ft) ❋❋❋ H5 ☼ ☼ ◊ ◐ ◯

Taxus baccata 'Fastigiata'

Irish yew has a narrow, upright habit, eventually forming a distinguished, columnar shape. This makes it useful as a focal point or accent plant in a border. Small red berries appear in summer. 'Fastigiata Aurea' is similar but has variegated yellow-green leaves. All parts are poisonous.

↕10m (30ft) ↔2m (6ft) ❋❋❋ H6 ☼ ☼ ◊ ◐ ♦

Tsuga canadensis 'Aurea'

A graceful species of conifer, there are many varieties of Eastern hemlock available. 'Aurea' is an elegant, compact, and fairly slow-growing tree with golden yellow juvenile foliage, which darkens to green with age. It is useful for evergreen interest in partially shaded areas.

↕8m (25ft) ↔4m (12ft) ❋❋❋ H6 ☼ ☼ ◊ ◐ ◯

TREES FOR AUTUMN COLOUR

- *Acer campestre* p.288
- *Acer griseum* p.292
- *Acer japonicum* 'Vitifolium' p.292
- *Acer palmatum* 'Bloodgood' p.292
- *Acer palmatum* 'Ōsakazuki' p.292
- *Acer palmatum* 'Sango-kaku' p.292
- *Acer platanoides* 'Crimson King' p.288
- *Acer rubrum* 'October Glory' p.288
- *Amelanchier lamarckii* p.292
- *Cercidiphyllum japonicum* p.289
- *Crataegus persimilis* 'Prunifolia' p.293
- *Gleditsia triacanthos* f. *inermis* 'Sunburst' p.290
- *Liquidambar styraciflua* 'Worplesdon' p.290
- *Malus* 'Evereste' p.294
- *Nyssa sinensis* p.290
- *Prunus* 'Mount Fuji' p.294
- *Prunus padus* 'Watereri' p.291
- *Prunus* 'Spire' p.294
- *Rhus typhina* p.295
- *Sorbus commixta* p.295
- *Stewartia sinensis* p.295

Large shrubs

Aralia elata 'Variegata'

The Japanese angelica tree, *A. elata,* is an elegant, deciduous shrub with striking grey-green leaves that turn many shades of yellow, orange, or purple in autumn. Large heads of small white flowers appear in late summer. The leaves of 'Variegata' have creamy-white margins that shine out in a shady border.

↕5m (15ft) ↔5m (15ft) ❋ ❋ ❋ H5 ☼ ◐ ◊ ♦

Azara microphylla

An attractive evergreen shrub or small tree with large sprays of small, glossy, dark green leaves. Small clusters of vanilla-scented, deep yellow flowers are borne in late winter and early spring, making it a useful shrub for winter interest. It will tolerate part-shade and grows well against a wall.

↕7m (22ft) ↔4m (12ft) ❋ ❋ ❋ (boderline)H4 ☼ ◐ ◊ ♦

Buddleja alternifolia 'Argentea'

The slender, arching branches of this robust deciduous shrub have narrow grey-green leaves and carry dense clusters of very fragrant lilac flowers in summer. Its weeping habit makes it suitable for training as a standard. Prune after flowering to prevent branches from becoming tangled.

↕4m (12ft) ↔4m (12ft) ❋ ❋ ❋ H6 ☼ ◐ ◊ ♦

Calycanthus 'Aphrodite'

A large-growing, deciduous shrub that bears impressively showy, yellow-marked, red-purple, scented flowers up to 10cm (4in) across in summer. Flowers are held clear of the oval glossy green foliage, which turns yellowish before falling in autumn. This spreading, bushy plant is best in a sheltered site.

↕4m (12ft) ↔4m (12ft) ❋ ❋ ❋ H5 ☼ ◊

Camellia reticulata 'Leonard Messel'

Camellias are invaluable evergreen spring-flowering shrubs for acid soils in sheltered sites. 'Leonard Messel' produces a profusion of large, semi-double, pink flowers in spring that stand out vividly against a background of matt, dark green leaves. It is ideal as a specimen or in a woodland setting.

↕4m (12ft) ↔3m (10ft) ❋ ❋ ❋ H5 ☼ ◐ ◊

Chimonanthus praecox 'Grandiflorus'

Known as wintersweet, this deciduous shrub produces pale yellow flowers that hang from its bare stems throughout winter, perfuming the air with intoxicating scent. Grow it as a specimen shrub, as part of a border planting, or train it on a sunny wall. The stems can be cut for indoor displays.

↕4m (12ft) ↔3m (10ft) ❋ ❋ ❋ H5 ☼ ◊

Clerodendrum trichotomum var. *fargesii*

This spectacular deciduous shrub has an upright habit and attractive bronze young leaves. Fragrant, white, star-shaped flowers with green sepals open from pink and greenish-white buds in late summer. Jewel-like, bright blue berries, surrounded by pronounced maroon calyxes, follow the flowers.

↕6m (20ft) ↔6m (20ft) ❋ ❋ ❋ H5 ☼ ◊

Cordyline australis 'Red Star'

The New Zealand cabbage palm is a popular evergreen shrub grown for its striking foliage. In warm regions, it makes an eye-catching architectural plant for a sheltered courtyard garden; in frost-prone areas, keep it in a pot in a cool greenhouse during winter. 'Red Star' has rich red-bronze, sword-like leaves.

↕3–10m (10–30ft) ↔1–4m (3–12ft) ❋ ❋ ❋ H3 ☼ ◐ ◊

Cornus mas

Shrubs that flower in winter, such as this Cornelian cherry, are a valuable asset to the designer. It bears little clusters of tiny yellow flowers on bare branches in late winter, before the leaves appear. Bright red fruits are produced in late summer, and the leaves turn red-purple in autumn.

↕5m (15ft) ↔5m (15ft) ❋ ❋ ❋ H6 ☼ ◐ ◊

PLANT GUIDE

❅❅❅ H7–H5 fully hardy ❅❅ H4–H3 hardy in mild regions/sheltered sites ❅ H2 protect from frost over winter
❆ H1c–H1a no tolerance to frost ☀ full sun ☼ partial sun ⬤ full shade ◊ well-drained soil ◐ moist soil ● wet soil

Corylus maxima 'Purpurea'
The intense colour of this deciduous, deep purple-leaved hazel makes an immediate impact in a garden. Attractive purple-tinged catkins appear in late winter, and edible nuts ripen in autumn. Grow as a specimen plant or as a focal point in a shrub border. The best colour is produced in full sun.

↕6m (20ft) ↔5m (15ft) ❅❅❅ H6 ☀ ☼ ◊

Cotinus 'Grace'
A vigorous smoke bush cultivar that can be grown as a small bushy tree or as a tall multi-stemmed shrub. Large, dark pink flower clusters appear above the foliage in summer, and the soft purple-red leaves turn a brilliant orange-red before falling. An excellent choice for autumn colour.

↕6m (20ft) ↔5m (15ft) ❅❅❅ H5 ☀ ☼ ◊ ◐

Cotoneaster frigidus 'Cornubia'
A large, arching, semi-evergreen shrub, this cotoneaster has narrow green leaves that are tinted bronze in autumn. Creamy-white, early summer flowers are produced in profusion, followed by heavy clusters of bright red fruit that are attractive to birds. It can be trained as a single-stemmed tree.

↕10m (30ft) ↔10m (30ft) ❅❅❅ H6 ☀ ◊

Cotoneaster lacteus
This dense, evergreen shrub sports distinctive, dark green, leathery leaves. Cup-shaped, milky-white flowers appear in summer, followed by clusters of dark red fruit that persist well into winter. It makes an attractive hedge or screen, and it can also be grown as a small tree.

↕4m (12ft) ↔4m (12ft) ❅❅❅ H6 ☀ ☼ ◊

Dipelta floribunda
This handsome deciduous shrub offers interest through the seasons. Masses of fragrant pale pink flowers with yellow markings appear in late spring, its light green leaves turn yellow in autumn, and it has attractive peeling bark in winter. Grow it as a specimen plant or in a shrub border.

↕4m (12ft) ↔4m (12ft) ❅❅❅ H5 ☀ ◊

Elaeagnus x *ebbingei* 'Gilt Edge'
A hardy, evergreen, dense shrub, 'Gilt Edge' has brown scaly stems and glossy leaves with green centres and golden yellow margins. Small, lightly-scented flowers are produced from mid- to late autumn. The plant's hardiness makes it a good choice for a shelter belt or hedge, especially in coastal areas.

↕4m (12ft) ↔4m (12ft) ❅❅❅ H5 ☀ ◊

Elaeagnus 'Quicksilver'
With silvery shoots and narrow, silver-grey leaves, this fast-growing shrub makes a great foil for dark-leaved plants. Although bushy, with a loose, spreading crown, it can be trained as a small tree. Star-shaped, fragrant, creamy-yellow flowers open from silvery buds in late spring or summer.

↕4m (12ft) ↔4m (12ft) ❅❅❅ H5 ☀ ◊

Hamamelis x *intermedia* 'Pallida'
Witch hazel is a handsome shrub that produces spider-like scented flowers on bare branches in winter. There are many cultivars. 'Jelena' has large, coppery-orange flowers and orange and red autumn foliage. 'Pallida' bears large, fragrant, yellow flowers and has golden autumn leaves.

↕4m (12ft) ↔4m (12ft) ❅❅❅ H5 ☀ ☼ ◊ ◐

SHRUBS FOR FOCAL POINTS

- *Acer palmatum* p.300
- *Cordyline australis* 'Red Star' p.296
- *Cotinus* 'Grace' p.297
- *Euphorbia characias* subsp. *wulfenii* 'John Tomlinson' p.308
- *Fatsia japonica* p.302
- *Fothergilla* species p.211
- *Hamamelis* x *intermedia* 'Pallida' p.297
- *Hydrangea* RUNAWAY BRIDE SNOW WHITE ('Ushyd0405') p.309
- *Juniperus communis* 'Hibernica' p.298
- *Magnolia liliiflora* 'Nigra' p.303
- *Magnolia stellata* p.303
- *Photinia* x *fraseri* 'Red Robin' p.298
- *Rosa* FOR YOUR EYES ONLY ('Cheweyesup') p.311
- *Sambucus nigra* f. *porphyrophylla* 'Eva' p.305
- *Schefflera taiwaniana* p.299
- *Trithrinax campestris* p.299
- *Viburnum plicatum* f. *tomentosum* 'Mariesii' p.305
- *Yucca filamentosa* 'Bright Edge' p.313

Large shrubs

Hippophae rhamnoides
Sea buckthorn thrives in harsh conditions and makes an excellent screening plant for a coastal garden. It has a bushy habit, but can be trained to make a small tree, and has thorny stems with narrow, silver-grey leaves. Grow male and female plants together to produce brilliantly orange-coloured berries.

↕6m (20ft) ↔6m (20ft) ❄❄❄ H7 ☀ ◊

Hydrangea paniculata 'Unique'
Hydrangeas are mainly grown for their showy flowerheads, but some have pretty bark and others develop good autumn colour. *H. paniculata* 'Unique' bears large, creamy-white flowerheads from midsummer to early autumn, and its leaves turn yellow before falling. It's best planted singly or in a shrub border.

↕3–7m (10–22ft) ↔2.5m (8ft) ❄❄❄ H5 ☀ ☼ ◊ ◊

Ilex aquifolium 'Silver Queen'
Common holly has dark green leaves, but there are many cultivars with white, cream, or yellow variegation. 'Silver Queen' is a male variety (it does not bear berries); it forms an upright evergreen, with purple stems and striking leaves with broad, creamy-white margins. It is ideal for hedges and screens.

↕10m (30ft) ↔4m (12ft) ❄❄❄ H6 ☀ ☼ ◊ ◊

Itea ilicifolia
This spectacular evergreen shrub has holly-like, shiny, dark green leaves. Long catkins made up of small, greenish-white flowers appear in late summer, and a honey-like scent is discernible on warm evenings. A fine freestanding shrub for mild areas, but plant it against a wall in more exposed sites.

↕3–5m (20–15ft) ↔3m (10ft) ❄❄❄ H5 ☀ ◊

Juniperus communis 'Hibernica'
Junipers tolerate a wide range of soils and growing conditions, are tough enough for hot, sunny sites, and need little pruning. 'Hibernica', also known as the Irish juniper, forms a slender column of crowded, needle-like leaves, each with a silver line, and makes an excellent structural plant for formal schemes.

↕3–5m (10–15ft) ↔30cm (12in) ❄❄❄ H7 ☀ ◊

Ligustrum ovalifolium 'Aureum'
A vigorous, semi-evergreen shrub, golden privet has variegated leaves with bright yellow margins and bears dense clusters of white flowers in midsummer, followed by black berries. It clips easily and is ideal for hedging and topiary. Shade tolerant, it can be planted to brighten a shady corner of the garden.

↕4m (12ft) ↔4m (12ft) ❄❄❄ H5 ☀ ☼ ◊

Mahonia x media 'Charity'
With their attractive foliage, bright yellow flowers, and decorative fruits, mahonias make magnificent architectural features in a winter garden. 'Charity' is fast-growing and has spiny holly-like leaves. Bright yellow to lemon yellow flowers are produced in spikes from late autumn to late winter.

↕5m (15ft) ↔4m (12ft) ❄❄❄ H5 ☀ ◊ ◊

Olearia macrodonta
New Zealand holly is a vigorous evergreen shrub with sharply-toothed, sage green leaves, which provide mellow colour all year. Fragrant, white, daisy-like flowers are borne in early summer. A handsome freestanding shrub in mild areas, it also makes an excellent screen for exposed coastal gardens.

↕6m (20ft) ↔5m (15ft) ❄❄❄ (borderline) H4 ☀ ◊

Photinia x fraseri 'Red Robin'
This hardy evergreen shrub is grown for its conspicuous, deep red young foliage, which is produced in spring on the tips of the branches. It looks good in a woodland garden or in a shrub border, and can also be used for hedging. 'Red Robin' is a compact cultivar, with especially bright red young leaves.

↕5m (15ft) ↔5m (15ft) ❄❄❄ H5 ☀ ☼ ◊ ◊

PLANT GUIDE

✽✽✽ H7–H5 fully hardy   ✽✽ H4–H3 hardy in mild regions/sheltered sites   ✽ H2 protect from frost over winter
❀ H1c–H1a no tolerance to frost   ☼ full sun   ☼ partial sun   ☼ full shade   ◊ well-drained soil   ◗ moist soil   ◆ wet soil

Pittosporum tenuifolium 'Silver Queen'
A marvellous evergreen with sparkling, undulating grey-green leaves on black stems. Foliage is variegated, each leaf edged in creamy white. It forms a neat and dense, cone-shaped shrub, at its best in winter. It is great free-standing or against a wall in colder areas; it can also be trimmed to keep it more compact.

↕to 6m (20ft) ↔4m (12ft) ✽✽ H4 ☼ ◊

Rhamnus alaternus 'Argenteovariegata'
This handsome, bushy, evergreen shrub bears glossy grey-green leaves with creamy-white margins. Small yellow-green flowers appear in early summer, followed by spherical red berries in a warm summer, which ripen to black. It does well in coastal and city gardens, but needs shelter in colder areas.

↕5m (15ft) ↔4m (12ft) ✽✽✽ H5 ☼ ◊

Rhododendron luteum
An elegant deciduous azalea, *R. luteum* bears rounded clusters of funnel-shaped yellow flowers in late spring, which have a delightful scent. The rich green leaves turn shades of crimson, purple, and orange in autumn, making it a valuable garden plant over a long season. It requires acid soil.

↕4m (12ft) ↔4m (12ft) ✽✽✽ H5 ☼ ◊ ◗

Schefflera taiwaniana
Evergreen shrub or small tree with glossy palmate leaves, each divided into 7–11 radiating leaflets. Silvery new growth is especially dramatic in spring. Plants can be single- or multi-stemmed, forming a rounded canopy, and may bear small white flowers and rounded black fruit. It needs shelter.

↕4m (12ft) ↔2.5m (8ft) ✽✽ H4 ☼ ☼ ◗

Syringa vulgaris 'Mrs Edward Harding'
Lilacs form spreading deciduous shrubs with pretty heart-shaped leaves and make useful screening plants. Sweetly scented flowerheads appear from spring to early summer. There are over 500 cultivars of common lilac to choose from; 'Mrs Edward Harding' has double purple-red flowers.

↕to 7m (22ft) ↔to 7m (22ft) ✽✽✽ H6 ☼ ◊

Tamarix ramosissima 'Pink Cascade'
Tamarisks are excellent shrubs for exposed coastal gardens where they can make an effective screen. They have attractive, feathery foliage, formed of needle-like leaves. *T. ramosissima* is deciduous, with arching branches and upright plumes of small, pink flowers; 'Pink Cascade' has rich pink flowers.

↕5m (15ft) ↔5m (15ft) ✽✽✽ H5 ☼ ◊

Trithrinax campestris
Unusual and slow growing, this hardy palm bears rigid blue-grey, spine-tipped foliage, held atop a spiny, fibre-covered rather stout trunk. It makes a wonderful drought-tolerant architectural plant for the sunniest, driest sites, perhaps in a gravel garden. Grow well away from paths due to sharp spines.

↕4m (12ft) ↔2m (6ft) ✽✽ H3 ☼ ◊

Viburnum opulus
The guelder rose is a good choice for a wildlife garden as birds love the translucent red berries; as a bonus, the leaves also turn a rich red in autumn. The late spring blooms are attractive, too, forming lacecap-like heads of white flowers. This deciduous plant is vigorous and is commonly seen in hedgerows.

↕5m (15ft) ↔4m (12ft) ✽✽✽ H6 ☼ ☼ ◊ ◗

SHRUBS FOR HOT, DRY SITES

- *Artemisia arborescens* p.306
- *Ceanothus thyrsiflorus* var. *repens* p.307
- *Choisya* x *dewitteana* 'Aztec Pearl' p.301
- *Cistus* cultivars p.307
- *Convolvulus cneorum* p.307
- *Helianthemum* 'Wisley Primrose' p.309
- *Lavandula angustifolia* 'Munstead' p.309
- *Origanum* 'Kent Beauty' p.310
- *Pinus mugo* 'Mops' p.310
- *Pittosporum tenuifolium* 'Golf Ball' p.310
- *Pittosporum tenuifolium* 'Silver Queen' p.299
- *Pittosporum tenuifolium* 'Tom Thumb' p.310
- *Pittosporum tobira* 'Nanum' p.310
- *Potentilla fruticosa* cultivars p.311
- *Ribes sanguineum* 'Pulborough Scarlet' p.304
- *Salvia officinalis* cultivars p.312
- *Salvia rosmarinus* p.312
- *Santolina pinnata* subsp. *neapolitana* 'Sulphurea' p.313
- *Trithrinax campestris* p.299

Medium-sized shrubs

Abelia × *grandiflora*
A vigorous, semi-evergreen shrub with glossy dark green foliage and an abundance of fragrant, pink-flushed white flowers from midsummer to mid-autumn. Plant either as a freestanding shrub, or as an informal hedge. It is best fan-trained against a sunny wall in colder areas.

↕3m (10ft) ↔4m (12ft) ❄❄❄ H5 ☼ ◊

Acer palmatum
Most Japanese maples are low-growing and shrubby, and look their best at the front of a border; many have beautiful foliage and fiery autumn colour. *A. palmatum* forms a mound of narrow, very finely toothed red-purple leaves, turning deep yellow or orange in autumn.

↕2m (6ft) ↔3m (10ft) ❄❄❄ H6 ☼ ☼ ◊ ◊

Aucuba japonica 'Crotonifolia'
Hardy evergreen shrubs, spotted laurels are easy to grow and tolerant of a wide range of growing conditions – shade, dry sites, and even areas with polluted air. 'Crotonifolia' has large, glossy green leaves speckled with yellow marks. In mid-spring, small red-purple flowers appear, followed by red berries.

↕3m (10ft) ↔3m (10ft) ❄❄❄ H5 ☼ ☼ ◊ ◊

Berberis darwinii
This vigorous, dense, mounded evergreen shrub has glossy dark green foliage on prickly stems. During spring, it bears drooping clusters of bright orange flowers, which are followed by round blue-black fruit. It makes an attractive informal hedge, and tolerates heavy clay soils.

↕3m (10ft) ↔3m (10ft) ❄❄❄ H5 ☼ ☼ ◊

Berberis julianae
A handsome evergreen shrub with spiny-margined, glossy deep green leaves, this plant is often used as a screen. From spring to early summer, clusters of scented yellow or red-tinged flowers are produced, followed by egg-shaped, blue-black fruits. It is best planted where its scent will be appreciated.

↕3m (10ft) ↔3m (10ft) ❄❄❄ H5 ☼ ☼ ◊

Buddleja davidii 'Dartmoor'
An outstanding butterfly bush cultivar, 'Dartmoor' has arching stems and soft green leaves that are white beneath. In late summer and autumn, it bears broad, open-branched plumes of highly scented, pinkish purple flowers. Loved by butterflies and ideally suited to wildlife gardens.

↕2.5m (8ft) ↔2.5m (8ft) ❄❄❄ H6 ☼ ☼ ◊

Buddleja × *weyeriana* 'Sun Gold'
One of the most reliable late-flowering deciduous shrubs, this splendid plant bears racemes of scented pink-tinged orange flowers, produced freely from summer until the first frosts if deadheaded. It is also attractive to a wide range of pollinators. It is easily grown in a sunny site and best pruned annually.

↕3m (10ft) ↔3m (10ft) ❄❄❄ H6 ☼ ◊

Camellia japonica 'Bob's Tinsie'
Camellias make elegant evergreen flowering plants for gardens with acid soil. New variations of *C. japonica* appear every year and there is a huge range of cultivars to choose from. 'Bob's Tinsie' has an upright habit, and bears small, clear red flowers from early to late spring. Shelter from cold, drying winds.

↕2m (6ft) ↔1m (3ft) ❄❄❄ H5 ☼ ◊ ◊

Ceanothus 'Concha'
Ceanothus are cultivated for their flowers, which may be blue, white, or pink. *C.* 'Concha' is a good choice for a warm, sunny wall or fence. It forms a dense evergreen shrub with finely toothed, dark green leaves and produces masses of reddish-purple buds in late spring that open up to dark blue flowers.

↕3m (10ft) ↔3m (10ft) ❄❄ H4 ☼ ◊

PLANT GUIDE

❋ ❋ ❋ H7–H5 fully hardy ❋ ❋ H4–H3 hardy in mild regions/sheltered sites ❋ H2 protect from frost over winter
❋ H1c–H1a no tolerance to frost ☼ full sun ☼ partial sun ☀ full shade ◊ well-drained soil ◐ moist soil ● wet soil

Chaenomeles speciosa 'Moerloosei'
Ornamental quinces make reliable garden shrubs, and can even be trained against a shaded wall or fence. This variety (also sold as 'Apple Blossom') bears large clusters of white flowers, flushed dark pink, in spring and early summer, followed by aromatic fruits. Prune after flowering.

↕2.5m (8ft) ↔5m (15ft) ❋❋❋ H6 ☼ ☼ ◊

Choisya × *dewitteana* 'Aztec Pearl'
A compact, elegant example of Mexican orange blossom, this pretty evergreen shrub with slim dark green leaves is suitable for a small garden or container. Fragrant clusters of white star-shaped flowers emerge from pink buds in late spring, and appear again in smaller numbers in late summer and autumn.

↕2.5m (8ft) ↔2.5m (8ft) ❋❋ H4 ☼ ☼ ◊

Cornus alba 'Aurea'
This golden-leaved, vigorous dogwood offers a combination of summer and winter interest. Throughout summer it forms a mound of broad greenish-yellow leaves and, after these fall in late autumn, the dark red stems create a stunning display. Cut down a third of the stems in spring to rejuvenate the plant.

↕3m (10ft) ↔3m (10ft) ❋❋❋ H7 ☼ ☼ ◊

Cornus alba 'Sibirica'
A deciduous, upright dogwood, 'Sibirica' forms a dense thicket of young scarlet stems. These are seen at their best in sunshine, and set a dull winter garden ablaze with their fiery colours. It is one of the best cultivars for autumn colour, its dark green leaves turning red before falling.

↕3m (10ft) ↔3m (10ft) ❋❋❋ H7 ☼ ☼ ◊

Cornus sericea 'Flaviramea'
The winter shoots of this vigorous dogwood display their most vivid colour when grown in a sunny site. The plant bears white flowers from late spring to early summer, and the dark green leaves turn red and orange in autumn. The form 'Flaviramea' produces bright yellow-green winter stems.

↕2m (6ft) ↔4m (12ft) ❋❋❋ H7 ☼ ☼ ◊

Daphne bholua 'Jacqueline Postill'
A shrub for a border or rock garden, *D. bholua* is best planted in a sheltered position where the richly fragrant flowers will be appreciated. This cultivar is vigorous, evergreen, and bears clusters of deep purple-pink flowers, white inside, over a long flowering season in late winter. Mulch to retain moisture.

↕2m (6ft) ↔1.5m (5ft) ❋❋ H4 ☼ ☼ ◊ ◐

Erica arborea var. *alpina*
This tree heath makes a dense, compact, upright shrub, crowded with needle-shaped, bright green evergreen leaves. Masses of tiny, fragrant, bell-shaped white flowers appear in spring. Grow it in acid soil for the best results, and prune hard after flowering to keep it in shape and encourage new growth.

↕2m (6ft) ↔90cm (36in) ❋❋ H4 ☼ ◊

Exochorda × *macrantha* 'The Bride'
Pure white, showy, saucer-shaped flowers on arching branches cover this spreading evergreen shrub in late spring, making a beautiful display. Mound-forming and wider than it is tall, it is suitable for growing as a specimen plant, although it can also be grown in a shrub border.

↕2m (6ft) ↔3m (10ft) ❋❋❋ H6 ☼ ☼ ◊ ◐

SHRUBS FOR SHADE

- *Aucuba japonica* 'Crotonifolia' p.300 (dry shade)
- *Azara microphylla* p.296 (dry shade)
- *Buxus sempervirens* 'Suffruticosa' p.306 (dry shade)
- *Chaenomeles speciosa* 'Moerloosei' p.301 (dry shade)
- *Cornus alba* 'Aurea' p.301 (damp conditions)
- *Cornus sericea* 'Flaviramea' p.301 (damp conditions)
- *Mahonia japonica* p.303 (dry shade)
- *Rhododendron* 'Kure-no-yuki' (Kurume) p.311 (dry shade)
- *Sarcococca hookeriana* var. *digyna* p.313 (dry shade)
- *Sarcococca hookeriana* WINTER GEM ('Pmoore03') p.313
- *Viburnum opulus* p.299 (damp conditions)

Medium-sized shrubs

Euphorbia × *pasteurii*
Broad evergreen shrub with spreading then upright stems, often bare lower down. Leaves are long and lanceolate with a silvery midrib, and develop fiery tints during autumn and winter. Impressive terminal lime-green heads of brown flowers appear in late spring. It needs a sheltered site.

↕3m (10ft) ↔3m (10ft) ❄❄ H4 ☼ ◊

Fatsia japonica
The castor oil plant is valued for its bold evergreen foliage and architectural habit. Its long-stalked, palmate, shiny dark green leaves give a subtropical effect, while striking branched clusters of creamy-white flowers emerge in autumn, followed by small black berries. It is tolerant of coastal exposure.

↕↔1.5–4m (5–12ft) ❄❄❄ H6 ☼ ☼ ◊ ◊

Fuchsia magellanica
In frost-free regions, this deciduous shrub, the hardiest of the fuchsia species, can be grown on its own or as informal hedging. It carries small, lantern-like flowers with red tubes, long red sepals and purple petals, from midsummer through into autumn. The flowers are followed by black fruits.

↕to 3m (10ft) ↔to 3m (10ft) ❄❄ H4 ☼ ☼ ◊ ◊

Hebe 'Midsummer Beauty'
Hebes are adaptable evergreen shrubs that suit a wide range of growing conditions, including containers. 'Midsummer Beauty', an upright, rounded shrub with purplish-brown stems and bright green leaves, bears tapering plumes of medium-sized, lilac-purple flowers from midsummer to late autumn.

↕2m (6ft) ↔1.5m (5ft) ❄❄ H4 ☼ ◊ ◊

Hibiscus syriacus 'Diana'
Large showy flowers are the main allure of hibiscus cultivars. They thrive in a sunny border and flower over a long period. 'Diana' is an erect, deciduous shrub with toothed, dark green leaves that produces trumpet-shaped, white flowers with wavy-margined petals, from late summer to mid-autumn.

↕3m (10ft) ↔2m (6ft) ❄❄❄ H5 ☼ ◊ ◊

Hydrangea arborescens 'Annabelle'
Excellent as specimen plants or in groups, in a mixed border or in containers, hydrangeas are versatile garden shrubs. 'Annabelle', one of the most elegant cultivars, is deciduous and, from summer to early autumn, bears large, spherical flowerheads, crowded with creamy-white flowers.

↕2.5m (8ft) ↔2.5m (8ft) ❄❄❄ H6 ☼ ☼ ◊ ◊

Hydrangea aspera HOT CHOCOLATE ('Haopr012')
This deciduous and rather bushy shrub has wide-spreading branches and impressive dark purple young foliage, which slowly ages to bronze-green. The leaves have a velvety feel and are covered thickly with soft hairs. In late summer, large, showy lacecap flowerheads are produced in blue-purple and pink.

↕3m (10ft) ↔2m (6ft) ❄❄ H4 ☼ ☼ ◊

Hydrangea macrophylla 'Mariesii Lilacina'
This rounded, deciduous shrub is grown for its mauve-pink to blue, showy lacecap flowers, which appear from mid- to late summer. It makes a fine freestanding shrub, and is also useful for mass planting in shady areas. Leave the flowerheads on over winter to protect the plant from frost damage.

↕2m (6ft) ↔2.5m (8ft) ❄❄❄ H5 ☼ ☼ ◊ ◊

Hydrangea quercifolia SNOW QUEEN ('Flemygea')
The oak-leaved hydrangea is grown chiefly for its deeply lobed, dark green leaves, which turn magnificent tints of bronze and purple in autumn before falling. From midsummer to autumn, SNOW QUEEN, also known as 'Flemygea', produces large, white, conical flowerheads, which fade to pink as they age.

↕2m (6ft) ↔2.5m (8ft) ❄❄❄ H5 ☼ ☼ ◊ ◊

❄❄❄ H7–H5 fully hardy   ❄❄ H4–H3 hardy in mild regions/sheltered sites   ❄ H2 protect from frost over winter
❄ H1c–H1a no tolerance to frost   ☀ full sun   ☀ partial sun   ☀ full shade   ◊ well-drained soil   ◊ moist soil   ◆ wet soil

Ilex crenata
With its dense growth and small, glossy, dark green foliage, this slow-growing evergreen is a possible replacement for disease-blighted box (*Buxus*). It is tolerant of regular trimming, making it highly useful as a hedging or topiary plant. It may bear small white flowers in summer, followed by black fruits if unpruned.
↕6–8m (20–25ft) ↔1.5m (5ft) ❄❄❄ H6 ☀ ☀ ◊

Indigofera heterantha
Elegant, fern-like, grey-green leaves clothe the arching branches of this spreading, multi-stemmed, deciduous shrub. From early summer through to autumn, small, purple-pink, pea-like flowers are carried in dense spikes. It thrives when fan-trained against a sunny wall, especially in colder areas.
↕2–3m (6–10ft) ↔2–3m (6–10ft) ❄❄❄ H5 ☀ ◊ ◆

Jasminum nudiflorum
Winter jasmine has long, slender, arching, leafless shoots bearing bright yellow flowers from winter to early spring. Oval, dark green leaves emerge after flowering. It is ideal for training on a low wall or trellis. Prune once flowering has finished to maintain a neat shape.
↕3m (10ft) ↔3m (10ft) ❄❄❄ H5 ☀ ☀ ◊

Kolkwitzia amabilis 'Pink Cloud'
A hardy, deciduous shrub, the beauty bush forms a dense twiggy shape. Bell-shaped pink flowers, with yellow-flushed throats, are borne in profusion from late spring to early summer. Pale, bristly seed clusters follow. It makes a fine freestanding shrub, but can be planted as an informal hedge.
↕3m (10ft) ↔4m (12ft) ❄❄❄ H6 ☀ ◊

Magnolia liliiflora 'Nigra'
One of the most reliable of all magnolias, this cultivar produces beautiful large, dark purple-red upright flowers in early summer and intermittently into the autumn. It is compact and deciduous, with glossy dark green leaves that provide a foil to the flowers. Grow as a specimen plant for the best effect.
↕3m (10ft) ↔2.5m (8ft) ❄❄❄ H6 ☀ ☀ ◊ ◆

Magnolia stellata
This graceful, deciduous shrub is slow-growing but well worth the wait. The star magnolia bears pure white, sometimes pink-flushed, star-shaped flowers in early spring, before the leaves emerge. A compact shrub, it is initially bushy and then spreading. Spring frosts may damage early blooms.
↕3m (10ft) ↔4m (12ft) ❄❄❄ H6 ☀ ☀ ◊ ◆

Mahonia japonica
Invaluable in a winter garden, this handsome evergreen shrub thrives in shady spots. It has spectacular, sharply toothed, dark green leaves. Arching spikes of fragrant, pale yellow flowers appear from late autumn to early spring, followed by blue-purple berries.
↕2m (6ft) ↔3m (10ft) ❄❄❄ H5 ☀ ☀ ◊ ◆

Nandina domestica
The fruit, flowers, and foliage of this evergreen shrub give it a long season of interest. The leaves have warm red tints in spring and autumn, and small star-shaped white flowers emerge in midsummer, followed by bright red berries. The cultivar 'Fire Power' is a compact form with bright red leaves.
↕2m (6ft) ↔1.5m (5ft) ❄❄❄ H5 ☀ ◊ ◆

SHRUBS FOR FOLIAGE INTEREST

- *Acer palmatum* p.300
- *Aralia elata* 'Variegata' p.296
- *Artemisia arborescens* p.306
- *Aucuba japonica* 'Crotonifolia' p.300
- *Cordyline australis* 'Red Star' p.296
- *Corylus maxima* 'Purpurea' p.297
- *Cotinus* 'Grace' p.297
- *Elaeagnus* x *ebbingei* 'Gilt Edge' p.297
- *Elaeagnus* 'Quicksilver' p.297
- *Fatsia japonica* p.302
- *Hydrangea aspera* HOT CHOCOLATE ('Haopr012') p.302
- *Hydrangea quercifolia* SNOW QUEEN ('Flemygea') p.302
- *Ilex aquifolium* 'Silver Queen' p.298
- *Photinia* x *fraseri* 'Red Robin' p.298
- *Physocarpus opulifolius* 'Diabolo' p.304
- *Pittosporum tenuifolium* 'Silver Queen' p.299
- *Sambucus racemosa* 'Plumosa Aurea' p.305
- *Schefflera taiwaniana* p.299

Medium-sized shrubs

Osmanthus × *burkwoodii*
This hardy evergreen shrub is grown for its glossy dark green leaves, and clusters of tiny, creamy-white trumpet-shaped flowers, which are sweetly scented and appear from mid- to late spring. Its dense habit makes it useful for hedging and topiary. Trim into shape after flowering.

↕3m (10ft) ↔3m (10ft) ❋ ❋ ❋ H5 ☼ ☼ ◊

Paeonia delavayi
In early summer, this magnificent tree peony produces single, cup-shaped, dark crimson flowers on long lax stems. The handsome, deeply cut, dark green leaves are tinged burgundy in spring. A stunning deciduous shrub for a mixed border; it does not tolerate being moved.

↕2m (6ft) ↔1.2m (4ft) ❋ ❋ ❋ H5 ☼ ☼ ◊ ◊

Philadelphus 'Belle Étoile'
Mock oranges are grown for their beautiful flowers, which are often scented and usually white. 'Belle Étoile' makes an arching, deciduous shrub with tapering leaves. Its fragrant white flowers are single with a maroon flush at the centre, and are freely produced from late spring to early summer.

↕1.2m (4ft) ↔2.5m (8ft) ❋ ❋ ❋ H6 ☼ ☼ ◊

Physocarpus opulifolius 'Diabolo'
Grown chiefly for its attractive purple foliage and upright red stems, this spreading deciduous shrub also produces clusters of small pinkish-white flowers in late spring, followed by maroon fruit. The peeling bark gives additional winter interest. Cut down to the ground in spring to rejuvenate.

↕2m (6ft) ↔2.5m (8ft) ❋ ❋ ❋ H7 ☼ ☼ ◊

Pieris japonica 'Blush'
A versatile evergreen shrub for acid soils, *P. japonica* has narrow, glossy leaves, which are an attractive coppery-red when young. Tassels of white flowers appear from early to mid-spring. The compact cultivar 'Blush' has dark green leaves and its pink-flushed white flowers open from dark pink buds.

↕2m (6ft) ↔2m (6ft) ❋ ❋ ❋ H5 ☼ ☼ ◊ ◊

Pyracantha SAPHYR JAUNE ('Cadaune')
Firethorns can be grown as a freestanding feature, against a wall, or for hedging. This cultivar, also known as 'Cadaune', is an upright, evergreen shrub with spiny branches, dark green leaves, and small, white late-spring flowers. Its bright yellow autumn fruits provide a flash of colour as winter approaches.

↕4m (12ft) ↔3m (10ft) ❋ ❋ ❋ H6 ☼ ☼ ◊

Ribes sanguineum 'Pulborough Scarlet'
Fairly upright when young, this flowering currant becomes spreading with maturity. It is a vigorous, deciduous shrub, with aromatic leaves and clusters of dark red, white-centred tubular flowers in spring, followed by round, blue-black berries. It is ideal for the back of a mixed border.

↕3m (10ft) ↔2.5m (8ft) ❋ ❋ ❋ H6 ☼ ◊

Rosa 'Geranium' (*moyesii* hybrid)
This spectacular shrub rose has arching branches and small, dark green leaves. A profusion of open, scarlet flowers, with prominent cream stamens is produced in summer, followed by blazing orange-red, bottle-shaped hips in autumn, which extend the season of visual interest.

↕2.5m (8ft) ↔1.5m (5ft) ❋ ❋ ❋ H6 ☼ ◊ ◊

Rosa xanthina 'Canary Bird'
This large, spreading, early-flowering rose has thorny, arching stems and dainty, almost fern-like foliage. In spring, the branches are wreathed with masses of upward-facing bright yellow, single, slightly scented, five-petalled flowers. It is a glorious sight, and one that attracts many pollinating insects.

↕2.5m (8ft) ↔4m (12ft) ❋ ❋ ❋ H6 ☼ ◊

PLANT GUIDE

✤✤✤ H7–H5 fully hardy   ✤✤ H4–H3 hardy in mild regions/sheltered sites   ✤ H2 protect from frost over winter
❀ H1c–H1a no tolerance to frost   ☼ full sun   ◐ partial sun   ● full shade   ◊ well-drained soil   ◐ moist soil   ● wet soil

Sambucus nigra f. porphyrophylla 'Eva'
This graceful elder is attractive for most of the year. The dark purple lacy foliage provides colour contrast in a mixed border. Showy pale pink, lemon-scented, flattened flowerheads appear in midsummer, followed by dark red elderberries. Full sun is best for foliage colour. It is also sold as 'Black Lace'.

↕3m (10ft) ↔2m (6ft) ✤✤✤ H6 ☼ ◐ ◊ ◐

Sambucus racemosa 'Plumosa Aurea'
A bushy plant with arching shoots; the deeply cut leaves, which are bronze in youth and mature to golden yellow, provide a bright splash of colour in a border. Small, creamy yellow flowers appear in mid-spring, followed by round, glossy red fruits in summer. The foliage may scorch in hot sun.

↕3m (10ft) ↔3m (10ft) ✤✤✤ H7 ☼ ◐ ◊ ◐

Skimmia x confusa 'Kew Green'
In spring, this compact, mounded, evergreen shrub produces dense, conical heads of fragrant, creamy-white flowers above deep green, pointed, aromatic leaves. Suitable for a shady border or woodland garden, it also looks attractive in a container. An adaptable shrub, it can cope with polluted air.

↕3m (10ft) ↔1.5m (5ft) ✤✤✤ H5 ◐ ● ◊ ◐

Spiraea nipponica 'Snowmound'
At its peak in early summer, this spiraea presents a marvellous display, with clusters of bowl-shaped white flowers carried all along the upper sides of the arching stems. Deciduous, fast-growing, and densely leaved, it forms a spreading shape and is perfect for growing near the back of a sunny mixed border.

↕2.5m (8ft) ↔2.5m (8ft) ✤✤✤ H6 ☼ ◊ ◐

Viburnum x bodnantense 'Dawn'
Useful for providing winter interest in a garden, this shrub produces clusters of scented, tubular, rose-tinted flowers on bare stems over a long season, from late autumn to spring. It is upright and deciduous, with toothed, dark green leaves. A range of cultivars is available; 'Deben' has white flowers.

↕3m (10ft) ↔2m (6ft) ✤✤✤ H6 ☼ ◐ ◊ ◐

Viburnum carlesii 'Aurora'
Suitable for a border or woodland garden, this deciduous shrub is densely bushy with irregularly toothed, dark green leaves. 'Aurora' is mainly grown for its clusters of perfumed flowers, which emerge in mid-spring. The buds are initially red and then open up to the pink tubular blooms.

↕2m (6ft) ↔2m (6ft) ✤✤✤ H6 ☼ ◐ ◊ ◐

Viburnum plicatum f. tomentosum KILIMANJARO ('Jwwl')
A spreading, deciduous shrub with a pyramid-shaped habit, it has branches arranged in layers. Dark green oval foliage contrasts with flat, lacy, white then pinkish flowerheads freely produced in early summer. Leaves turn purple in autumn and a second flush of flowers may appear. Red then black berries may form.

↕4m (12ft) ↔4m (12ft) ✤✤✤ H6 ☼ ◐ ◊

Viburnum plicatum f. tomentosum 'Mariesii'
This viburnum has distinctive extended horizontal branches that create a striking architectural effect, which is best appreciated when the shrub is grown as a specimen plant in a lawn. The flowers are white and the heart-shaped, dark green leaves turn red-purple in autumn.

↕3m (10ft) ↔4m (12ft) ✤✤✤ H5 ☼ ◐ ◊ ◐

SHRUBS FOR GROUND COVER

- *Calluna vulgaris* 'Gold Haze' p.306
- *Ceanothus thyrsiflorus* var. *repens* p.307
- *Cotoneaster dammeri* p.308
- *Cotoneaster salicifolius* 'Gnom' p.308
- *Euonymus fortunei* 'Emerald Gaiety' p.308
- *Hebe pinguifolia* 'Pagei' p.308
- *Helianthemum* 'Wisley Primrose' p.309
- *Juniperus procumbens* p.309
- *Juniperus squamata* 'Blue Carpet' p.309
- *Lonicera pileata* p.310
- *Potentilla fruticosa* 'Dart's Golddigger' p.208
- *Prunus laurocerasus* 'Zabeliana' p.311
- *Santolina chamaecyparissus* p.208
- *Vinca major* p.209
- *Vinca minor* 'La Grave' p.313

Small shrubs

Artemisia arborescens
Grown for its silver-grey, feathery foliage, this evergreen shrub is tolerant of exposed sites and is useful in a coastal garden. It carries clusters of small yellow flowers in summer and autumn, but is most valued for its elegant leaves. It is also suitable for a herb or rock garden.

↕1m (3ft) ↔1.5m (5ft) ❄ ❄ H4 ☼ ◊

Ballota 'All Hallows Green'
Originally from the Mediterranean, ballota thrives in dry, free-draining, sunny sites and makes an attractive edging plant. This cultivar forms a bushy evergreen subshrub with heart-shaped, lime green leaves. Small, pale green flowers appear in midsummer. Trim in spring to keep the shrub compact.

↕60cm (24in) ↔75cm (30in) ❄ ❄ ❄ H5 ☼ ◊

Berberis x *stenophylla* 'Corallina Compacta'
This is a compact cultivar of the much larger evergreen shrub, *B.* x *stenophylla*, which can be grown as an informal hedge. Like its parent, it has arching stems and narrow, spine-tipped, dark green leaves. In late spring, small clusters of pale orange flowers open from red buds along the branches.

↕30cm (12in) ↔30cm (12in) ❄ ❄ ❄ H5 ☼ ☼ ◊

Berberis thunbergii 'Aurea'
Create a splash of colour in the garden with this compact, deciduous berberis, which has vivid yellow young foliage, maturing to yellow-green. Pale yellow flowers are produced along the branches in mid-spring, followed by glossy red fruit. Suitable for hedging, but the leaves may scorch in full sun.

↕1.5m (5ft) ↔2m (6ft) ❄ ❄ ❄ H7 ☼ ◊

Berberis thunbergii f. *atropurpurea* 'Atropurpurea Nana'
A dwarf, dome-shaped berberis with rounded, red-purple leaves, a dense, twiggy habit, and small bright red berries that are attractive to birds. It tolerates polluted air and is a very adaptable shrub, ideal for a border or a rock garden.

↕60cm (24in) ↔75cm (30in) ❄ ❄ ❄ H7 ☼ ◊

Berberis thunbergii f. *atropurpurea* 'Helmond Pillar'
This deciduous barberry has distinctive columnar stems and dark wine-red leaves, which turn bright red in autumn. Tiny yellow flowers appear in spring, followed by red berries. Its upright habit makes it useful for filling gaps in a border.

↕1.2m (4ft) ↔60cm (24in) ❄ ❄ ❄ H7 ☼ ☼ ◊

Buxus sempervirens 'Suffruticosa'
This compact, very slow-growing selection of box is good for hedging or screens, and is one of the best types for the structure of a knot garden or parterre. Its dense habit makes it easy to trim into different shapes. It prefers partial shade, but can tolerate full sun if it is not allowed to get too dry.

↕1m (3ft) ↔1.5m (5ft) ❄ ❄ ❄ H6 ☼ ◊

Calluna vulgaris 'Gold Haze'
Heathers are robust plants and make good low-maintenance ground cover. There are many cultivars to choose from, all derived from *C. vulgaris*, a hardy, bushy, evergreen shrub that grows on acid soils in the wild. 'Gold Haze' has pale yellow leaves and short spikes of white bell-shaped flowers.

↕60cm (24in) ↔45cm (18in) ❄ ❄ ❄ H7 ☼ ◊

Calluna vulgaris 'Spring Cream'
A compact heather with mid-green leaves, which are tipped with cream in spring, this cultivar produces short spikes of white bell-shaped flowers that remain from midsummer until late autumn. Along with other heathers, it is attractive to bees. Grow on a moist, but free-draining sunny bank.

↕35cm (14in) ↔45cm (18in) ❄ ❄ ❄ H7 ☼ ◊

PLANT GUIDE

❋❋❋ H7–H5 fully hardy   ❋❋ H4–H3 hardy in mild regions/sheltered sites   ❋ H2 protect from frost over winter
❋ H1c–H1a no tolerance to frost   ☼ full sun   ☼ partial sun   ☀ full shade   ◊ well-drained soil   ◐ moist soil   ◆ wet soil

Caryopteris x *clandonensis* 'Worcester Gold'
The small but vivid blue flowers are the main attraction of *Caryopteris*. The cultivar 'Worcester Gold' has lavender-blue flowers, which are produced from late summer to early autumn on the current year's shoots. They stand out against a dense mound of warm yellow, deciduous foliage.

↕1m (3ft) ↔1.5m (5ft) ❋❋ H4 ☼ ◊

Ceanothus x *delilianus* 'Gloire de Versailles'
Also known as California lilac, ceanothus are grown for their abundant blue, pink, or white flowers. 'Gloire de Versailles' is a fast-growing, bushy, deciduous shrub with finely toothed, mid-green leaves. From midsummer to autumn, it produces loose bunches of scented, powder-blue flowers.

↕1.5m (5ft) ↔1.5m (5ft) ❋❋ H4 ☼ ◊

Ceanothus thyrsiflorus var. *repens*
Also known as creeping blue blossom, this is a useful, low-growing, evergreen ceanothus. It forms a natural mound of glossy mid-green foliage and, in late spring, produces an abundance of fluffy, pale to dark blue flowers. A perfect shrub for the front of a border or to clothe a sunny bank.

↕1m (3ft) ↔2.5m (8ft) ❋❋❋ (borderline) H4 ☼ ◊

Ceratostigma willmottianum
This loosely-domed, deciduous shrub produces clusters of pale to mid-blue flowers from late summer through to autumn. The pointed, bristly leaves are initially mid- to dark green with purple margins and then turn red in autumn. It needs a warm, sunny sheltered site to thrive.

↕1m (3ft) ↔1.5m (5ft) ❋❋❋ (borderline) H4 ☼ ◊ ◐

Cistus x *dansereaui* 'Decumbens'
Rock roses prefer a sunny site and can be grown in beds or containers. The flowers, usually white or pink, only last a day but are carried in profusion. 'Decumbens' is a low-growing, spreading, evergreen shrub that bears large white flowers with a crimson blotch at the base of each petal.

↕60cm (24in) ↔1m (3ft) ❋❋ H4 ☼ ◊

Cistus x *purpureus*
The narrow, green leaves of this rounded, evergreen shrub make a good foil for the single, crinkled, dark pink flowers, which appear in succession throughout summer. Each petal has a crimson mark at the base. The stems are upright and red-flushed. It is drought-tolerant and needs a sunny site.

↕1m (3ft) ↔1m (3ft) ❋❋ H4 ☼ ◊

Convolvulus cneorum
With its silky, silvery leaves and stems, this convolvulus is an asset even when not in bloom. The delicate flowers emerge from pink buds from late spring to summer, and are white and funnel-shaped with yellow centres. In colder areas, grow in a pot and move into a conservatory or greenhouse over winter.

↕60cm (24in) ↔90cm (36in) ❋❋ H4 ☼ ◊

Coronilla valentina subsp. *glauca*
The leaves of this bushy, rounded evergreen shrub are an attractive blue-green and fleshy. From late winter to early spring, and again in late summer, fragrant, yellow, pea-like flowers appear, followed by slim pods. Either grow it in a shrub border or at the base of a warm, sunny wall.

↕80cm (32in) ↔80cm (32in) ❋❋ H4 ☼ ◊

SHRUBS FOR SPRING INTEREST

- *Berberis darwinii* p.300
- *Camellia japonica* 'Bob's Tinsie' p.300
- *Camellia reticulata* 'Leonard Messel' p.296
- *Ceanothus* 'Concha' p.300
- *Ceanothus thyrsiflorus* var. *repens* p.307
- *Choisya* x *dewitteana* 'Aztec Pearl' p.301
- *Euphorbia characias* subsp. *wulfenii* 'John Tomlinson' p.308
- *Exochorda* x *macrantha* 'The Bride' p.301
- *Magnolia stellata* p.303
- *Photinia* x *fraseri* 'Red Robin' p.298
- *Prunus laurocerasus* 'Zabeliana' p.311
- *Rhododendron* 'Kure-no-yuki' (Kurume) p.311
- *Ribes sanguineum* 'Pulborough Scarlet' p.304
- *Rosa xanthina* 'Canary Bird' p.304
- *Viburnum* x *burkwoodii* 'Anne Russell' p.313
- *Viburnum carlesii* 'Aurora' p.305
- *Viburnum opulus* p.299
- *Viburnum plicatum* f. *tomentosum* 'Mariesii' p.305

Small shrubs

Cotoneaster dammeri

Evergreen cotoneasters offer colour and texture all year round, and are at their best in autumn when the berries develop. *C. dammeri* is vigorous and spreading with long arching stems, and makes excellent ground cover. Small, white flowers are borne in early summer, followed in autumn by round red berries.

↕20cm (8in) ↔2m (6ft) ❋ ❋ ❋ H6 ☼ ☼ ◊

Cotoneaster salicifolius 'Gnom'

This dwarf, evergreen shrub makes a prostrate, dense dome, with wide-spreading branches bearing small, slender, dark green leaves. In early summer, white flowers are produced and these are followed by clusters of bright red fruits in the autumn. It is a good choice for ground cover.

↕30cm (1ft) ↔2m (6ft) ❋ ❋ ❋ H6 ☼ ◊

Daphne odora 'Aureomarginata'

This evergreen species of daphne is one of the most fragrant flowering shrubs for a winter garden. The variegated cultivar 'Aureomarginata' has leaves with narrow yellow margins. Clusters of pink trumpet-shaped flowers appear from midwinter to early spring, followed by red fruit.

↕1.5m (5ft) ↔1.5m (5ft) ❋ ❋ H4 ☼ ☼ ◊

Daphne × *transatlantica* ETERNAL FRAGRANCE ('Blafra')

Small and slow growing, and more or less evergreen, this plant combines dense, neat, rounded growth and small, glossy, dark green leaves with a continuous display of sweetly perfumed, palest pink, four-petalled flowers from spring until autumn. Neat and easily grown, it is one of the best shrubs for scent.

↕1m (3ft) ↔1m (3ft) ❋ ❋ ❋ H5 ☼ ◊

Euonymus fortunei 'Emerald Gaiety'

Poor soil and full sun suit many *E. fortunei* cultivars, making them useful shrubs for difficult sites. They make good ground cover, and can be fan-trained against a wall if supported. The evergreen 'Emerald Gaiety' is compact and bushy, with bright green leaves with white margins, tinged pink in winter.

↕1m (3ft) ↔1.5m (5ft) ❋ ❋ ❋ H5 ☼ ☼ ◊

Euphorbia characias subsp. *wulfenii* 'John Tomlinson'

This striking evergreen shrub produces upright stems with grey-green leaves one year, followed the next spring by large showy heads of small, bright, yellow-green cup-shaped flowers, which last from early spring to early summer.

↕1.2m (4ft) ↔1.2m (4ft) ❋ ❋ H4 ☼ ◊

Hebe 'Great Orme'

Adaptable shrubs, hebes will grow in a wide range of garden situations, from a mixed border to a rock garden. 'Great Orme' is an open, rounded, evergreen shrub with deep purplish shoots and glossy, dark green leaves. Spikes of deep pink flowers, fading to white, appear from midsummer to mid-autumn.

↕1.2m (4ft) ↔1.2m (4ft) ❋ ❋ H4 ☼ ◊ ◊

Hebe macrantha

This evergreen is bushy, initially open-branched and then later spreading, with oval, fleshy, bright green leaves. In early summer, large white flowers are produced in clusters of three. It is suitable for a container or rock garden, and needs little or no pruning.

↕60cm (24in) ↔90cm (36in) ❋ ❋ H4 ☼ ☼ ◊ ◊

Hebe pinguifolia 'Pagei'

An evergreen, semi-prostrate shrub, 'Pagei' has small, slightly cupped blue-green leaves. Short spikes of delicate pure white flowers emerge in profusion in late spring or early summer. It is an excellent plant for a rock garden or for ground cover, and needs little or no pruning. It flowers best in full sun.

↕30cm (12in) ↔90cm (36in) ❋ ❋ ❋ H5 ☼ ☼ ◊ ◊

❋❋❋ H7–H5 fully hardy   ❋❋ H4–H3 hardy in mild regions/sheltered sites   ❋ H2 protect from frost over winter
❋ H1c–H1a no tolerance to frost   ☼ full sun   ☼ partial sun   ☼ full shade   ◊ well-drained soil   ◑ moist soil   ● wet soil

Hebe 'Red Edge'
A decorative small shrub, 'Red Edge' has grey-green leaves that have narrow red margins and veins when the foliage is young. Lilac-blue flowers, which fade to white, are produced in spikes in summer. It is mound-forming and makes an attractive plant for edging, or for the front of a border.

↕ 45cm (18in) ↔ 60cm (24in) ❋❋ H4 ☼ ◊

Helianthemum 'Wisley Primrose'
Also known as rock roses, helianthemums are sun-loving, carpeting plants that thrive in a rock garden or on a sunny bank. 'Wisley Primrose' forms low hummocks of evergreen, grey-green foliage, and bears plenty of saucer-shaped, pale yellow flowers with deep yellow centres, throughout summer.

↕ to 30cm (12in) ↔ to 45cm (18in) ❋❋ H4 ☼ ◊

Helichrysum italicum subsp. serotinum
The curry plant is a low-growing, evergreen subshrub with woolly stems and intensely aromatic, slim, silver-grey leaves. From summer to autumn, it produces dark yellow flowers, which many designers remove if using the plant for its foliage. One of the best silver shrubs for a dry, sunny site.

↕ 60cm (24in) ↔ 1m (3ft) ❋❋ H4 ☼ ◊

Hydrangea RUNAWAY BRIDE SNOW WHITE ('Ushyd0405')
A rounded deciduous shrub with rather lax, almost trailing stems, it is grown for its remarkably continuous display of white or pink-tinged lacy flowerheads throughout the growing season. The show starts in spring and ends with the frosts. The blooms stand out well against the dark green oval foliage.

↕ 1.5m (5ft) ↔ 1.5m (5ft) ❋❋❋ H5 ☼ ◊

Juniperus x pfitzeriana 'Pfitzeriana Aurea'
Junipers are hardy conifers, tolerant of a wide range of soils and growing conditions. *J.* x *pfitzeriana* is a spreading shrub, eventually forming a flat-topped bush with tiered foliage. 'Pfitzeriana Aurea' has golden yellow leaves, which turn yellowish-green over winter. Junipers need little pruning.

↕ 90cm (36in) ↔ 2m (6ft) ❋❋❋ H6 ☼ ◊

Juniperus procumbens
Creeping juniper is a dwarf species with long, stiff branches that intertwine to form a mat, making it excellent as ground cover and in rock gardens. It has needle-like, bluish-green leaves, and small brown or black berry-like cones. It grows best in a sunny, open position.

↕ to 50cm (20in) ↔ to 2m (6ft) ❋❋❋ H7 ☼ ◊

Juniperus squamata 'Blue Carpet'
The wide-spreading stems of this vigorous, prostrate juniper create a wide, undulating, low mat of prickly foliage, making it an excellent plant for ground cover. The cultivar 'Blue Carpet' is fast-growing, with needle-like, aromatic leaves that are a bright steely blue.

↕ 30cm (12in) ↔ 2–3m (6–10ft) ❋❋❋ H7 ☼ ◊

Lavandula angustifolia 'Munstead'
This evergreen, compact, bushy lavender has narrow, aromatic, grey-green leaves. From mid- to late summer, dense spikes of small, fragrant blue-purple flowers are produced on long stalks. Lavenders prefer warm conditions but suit a variety of situations, from a shrub border to a rock garden.

↕ 45cm (18in) ↔ 60cm (24in) ❋❋❋ H5 ☼ ◊

SHRUBS FOR SUMMER COLOUR

- *Abelia* x *grandiflora* p.300
- *Buddleja davidii* 'Dartmoor' p.300
- *Buddleja* x *weyeriana* 'Sun Gold' p.300
- *Calycanthus* 'Aphrodite' p.296
- *Caryopteris* x *clandonensis* 'Worcester Gold' p.307
- *Cistus* x *purpureus* p.307
- *Helianthemum* 'Wisley Primrose' p.309
- *Kolkwitzia amabilis* 'Pink Cloud' p.303
- *Lavandula angustifolia* 'Munstead' p.309
- *Lavandula stoechas* p.310
- *Paeonia delavayi* p.304
- *Phygelius* x *rectus* 'African Queen' p.310
- *Potentilla fruticosa* 'Goldfinger' p.311
- *Rhododendron* 'Golden Torch' p.311
- *Rhododendron luteum* p.299
- *Rosa* FOR YOUR EYES ONLY ('Cheweyesup') p.311
- *Rosa* 'Geranium' (moyesii hybrid) p.304
- *Salvia* 'Blue Spire' p.312

Small shrubs

Lavandula stoechas
French lavender is a compact shrub that blooms from late spring to summer. Dense spikes of fragrant dark purple flowers, topped by distinctive rose-purple bracts, are carried on long stalks above the silvery-grey leaves. It grows best in a warm, sunny site, and also makes a good container plant.

↕60cm (24in) ↔60cm (24in) ❈ H4 ☼ ◊

Lonicera pileata
With its wide-spreading habit, the shrubby honeysuckle is a good plant for ground cover. It is a low-growing evergreen with narrow dark green leaves, and in late spring it produces tiny, funnel-shaped, creamy-white flowers, which are occasionally followed by purple fruits.

↕60cm (24in) ↔2.5m (8ft) ❈❈❈ H6 ☼ ◊

Origanum 'Kent Beauty'
A pretty addition to a rock garden or container, this decorative subshrub (a cultivar of the herb oregano) has slender trailing stems and smooth aromatic leaves. In late summer, hop-like clusters of pale pink flowers appear above rose-tinted green bracts. It prefers a sunny position.

↕10cm (4in) ↔to 20cm (8in) ❈❈ H4 ☼ ◊

Phlomis fruticosa
A mound-forming evergreen shrub, Jerusalem sage has aromatic, wrinkled, grey-green leaves, which are woolly underneath, and produces short spikes of hooded dark yellow flowers from early to midsummer. It looks effective when massed in a border, and also suits a sunny gravel garden.

↕1m (3ft) ↔1.5m (5ft) ❈❈❈ H5 ☼ ◊

Phygelius × rectus 'African Queen'
This upright evergreen shrub has dark green leaves and graceful upward-curving branches. The pendent tubular flowers produced by the cultivar 'African Queen' are brightly coloured: pale red with orange-red lobes and yellow mouths. Deadhead regularly to encourage further flowering.

↕1m (3ft) ↔1.2m (4ft) ❈❈❈ H5 ☼ ◊ ◐

Pinus mugo 'Mops'
The evergreen dwarf mountain pine forms a spherical mound of thick branches bearing dark green needles and brown cones. It grows best in a sunny position and would suit a rock garden or large container; the shrub's rounded shape also creates a cloud-like effect when it is planted en masse.

↕1m (3ft) ↔1m (3ft) ❈❈❈ H7 ☼ ◊

Pittosporum tenuifolium 'Golf Ball'
This compact, evergreen shrub has a dense, naturally low-growing, ball-shaped habit. The bright green shiny foliage looks fresh year round, making it ideal for adding year-round structure to plantings or for growing in containers. It is not suitable for the coldest sites, but is a good choice in coastal gardens.

↕1m (3ft) ↔1m (3ft) ❈❈ H4 ☼ ◊

Pittosporum tenuifolium 'Tom Thumb'
This popular and compact evergreen is distinctive for its glossy, red-purple, rather corrugated foliage, forming a dense, rounded mound. It is especially striking in spring, when fresh green young shoots contrast vividly with the mature growth. It is not suitable for cold sites, but is a good choice in coastal gardens.

↕1m (3ft) ↔1m (3ft) ❈❈ H4 ☼ ◊

Pittosporum tobira 'Nanum'
With young foliage that shines like patent leather, this evergreen makes a great choice for a sheltered garden. Slow growing, it forms dense, low mounds of broad, bright green foliage. Clusters of sweetly scented, creamy white flowers appear in early summer. Grow it in a pot in cold areas.

↕1m (3ft) ↔1.2m (4ft) ❈ H3 ☼ ◊

PLANT GUIDE

✽✽✽ H7–H5 fully hardy   ✽✽ H4–H3 hardy in mild regions/sheltered sites   ✽ H2 protect from frost over winter
❀ H1c–H1a no tolerance to frost   ☼ full sun   ☼ partial sun   ☀ full shade   ◊ well-drained soil   ◐ moist soil   ● wet soil

Potentilla fruticosa 'Abbotswood'
In summer and early autumn, this low, domed shrub is covered with small white flowers, set against a background of divided, dark blue-green leaves. Shrubby potentillas are compact, bushy, deciduous plants and their long flowering season makes them ideal for a mixed border or a low hedge.

↕75cm (30in) ↔1.2m (4ft) ✽✽✽ H7 ☼ ◊

Potentilla fruticosa 'Goldfinger'
There are numerous cultivars of shrubby potentilla to choose from, with flower colours ranging from white, yellow, and orange to shades of pink and red. 'Goldfinger' is covered in large, saucer-shaped, rich yellow flowers, from late spring to autumn, and has small deep green leaves.

↕1m (3ft) ↔1.5m (5ft) ✽✽✽ H7 ☼ ◊

Prunus laurocerasus 'Zabeliana'
The cherry laurel is an evergreen bushy shrub, which looks its best in spring when long spikes of cup-shaped, fragrant white flowers appear. 'Zabeliana' has a low, wide-spreading habit, making it suitable for ground cover. The flowers are followed by red, cherry-like fruits, which later turn black.

↕1m (3ft) ↔2.5m (8ft) ✽✽✽ H5 ☼ ☼ ◊ ◐

Rhododendron 'Golden Torch'
This small evergreen shrub has medium-sized leaves and is popularly grown for its trusses of flowers, which emerge as salmon-pink buds and open to funnel-shaped, pale creamy-yellow blooms in late spring and early summer. Rhododendrons need acid soil and some shade to thrive.

↕1.5m (5ft) ↔1.5m (5ft) ✽✽✽ H6 ☼ ◊ ◐

Rhododendron 'Kure-no-yuki' (Kurume)
A dwarf azalea with a compact habit, 'Kure-no-yuki' has small leaves and produces clusters of pure white flowers in mid-spring. Azaleas prefer sheltered conditions in deep, acid soil and do best in a woodland garden in dappled shade. This cultivar would make a pretty feature in a Japanese garden.

↕1m (3ft) ↔1m (3ft) ✽✽✽ H5 ☼ ◊ ◐

Rosa ANNA FORD ('Harpiccolo')
There are roses for virtually every situation, but whether they are grown in pots, against a wall, or in a border, most prefer a sunny site. This is a compact, dwarf floribunda rose with dark green leaves and semi-double, orange-red blooms that appear over a long season from summer to autumn.

↕45cm (18in) ↔40cm (16in) ✽✽✽ H6 ☼ ◊ ◐

Rosa FOR YOUR EYES ONLY ('Cheweyesup')
This is a marvellous, vigorous, rather thorny rose bearing multitudes of pale pink flowers, each with a contrasting darker pink heart. Flowering starts in late spring until the first hard frosts of winter, if deadheaded. It is remarkably easy to grow and disease free, and best pruned annually to keep it compact.

↕1.4m (4½ft) ↔1.4m (4½ft) ✽✽✽ H6 ☼ ◊

Rosa 'Golden Wings'
This bushy, spreading shrub rose is suitable for hedging or a border. It has prickly stems and light green leaves, and bears cupped, fragrant, single pale yellow flowers from summer to autumn. A position in full sun will encourage repeat flowering. Apple green hips follow the flowers.

↕1.1m (3.5ft) ↔1.3m (4.5ft) ✽✽✽ H6 ☼ ◊ ◐

AUTUMN- AND WINTER-FLOWERING SHRUBS

- *Azara microphylla* (late winter to early spring) p.296
- *Chimonanthus praecox* 'Grandiflorus' (winter) p.296
- *Cornus mas* (winter) p.296
- *Coronilla valentina* subsp. *glauca* (late winter to early spring) p.307
- *Elaeagnus* x *ebbingei* 'Gilt Edge' (mid- to late autumn) p.297
- *Hamamelis* x *intermedia* 'Pallida' (mid- and late winter) p.297
- *Jasminum nudiflorum* (winter to early spring) p.303
- *Mahonia japonica* (late autumn to early spring) p.303
- *Mahonia* x *media* 'Charity' (late autumn to late winter) p.298
- *Sarcococca* selections p.313 (winter)
- *Viburnum* x *bodnantense* 'Dawn' (late autumn to spring) p.305

Small shrubs

Rosa PEARL DRIFT ('Leggab')
A vigorous shrub rose, spreading in habit, PEARL DRIFT produces clusters of lightly scented, semi-double, pale pink flowers from summer to autumn, against a background of glossy dark green leaves. It is ideal for a mixed cottage-style border, and is also sold under the official cultivar name of 'Leggab'.

↕1m (3ft) ↔1.2m (4ft) ❀ ❀ ❀ H6 ☼ ◊ ◊

Rosa 'The Fairy' (Poly)
Suited to a border or a container, 'The Fairy' is a small shrub rose with a dense cushion-forming habit. The thorny stems are covered with small, glossy leaves, and from late summer to autumn it produces sprays of small, double, pink flowers.

↕60–90cm (24–36in) ↔60–90cm (24–36in)
❀ ❀ ❀ H6 ☼ ◊ ◊

Rosa WILDEVE ('Ausbonny')
This robust rose has long, arching stems and forms a bushy shrub. The flower buds are pink, and open to fully-double, apricot-flushed pink fragrant blooms, which appear from late spring to early summer. Grow WILDEVE in a mixed border, or use for hedging. Its official cultivar name is 'Ausbonny'.

↕1.1m (3.5ft) ↔1.25m (2.5ft) ❀ ❀ ❀ H7 ☼ ◊ ◊

Ruta graveolens
This evergreen subshrub, also known as common rue, is grown for its aromatic, deeply divided blue-green leaves and is sometimes used as a medicinal herb. Cup-shaped yellow flowers appear in summer. 'Jackman's Blue' is a compact cultivar with intensely glaucous foliage.

↕1m (3ft) ↔75cm (30in) ❀ ❀ ❀ H5 ☼ ◊

Salvia 'Blue Spire'
This sage forms a clump of grey-green toothed leaves. In late summer, grey-white upright stems carry elegant spires of small, tubular purple-blue flowers. An eyecatching plant for a border, it looks particularly effective when planted in groups. The frosty-looking stems are attractive in winter.

↕1.2m (4ft) ↔1m (3ft) ❀ ❀ ❀ H5 ☼ ◊

Salvia microphylla
From late summer to autumn this salvia bears crimson flowers among its mid- to deep green leaves. It makes a colourful addition to a late season border or herb garden, but needs a sunny site to produce its best flower display.

↕90–120cm (36–48in) ↔60–100cm (24–39in)
❀ ❀ H4 ☼ ◊

Salvia officinalis 'Purpurascens'
The aromatic downy leaves of this shrubby evergreen or semi-evergreen perennial are purple when young, and later greyish-green. Purple sage is used as a culinary herb but is also decorative in a gravel garden or mixed border. Blue purple flowers are borne on spikes in early and midsummer.

↕to 80cm (32in) ↔1m (3ft) ❀ ❀ H4 ☼ ◊

Salvia officinalis 'Tricolor'
This cultivar of the common sage has grey-green leaves with creamy-white margins, flushed pink when young. It makes a compact plant and colours best in a sunny site. The leaves are aromatic and can be used for culinary purposes, while the flowers are attractive to bees and butterflies.

↕to 80cm (32in) ↔1m (3ft) ❀ ❀ ❀ H5 ☼ ◊

Salvia rosmarinus
Rosemary is a tough evergreen Mediterranean shrub, grown for its aromatic leaves. It forms an attractive upright plant with slim, leathery leaves, and produces tubular, purple-blue to white flowers from mid-spring to early autumn. It needs a well-drained site and suits a rock or herb garden.

↕1.5m (5ft) ↔1.5m (5ft) ❀ ❀ H4 ☼ ◊

❄❄❄ H7–H5 fully hardy   ❄❄ H4–H3 hardy in mild regions/sheltered sites   ❄ H2 protect from frost over winter
❄ H1c–H1a no tolerance to frost   ☼ full sun   ◐ partial sun   ● full shade   ○ well-drained soil   ◐ moist soil   ● wet soil

Santolina pinnata subsp. *neapolitana* 'Sulphurea'
An evergreen shrub native to the Mediterranean, santolina forms a low, domed shape. The primrose-yellow, tubular flowers form button-like heads on long stems above narrow, feathery, grey-green leaves. It is useful as edging, and as part of a Mediterranean-style scheme.

↕75cm (30in) ↔1m (3ft) ❄❄❄ H5 ☼ ○

Sarcococca hookeriana var. *digyna*
The robustness of this winter-flowering evergreen makes it a useful shrub for difficult sites in the garden, as it will tolerate dry shade and air pollution, and needs very little attention. It has slender, tapered dark green leaves and is prized for its highly fragrant white flowers, followed by black fruit.

↕1.5m (5ft) ↔2m (6ft) ❄❄❄ H5 ◐ ● ○ ◐

Sarcococca hookeriana WINTER GEM ('Pmoore03')
A handsome, compact evergreen with upright leafy stems and a suckering habit, it forms broad, neat clumps of growth. The shiny, dark green oval leaves contrast with the small but powerfully fragrant white winter flowers, which open freely from distinctive rich red buds. The scent carries well on the air.

↕70cm (30in) ↔1m (3ft) ❄❄❄ H5 ☼ ◐ ○

Viburnum × *burkwoodii* 'Anne Russell'
This compact, deciduous or semi-evergreen shrub produces clusters of intensely fragrant white flowers from mid- to late spring. 'Anne Russell' is suited to growing in a shrub border or woodland garden; plant it close to a seating area or pathway to make the most of its spring scent.

↕1.5m (5ft) ↔1.5m (5ft) ❄❄❄ H6 ☼ ◐ ○ ◐

Viburnum davidii
This evergreen shrub forms a dome of dark green gleaming foliage on branching stems. The flowers appear above the deeply veined, oval leaves in late spring, producing flattened heads of small white blooms. Where male and female plants are grown together, metallic-blue fruits form on the female.

↕1–1.5m (3–5ft) ↔1–1.5m (3–5ft) ❄❄❄ H5 ☼ ◐ ○ ◐

Vinca minor 'La Grave'
Woodland plants in the wild, periwinkles bear decorative, star-shaped flowers on slender stems. The evergreen foliage and pretty flowers make attractive ground cover, although they can be invasive and may need cutting back regularly. 'La Grave' (also seen as 'Bowles's Blue') has lavender-blue flowers.

↕10–20cm (4–8in) ↔indefinite ❄❄❄ H6 ☼ ◐ ○ ◐

Weigela florida 'Foliis Purpureis'
This is a dark-leaved cultivar of the deciduous, arching shrub *Weigela florida*. Funnel-shaped flowers, deep pink on the outside and pale pink to white inside, are produced in late spring and early summer, and look striking against the tapered bronze-green foliage. Grow in a mixed or shrub border.

↕1m (3ft) ↔1.5m (5ft) ❄❄❄ H6 ☼ ○

Yucca filamentosa 'Bright Edge'
A dramatic architectural plant, the yucca suits a hot, dry site, making it a good specimen plant for a warm courtyard. *Yucca filamentosa* produces stems of bell-shaped white flowers, tinged green, from mid- to late summer. The leaves of 'Bright Edge' have broad yellow margins.

↕75cm (30in) ↔1.5m (5ft) ❄❄❄ H5 ☼ ○

EVERGREEN SHRUBS

- *Aucuba japonica* 'Crotonifolia' p.300
- *Azara microphylla* p.296
- *Berberis* selections p.300
- *Camellia* selections, pp.296, 300
- *Ceanothus* 'Concha' p.300
- *Choisya* × *dewitteana* 'Aztec Pearl' p.301
- *Cotoneaster lacteus* p.297
- *Daphne bholua* 'Jacqueline Postill' p.301
- *Elaeagnus* × *ebbingei* 'Gilt Edge' p.297
- *Euphorbia* × *pasteurii* p.302
- *Fatsia japonica* p.302
- *Ilex crenata* p.303
- *Itea ilicifolia* p.298
- *Olearia macrodonta* p.298
- *Osmanthus* × *burkwoodii* p.304
- *Pieris japonica* 'Blush' p.304
- *Pittosporum* selections pp.299, 310
- *Rhamnus alaternus* 'Argenteovariegata' p.299
- *Skimmia* × *confusa* 'Kew Green' p.305

Climbers

Actinidia kolomikta
This deciduous climber's main attraction is the masses of purple-tinged young leaves, which later turn dark green with distinctive pink and silver splashes. Small, slightly scented white flowers appear in early summer. Although it is slow to establish, it is well worth the wait.

↕5m (15ft) ❋ ❋ ❋ H5 ☼ ◊

Akebia quinata
Also known as the chocolate vine, *A. quinata* is a vigorous semi-evergreen with attractive leaves and strong, twining stems. Clusters of cup-shaped, purplish female flowers in spring are followed by unusual sausage-shaped fruits. Grow against a wall or train into a tree or pergola.

↕10m (30ft) ❋ ❋ ❋ H6 ☼ ☼ ◊ ◊

Ampelopsis brevipedunculata
This vigorous, deciduous climber is valued for its attractive foliage and ornamental berries. The small summer flowers are green, and are followed by eye-catching, round, pinkish purple berries, which later turn a clear blue. Ideal for a warm, sheltered wall since fruiting is best in a sunny site.

↕5m (15ft) ❋ ❋ ❋ H6 ☼ ☼ ◊ ◊

Campsis x *tagliabuana* 'Madame Galen'
The trumpet creeper is a fast-growing, deciduous climber, which clings by aerial roots. In late summer or early autumn, 'Madame Galen' bears clusters of tubular, reddish-orange flowers that look striking against the rich green divided leaves. It may take a few seasons to establish.

↕3–5m (10–15ft) ❋ ❋ H4 ☼ ◊ ◊

Clematis armandi
This popular clematis is a vigorous climber and one of the hardiest of the evergreen species, bearing glossy, dark green leaves and producing masses of small, white scented flowers in early spring. It prefers a sunny, sheltered site and will clothe a wall or shed with ease.

↕3–5m (10–15ft) ❋ ❋ H4 ☼ ☼ ◊ ◊

Clematis 'Bill MacKenzie'
A vigorous, scrambling clematis, 'Bill MacKenzie' has small, single, yellow lantern-like nodding flowers in late summer and autumn, followed by large silky seedheads. The plant needs support from wires or netting, or leave it to scramble through shrubs and trees.

↕7m (22ft) ❋ ❋ ❋ H6 ☼ ☼ ◊ ◊

Clematis 'Étoile Violette'
From midsummer to late autumn, this deciduous viticella clematis produces masses of small, nodding, deep violet flowers with cream stamens. Flowers are produced on the current year's growth. 'Étoile Violette' can be grown through other shrubs or on a wall or fence.

↕3–5m (10–15ft) ❋ ❋ ❋ H6 ☼ ☼ ◊ ◊

Clematis florida var. *florida* 'Sieboldiana'
This deciduous or semi-evergreen clematis bears showy, single creamy white flowers with a distinctive domed cluster of purple stamens in late spring or summer. It does best in a warm, sunny sheltered location where its roots are shaded and moist. It is also suitable for growing in large containers.

↕2–2.5m (6–8ft) ❋ ❋ H3 ☼ ☼ ◊ ◊

Clematis 'Huldine'
A vigorous, deciduous, summer-flowering clematis, well suited to walls and fences. The small, cup-shaped, almost translucent white flowers with pale mauve margins and a mauve stripe beneath appear in summer. They are particularly attractive in sunshine when the stripes are more evident.

↕3–5m (10–15ft) ❋ ❋ ❋ H6 ☼ ☼ ◊ ◊

PLANT GUIDE   **315**

❀❀❀ H7–H5 fully hardy   ❀❀ H4–H3 hardy in mild regions/sheltered sites   ❀ H2 protect from frost over winter
❀ H1c–H1a no tolerance to frost   ☼ full sun   ☼ partial sun   ☀ full shade   ○ well-drained soil   ◐ moist soil   ● wet soil

Clematis 'Markham's Pink'
This early-flowering macropetala clematis is vigorous and prolific, producing masses of bell-shaped, double, rich pink flowers from spring to early summer, followed by silky seedheads in autumn. Try growing through a shrub or small tree, or against a wall or fence.

↕ 2.5–3.5m (8–11ft) ❀❀❀ H6 ☼ ☼ ○ ◐

Clematis montana var. montana 'Tetrarose'
This vigorous, deciduous climber is ideal for covering a fence or garden building. In late spring, it is resplendent with cascades of large, bright pink, four-petalled flowers, each with a central crown of golden stamens in a display that lasts several weeks. The foliage is bronze tinted.

↕ 8m (25ft) ❀❀❀ H5 ☼ ○

Eccremocarpus scaber
The Chilean glory flower is an evergreen, perennial, fast-growing climber with attractive ferny leaves. In warmer areas it will quickly clothe a trellis or pergola, or scramble through a large shrub or small tree. From late spring to autumn, spikes of orange-red tubular flowers appear.

↕ 3–5m (10–15ft) ❀❀ H3 ☼ ○

Hardenbergia violacea
The purple coral pea is a strong-growing Australian native and does best in a sunny position outdoors, but is suitable for a greenhouse in cold regions. From late winter to early summer, clusters of violet pea-like flowers appear against the leathery rich green leaves.

↕ 2m (6ft) or more ❀❀ H3 ☼ ☼ ○

Hedera colchica 'Sulphur Heart'
The Persian ivy cultivars 'Sulphur Heart' and 'Dentata Variegata' have similar large light green leaves with cream splashes. 'Sulphur Heart' (also known as 'Paddy's Pride') grows more rapidly, however, and the slightly more elongated leaves are splashed with creamy yellow.

↕ 5m (15ft) ❀❀❀ H5 ☼ ○ ◐

Hedera helix 'Oro di Bogliasco'
This striking ivy, also known as 'Goldheart', has dark, glossy evergreen leaves with a gold central splash. A self-clinging climber, it makes an excellent wall ivy, slow to establish but then fast-growing. Unlike most variegated ivies, it will tolerate shade.

↕ 8m (25ft) ❀❀❀ H5 ☼ ☼ ○ ◐

Hedera helix 'Parsley Crested'
As its name suggests, this ivy has dark green leaves with waved and crested margins. A vigorous, evergreen self-clinging climber with thick upright stems, it is hardy, easy to grow, and ideal for garden walls and fences, although its aerial roots may damage old brickwork.

↕ 2m (6ft) ❀❀❀ H5 ☼ ○ ◐

Humulus lupulus 'Aureus'
Hops make a good choice for shady walls and fences, although H. lupulus 'Aureus' produces its best leaf colour in sun. This strong-growing, herbaceous perennial climber has yellow-green, boldly lobed leaves and hairy, twining stems; spikes of female flowers (hops) appear in late summer.

↕ 6m (20ft) ❀❀❀ H6 ☼ ☼ ○ ◐

CLIMBERS FOR SPRING AND SUMMER FLOWERS

- *Campsis* x *tagliabuana* 'Madame Galen' p.314
- *Clematis armandi* p.314
- *Clematis* 'Bill Mackenzie' p.314
- *Clematis* 'Étoile Violette' p.314
- *Clematis* 'Markham's Pink' p.315
- *Clematis montana* var. *montana* 'Tetrarose' p.315
- *Jasminum officinale* 'Argenteovariegatum' p.316
- *Lonicera periclymenum* 'Serotina' p.316
- *Passiflora caerulea* p.316
- *Rosa* selections pp.316, 317
- *Solanum crispum* 'Glasnevin' p.317
- *Solanum laxum* 'Album' p.317
- *Tropaeolum speciosum* p.317
- *Wisteria floribunda* 'Multijuga' p.317

Climbers

Hydrangea anomala subsp. *petiolaris*
The climbing hydrangea is vigorous and produces large, open lacecap heads of creamy-white flowers in summer, on a background of broad, rounded leaves. The stems have rich brown peeling bark. Young plants need support until they are established; they then climb by self-clinging aerial roots.

↕15m (50ft) ❄ ❄ ❄ H5 ☼ ☼ ◊ ◊

Jasminum officinale 'Argenteovariegatum'
Strong-growing and semi-evergreen, climbing jasmine has pretty, ferny foliage and bears clusters of strongly scented, white star-shaped flowers in summer. The variegated cultivar 'Argenteovariegatum' has finely divided, grey-green leaves with cream margins.

↕12m (40ft) ❄ ❄ ❄ H5 ☼ ☼ ◊

Lonicera periclymenum 'Serotina'
A twining, vigorous climber, the late Dutch honeysuckle can be grown alone or through a small tree or shrub. The spring foliage is lush and new shoots are purple when young. In summer, it produces long-tubed fragrant creamy white flowers streaked with dark red-purple.

↕7m (22ft) ❄ ❄ ❄ H6 ☼ ☼ ◊

Parthenocissus henryana
This deciduous ornamental vine, sometimes known as the Chinese Virginia creeper, clings to surfaces by the adhesive tips of its tendrils, making it a useful climber for growing on a wall. It produces the best colour in partial shade, its silver-veined leaves turning a rich red in autumn before they fall.

↕10m (30ft) ❄ ❄ ❄ (borderline) H4 ☼ ☼ ◊ ◊

Parthenocissus tricuspidata 'Veitchii'
Also known as Boston ivy, *P. tricuspidata* is vigorous and woody, and will clothe a wall or other support quite quickly, clinging without assistance. The cultivar 'Veitchii' is noted for its autumn colour, when the mid-green ivy-like leaves turn a deep red-purple before falling.

↕20m (70ft) ❄ ❄ ❄ H5 ☼ ◊

Passiflora caerulea
A good climber for a sunny, warm wall or fence, the blue passion flower is fast-growing, with rich green divided leaves. The striking flowers are usually white, with purple, blue, and white coronas. The orange-yellow fruits that follow are decorative but not edible.

↕10m (30ft) or more ❄ ❄ H4 ☼ ◊ ◊

Rosa 'Compassion'
A hybrid tea rose, 'Compassion' is an upright, freely branching climber with dark green leaves. The flowers are rounded and fully double, salmon pink tinged with apricot, and fragrant. They appear from summer to autumn; deadheading will prolong the flowering season. It is a good choice for a wall.

↕3m (10ft) ❄ ❄ ❄ H6 ☼ ◊ ◊

Rosa 'Félicité Perpétue'
This rambler is a semi-evergreen rose with long, slender stems and dark green leaves. The summer flowers are fully double, pale pink in bud and opening to faintly pink-tinged white. It is a beautiful rose for an arch or arbour, or it can be grown through a shrub or small tree.

↕to 5m (15ft) ❄ ❄ ❄ H6 ☼ ◊ ◊

Rosa 'Madame Alfred Carrière'
This marvellous old climbing rose is vigorous and tough enough for growing and flowering well, even in shaded sites. Its green stems are relatively thorn-free, and the large, scented double flowers, which open palest pink and age to white, nod elegantly all through summer until the frosts. It is easy and disease free.

↕8m (25ft) ↔2.5m (8ft) ❄ ❄ ❄ H5 ☼ ☼ ◊

PLANT GUIDE

❄❄❄ H7–H5 fully hardy ❄❄ H4–H3 hardy in mild regions/sheltered sites ❄ H2 protect from frost over winter
❄ H1c–H1a no tolerance to frost ☼ full sun ☼ partial sun ☀ full shade ◊ well-drained soil ◊ moist soil ◆ wet soil

Rosa THE GENEROUS GARDENER ('Ausdrawn')
A sturdy, first-rate climbing rose, it provides modern attributes of repeat flowering and disease-free growth with the beauty of an old-fashioned selection. It bears a succession of beautiful sweetly scented, semi-double, soft pink flowers through summer and autumn, held above glossy apple-green foliage.

↕3m (10ft) ↔1.5m (5ft) ❄❄❄ H6 ☼ ◊

Schizophragma integrifolium
Schizophragmas are slow-growing and mainly cultivated for their hydrangea-like blooms – flattened heads of creamy-white flowers with conspicuous, oval cream-coloured bracts, which appear in summer among the pointed green leaves. The plant will attach itself to a wall surface by aerial roots.

↕12m (40ft) ❄❄❄ H5 ☼ ☼ ◊

Solanum crispum 'Glasnevin'
Vigorous and scrambling, *S. crispum* is a good choice for a warm, sunny wall or fence. The cultivar 'Glasnevin' produces sprays of long-lasting, deep purple-blue, star-shaped flowers from summer to autumn, and is evergreen in warmer areas. It is ideal for training through a shrub or small tree.

↕6m (20ft) ❄❄ H3 ☼ ☼ ◊ ◊

Solanum laxum 'Album'
Known as the potato vine, *S. laxum* is a scrambling semi-evergreen or evergreen climber which produces clusters of lightly fragrant flowers over a long season from summer to autumn. The cultivar 'Album' is a white-flowered form of the normally blue-flowered plant.

↕6m (20ft) ❄❄ H3 ☼ ◊ ◊

Tropaeolum speciosum
The flame nasturtium has fleshy, twining stems and long-stalked divided leaves, and is an excellent plant to train into trees, shrubs or hedges, where its brilliant colour will contrast with the green foliage. Long-spurred scarlet flowers appear from summer into autumn, followed by spherical blue fruits.

↕to 3m (10ft) or more ❄❄❄ H5 ☼ ☼ ◊

Vitis coignetiae
This ornamental vine is grown for its decorative foliage and vivid autumn colour. It is a vigorous, deciduous climber with large, heart-shaped leaves, brown-felted beneath, that turn bright red in autumn. Small, inedible, blue-black grapes appear at the same time. Train into a tree or shrub, or over a pergola.

↕15m (50ft) ❄❄❄ H5 ☼ ☼ ◊

Vitis vinifera 'Purpurea'
An ideal climber for a warm, sunny wall or fence, the claret vine is a vigorous form of the grape vine, but is grown for its autumn foliage rather than the inedible grapes. It is a woody deciduous vine with toothed leaves which are grey at first, then mid-purple, turning a very deep purple in autumn.

↕7m (22ft) ❄❄❄ H5 ☼ ☼ ◊ ◊

Wisteria floribunda 'Multijuga'
Showy, pendent spikes of pea-like early summer flowers make wisterias popular with garden designers. *W. floribunda* (Japanese wisteria) is a vigorous, twining climber with pretty leaves, available as a range of cultivars: 'Multijuga' bears fragrant, lilac-blue blooms; 'Alba' has white flowers.

↕9m (28ft) or more ❄❄❄ H6 ☼ ☼ ◊ ◊

CLIMBERS FOR FOLIAGE INTEREST AND COLOUR

- *Actinidia kolomikta* p.314
- *Akebia quinata* p.314
- *Ampelopsis brevipedunculata* p.314
- *Hedera colchica* 'Sulphur Heart' p.315
- *Hedera helix* 'Oro di Bogliasco' p.315
- *Hedera helix* 'Parsley Crested' p.315
- *Humulus lupulus* 'Aureus' p.315
- *Hydrangea anomala* subsp. *petiolaris* p.316
- *Parthenocissus henryana* p.316
- *Parthenocissus tricuspidata* 'Veitchii' p.316
- *Vitis coignetiae* p.317
- *Vitis vinifera* 'Purpurea' p.317

Tall perennials

Acanthus spinosus

From late spring through to midsummer, majestic spikes of white flowers sheltered by purple bracts rise from a bed of prickly, dark green leaves. This clump-forming perennial prefers rich soil and makes a striking architectural plant. Cut stems last well in flower arrangements.

↕1.5m (5ft) ↔60–90cm (24–36in) ❊ ❊ ❊ H5 ☼ ☼ ◊

Aconitum 'Spark's Variety'

Upright stems bearing deep violet, hooded flowers, well above the dark green, deeply divided leaves, identify this as one of the monkshoods. The flowers appear from mid- to late summer and perform best in moist, fertile soil, in a woodland garden or border. Taller plants may need staking. All parts are poisonous.

↕1.2–1.5m (4–5ft) ↔45cm (18in) ❊ ❊ ❊ H7 ☼ ◊

Anemone × *hybrida*

The Japanese anemone bears semi-double, pink flowers on wiry stems from late summer to mid-autumn. The white-flowered 'Honorine Jobert' will shine in any border and like the other Japanese anemones, prefers rich soil. It dislikes cold, wet conditions during winter months.

↕1.2–1.5m (4–5ft) ↔indefinite ❊ ❊ ❊ H7 ☼ ☼ ◊

Asphodeline lutea

The yellow asphodel strikes a dominant pose in the border as its rocket-like spikes of star-shaped flowers stand above other late-spring perennials. Eye-catching blue-green leaves stud the length of each flower stem. Most well-drained soils will suit this clump-forming perennial.

↕1.5m (5ft) ↔30cm (12in) ❊ ❊ ❊ (borderline) H4 ☼ ◊

Baptisia australis

Flowering in late spring or early summer, this clump-forming perennial bears airy, lupin-like spires of glowing, usually blue or purple pea-like flowers. They are held atop sturdy, self-supporting upright stems, just above mounds of fresh, rather lush, three-lobed foliage. Pods of seeds follow the flowers.

↕1–1.5m (3–5ft) ↔1m (3ft) ❊ ❊ ❊ H7 ☼ ◊

Cephalaria gigantea

The giant scabious needs a sizeable border for the best display of its tall flower stems bearing pale yellow, ruffled blooms in summer. Make the most of them by planting at the back of a border against a dark background, such as a conifer hedge or fence, for contrast.

↕to 2.5m (8ft) ↔60cm (24in) ❊ ❊ ❊ H7 ☼ ☼ ◊ ◊

Cirsium rivulare 'Atropurpureum'

The deep crimson flowers of this clump-forming perennial, coupled with its prickly green leaves, should make thistles more popular border plants than they are. Suited to damp conditions in a wild garden, they attract insects during the flowering season from early to midsummer.

↕1.2m (4ft) ↔60cm (24in) ❊ ❊ ❊ H7 ☼ ◊ ◊

Crambe cordifolia

Looking like a mass of confetti, the tiny white flowers of this perennial appear suspended in mid-air. The coarseness of the rich green leaves is softened by a cloud of blooms from late spring to midsummer. Crambes are suited to a wild garden and will tolerate coastal conditions. The flowers attract bees.

↕to 2.5m (8ft) ↔1.5m (5ft) ❊ ❊ ❊ H5 ☼ ◊

Cynara cardunculus

Few plants produce such large flowerheads as the cardoon. Fierce-looking bracts sit below brush-like flowerheads of blue-purple florets to create a dazzling summer and early autumn display. Protect plants from strong winds and in cold areas, mulch around the plant base.

↕1.5m (5ft) ↔1.2m (4ft) ❊ ❊ ❊ H5 ☼ ◊

PLANT GUIDE   319

❀ ❀ ❀ H7–H5 fully hardy   ❀ ❀ H4–H3 hardy in mild regions/sheltered sites   ❀ H2 protect from frost over winter
❀ H1c–H1a no tolerance to frost   ☼ full sun   ☼ partial sun   ☼ full shade   ◊ well-drained soil   ◐ moist soil   ● wet soil

Delphinium Pacific Hybrids
A cottage garden favourite, this tall perennial comes in a range of colours, including blue, pink, white, and violet. After the midsummer flowering, cut back the stems to encourage another flush of double flowers in late summer and early autumn. Protect from strong winds.

↕1.2–2m (4–6½ft) ↔90cm (36in) ❀ ❀ ❀ H6–H5 ☼ ◊

Dierama pulcherrimum
The delightful name of angel's fishing rod perfectly suits this elegant perennial whose pendent, pink bells move gracefully in the slightest breeze against narrow, grass-like, green leaves. This combination looks good in the middle of a border or as edging alongside a pathway.

↕1–1.5m (3–5ft) ↔60cm (24in) ❀ ❀ H4 ☼ ◐

Dryopteris wallichiana
Wallich's wood fern, named after the Danish plant collector, Nathaniel Wallich, is a deciduous fern that sports a shuttlecock-like array of young, green fronds with rusty-brown, furry mid-ribs, in spring. Provide shelter, shade and a generous depth of rich, moist soil.

↕90cm (36in) or more ↔75cm (30in) ❀ ❀ ❀ H5 ☼ ◐

Echinops bannaticus
The globe thistle is a good plant for a wild garden; it is very attractive to bees, with its spherical, blue flowerheads held above a spiny mass of grey-green leaves from mid- to late summer. The dense clumps can be divided from autumn to spring. The variety 'Taplow Blue' has powder-blue flowers.

↕0.5–1.2m (1½–4ft) ↔75cm (30in) ❀ ❀ ❀ H7 ☼ ◊

Eryngium agavifolium
An Argentinian plant, sea holly makes a dramatic silhouette in a border. Long, sword-shaped leaves, sharply toothed along their length, form rosettes from which the flowering stems emerge. The stalkless, greenish-white flowers form cone-like structures. Stems can be dried for flower arrangements.

↕1–1.5m (3–5ft) ↔60cm (24in) ❀ ❀ H4 ☼ ◊ ◐

Foeniculum vulgare 'Purpureum'
This attractive perennial is well known for its aniseed-flavoured seeds and feathery mid-green leaves, which are used in cooking. Flat flowerheads of small yellow flowers appear from mid- to late summer. 'Purpureum' is hardier than the species and has striking bronze-purple foliage.

↕1.8m (6ft) ↔45cm (18in) ❀ ❀ ❀ H5 ☼ ◊

Helianthus 'Lemon Queen'
Sunflowers are always a good choice for the back of a border and this variety is no exception. Pale yellow flowers with a slightly darker eye mark this out as one of the more subtly coloured choices. Expect a long-lasting display from late summer to mid-autumn.

↕1.7m (5½ft) ↔1.2m (4ft) ❀ ❀ (borderline) H4 ☼ ◊ ◐

Inula magnifica
This fast-growing, clump-forming plant needs plenty of space in the garden. Large, frilly-petalled flowers are formed, up to 20 at a time, in late summer above a foil of dark green leaves with softly hairy undersides. Ideal for a wild garden, the plant likes sun but will tolerate damp soil.

↕to 1.8m (6ft) ↔1m (3ft) ❀ ❀ ❀ H6 ☼ ◊ ◐ ●

PERENNIALS FOR ARCHITECTURAL INTEREST

- *Acanthus spinosus* p.318
- *Asplenium scolopendrium* Crispum Group p.322
- *Astelia chathamica* p.322
- *Athyrium filix-femina* p.323
- *Cynara cardunculus* p.318
- *Dryopteris wallichiana* p.319
- *Echinops bannaticus* p.319
- *Eryngium agavifolium* p.319
- *Euphorbia* x *martinii* p.324
- *Foeniculum vulgare* 'Purpureum' p.319
- *Hosta* 'Sum and Substance' p.326
- *Melianthus major* p.320
- *Musa basjoo* p.320
- *Phormium cookianum* subsp. *hookeri* 'Tricolor' p.329
- *Phormium tenax* Purpureum Group p.320
- *Sisyrinchium striatum* 'Aunt May' p.329

Tall perennials

Leucanthemella serotina
This large-flowered daisy makes excellent cutting material, lasting well in the vase. It is a vigorous plant, with stout stems that should not need staking, and prefers a moist situation with full sun or partial shade. It is useful for illuminating darker areas of the garden.

↕ to 1.5m (5ft) ↔ 90cm (36in) ❋ ❋ ❋ H7 ☼ ☼ ◊ ♦

Macleaya microcarpa 'Kelway's Coral Plume'
This pink-flowered plume poppy is at its peak in early and midsummer, when large, open floral sprays sit above a sea of grey-green leaves. A tall, showy plant, it is best sited on its own, forming an eye-catching screen, or at the back of a large mixed border. Macleayas can be invasive.

↕ to 2.2m (7ft) ↔ 1m (3ft) or more ❋ ❋ ❋ H6 ☼ ◊ ◊

Melianthus major
Grown more for its grey-green, tooth-edged leaves than its flowers, the honey bush is tolerant of sea air and is a good choice for coastal gardens. Use as an architectural focus or place it in strategic positions around the garden where its angular features can be admired. It is not frost hardy.

↕ 2–3m (6–10ft) ↔ 1–3m (3–10ft) ❋ ❋ ❋ H3 ☼ ◊ ◊

Musa basjoo
The Japanese banana can grow to 5m (15ft) and even flower and produce fruit (unpalatable, however) in cooler climates. It is ideal as a specimen plant, or can be used as the centrepiece of a tropical display. Strong winds can shred the leaves, so try to provide some protection.

↕ to 5m (15ft) ↔ to 4m (12ft) ❋ H2 ☼ ◊

Phlox paniculata 'Blue Paradise'
With arresting cone-shaped heads of five-petalled, purple-blue flowers, each with a dark red central eye and producing a lovely sweet perfume, this perennial makes a great choice. It flowers for several weeks from a little after midsummer, the blooms held on stout maroon stems that are usually self-supporting.

↕ 1.2m (4ft) ↔ 60cm (24in) ❋ ❋ ❋ H7 ☼ ☼ ◊

Phormium tenax Purpureum Group
Long, fibrous, sword-shaped leaves burst forth from the base of the New Zealand flax. The red-purple foliage contrasts well with paler phormiums or grasses. Alternatively, use it on its own to dominate a border. The plant likes fertile soil in full sun; mulch the base in winter in frost-prone areas.

↕ 2.5–2.8m (8–9ft) ↔ 1m (3ft) ❋ ❋ ❋ H5 ☼ ◊ ◊

Romneya coulteri 'White Cloud'
This plant will eventually become a woody perennial once it becomes established. Large white petals with a bobble of yellow stamens in the centre create a winning display. Protect plants from cold, strong winds, and in frost-prone areas, choose a site against a warm wall.

↕ 1–2.5m (3–8ft) ↔ indefinite ❋ ❋ ❋ H5 ☼ ◊

Salvia uliginosa
Native to South America, the bog sage comes into its own from late summer to mid-autumn, when square stems bearing clear blue flowers emerge above mid-green, toothed leaves. As the name suggests, bog sage is a moisture-loving plant. It is tall and suited to the back of a sunny border.

↕ to 2m (6ft) ↔ 90cm (36in) ❋ ❋ ❋ H4 ☼ ◊

Selinum wallichianum
This refined member of the cow parsley family is a supremely elegant summer- and autumn-flowering perennial. It bears lace-like heads of white flowers held on stout, purple stems above beautifully divided fresh, green foliage. The seed heads that follow last into winter if allowed to stand.

↕ 1.5m (5ft) ↔ 60cm (24in) ❋ ❋ ❋ H6 ☼ ◊

❋❋❋ H7–H5 fully hardy   ❋❋ H4–H3 hardy in mild regions/sheltered sites   ❋ H2 protect from frost over winter
❋ H1c–H1a no tolerance to frost   ☼ full sun   ☼ partial sun   ☼ full shade   ◊ well-drained soil   ◐ moist soil   ● wet soil

Symphyotrichum 'Ochtendgloren'
The long-lasting, purple-pink, daisy-like flowers of this aster are held on branching stems in late summer. It is a strong-growing plant, producing neat clumps that do not need to be regularly divided. It brightens up borders, can be grown in containers, and is also good for cutting.

↕1.2m (4ft) ↔80cm (32in) ❋❋❋ H4 ☼ ◊

Thalictrum 'Elin'
This tall-growing herbaceous perennial flowers in late summer, bearing clouds of tiny pale mauve flowers with cream stamens atop lofty purple-coloured stems. The plant is clump forming with attractive, delicate-looking blue-green foliage and looks handsome all through the growing season. It may need support.

↕2.5m (8ft) ↔1m (3ft) ❋❋❋❋ H6 ☼ ◐

Thalictrum flavum subsp. *glaucum*
The yellow meadow rue is a clump-forming perennial that spreads by means of underground stems or rhizomes. Its blue-green foliage is offset by the pale sulphur-yellow flowers formed in summer. The variety 'Illuminator' is taller than the subspecies and has bright green leaves.

↕to 1m (3ft) ↔60cm (24in) ❋❋❋ H7 ☼ ◐

Valeriana phu 'Aurea'
The leaves of this plant are soft yellow when young, turning green to lime green by summer. The leaves at the base of the stem are scented. Small white flowers appear in early summer to complete the display. A woodland plant in the wild, valerian suits a cottage garden or any informal scheme.

↕to 1.5m (5ft) ↔60cm (24in) ❋❋❋ H5 ☼ ◐

Verbascum 'Cotswold Queen'
Synonymous with cottage gardens, this semi-evergreen perennial will brighten any summer border with its prominent spikes of yellow, saucer-shaped flowers. In a garden exposed to the elements, this tall plant will probably need staking. Many *Verbascum* species are short-lived.

↕1.2m (4ft) ↔30cm (12in) ❋❋❋ H7 ☼ ◊

Verbena bonariensis
A popular plant, this verbena comes into its own when grown with grasses, allowing its branched flowerheads to punctuate a border display. It can be grown at the back of beds, but its slim stems also look striking at the front. It flowers from midsummer to early autumn.

↕to 2m (6ft) ↔45cm (18in) ❋❋ H4 ☼ ◊ ◐

Veronicastrum virginicum
From summer to autumn, the dainty flower spikes of this perennial bring white, pink, and purple shades to border plantings. For a pure white-flowered variety, look for *V. virginicum* 'Album' and grow it with dark foliage plants to bring out its best attributes.

↕to 2m (6ft) ↔45cm (18in) ❋❋❋ H7 ☼ ☼ ◐

Veronicastrum virginicum 'Fascination'
This impressive and reliable summer-flowering perennial has upright stems supporting whorls of lance-shaped foliage, terminating in branched, appealingly slender spires of small mauve-blue flowers. The display lasts well into autumn and is highly attractive to pollinating insects. It seldom needs staking.

↕1.5m (5ft) ↔1m (3ft) ❋❋❋ H7 ☼ ☼ ◊

PERENNIALS FOR ATTRACTING WILDLIFE

- *Agastache* 'Blackadder' p.322
- *Aquilegia vulgaris* 'William Guiness' p.322
- *Centaurea dealbata* 'Steenbergii' p.323
- *Cirsium rivulare* 'Atropurpureum' p.318
- *Crambe cordifolia* p.318
- *Digitalis* x *mertonensis* p.323
- *Doronicum* 'Little Leo' p.332
- *Echinacea* selections p.324
- *Echinops bannaticus* p.319
- *Eryngium bourgatii* 'Picos Blue' p.324
- *Geranium* selections pp.324, 325
- *Helenium* 'Moerheim Beauty' p.325
- *Knautia macedonica* p.326
- *Monarda* 'Squaw' p.327
- *Nepeta grandiflora* 'Dawn to Dusk' p.327
- *Nepeta* 'Six Hills Giant' p.327
- *Pulmonaria* 'Diana Clare' p.335
- *Veronicastrum virginicum* 'Fascination' p.321

Medium-sized perennials

Achillea 'Lachsschönheit' (Galaxy Series)
Feathery foliage and large, flat heads of salmon-pink flowers (the plant is also seen labelled 'Salmon Beauty') make this clump-forming perennial a good choice to grow with wild flowers or in a mixed border. It is one of the Galaxy Hybrids series, which offers a wide range of colours.

↕75–90cm (30–36in) ↔60cm (24in) ❋ ❋ ❋ H7 ☼ ◊ ◊

Achillea 'Taygetea'
Large, creamy-yellow flowerheads appear in summer and autumn, providing perfect landing pads for summer-visiting insects looking for a source of nectar. Finely-cut, greyish-green leaves appear along the length of the stems, acting as a contrasting foil to the flowers.

↕60cm (24in) ↔45cm (18in) ❋ ❋ ❋ H7 ☼ ◊ ◊

Agastache 'Blackadder'
An arresting, upright-growing perennial that, from late summer into autumn, produces lovely violet-blue flowers. These emerge from near-black, finger-like flower spikes that appear above the aromatic, fresh green toothed foliage. It is good for attracting pollinators, and is best in a fairly open but sheltered spot.

↕80cm–1m (32in–3ft) ↔60cm (24in) ❋ ❋ H4 ☼ ◊

Anaphalis triplinervis
These are easy garden plants to grow and are very effective in a border where the emphasis is on white and silver. The clusters of flowers, borne from mid- to late summer, have papery white bracts, and make good cut flowers.

↕80–90cm (32–36in) ↔45–60cm (18–24in)
❋ ❋ ❋ H7 ☼ ◊

Aquilegia vulgaris 'William Guiness'
There are many granny's bonnets to choose from, but the exquisite colours of 'William Guiness' (here shown against a background of hosta leaves) make it a popular choice. Tall flower stems are carried above divided leaves; the plants are suited to cottage gardens or mixed borders.

↕90cm (36in) ↔45cm (18in) ❋ ❋ ❋ H7 ☼ ☼ ◊ ◊

Artemisia ludoviciana 'Silver Queen'
Grown predominantly for its downy silver leaves, this artemisia is good for contrast in a mixed border or as an element in a white and silver planting scheme. Brownish-yellow flowerheads emerge from midsummer to autumn. The variety 'Valerie Finnis' has more deeply cut leaf margins.

↕75cm (30in) ↔60cm (24in) ❋ ❋ ❋ H6 ☼ ◊

Asplenium scolopendrium Crispum Group
The Hart's tongue fern is evergreen, with wavy-edged fronds, making it a year-round decorative asset in the garden. For the lushest plants, choose a position in dappled shade with moist, rich soil to prevent sun scorching. A mixed woodland border would be ideal.

↕30–60cm (12–24in) ↔60cm (24in) ❋ ❋ ❋ H6 ☼ ◊ ◊

Astelia chathamica
Dense clumps of arching, silver scaly leaves make this an attractive plant for a border or container. Pale yellowish-green flowers appear on long stalks from mid- to late spring, followed, on female plants, by orange berries. Do not allow roots to become over-wet during the winter months.

↕1.2m (4ft) ↔to 2m (6ft) ❋ ❋ H3 ☼ ☼ ◊

Astrantia 'Hadspen Blood'
Astrantias are well suited to areas of dappled shade in the garden. The cultivar 'Hadspen Blood' is clump-forming, with deeply cut, mid-green leaves and clusters of dark red flowers surrounded by equally dark red bracts.

↕30–90cm (12–36in) ↔45cm (18in)
❋ ❋ ❋ H7 ☼ ☼ ◊

PLANT GUIDE

❀ ❀ ❀ H7–H5 fully hardy   ❀ ❀ H4–H3 hardy in mild regions/sheltered sites   ❀ H2 protect from frost over winter
❀ H1c–H1a no tolerance to frost   ☼ full sun   ☼ partial sun   ☀ full shade   ◊ well-drained soil   ◐ moist soil   ● wet soil

Athyrium filix-femina
It is clear why the Victorians found ferns so charming when you see the lady fern at its best. Its large, very finely cut fronds, sometimes with red-brown stalks, suit dappled corners of the garden. Shady, sheltered areas or a woodland setting provide the perfect growing conditions.

↕ to 1.2m (4ft)   ↔ to 90cm (36in) ❀ ❀ ❀ H6 ☼ ◐

Campanula 'Burghaltii'
In midsummer, pendent, lavender-coloured bells, opening from blue-purple buds, dangle from the stems of this mound-forming perennial, against a background of heart-shaped leaves. The plant prefers neutral to alkaline conditions to thrive. Alternatively, grow it in a large container.

↕ 60cm (24in)   ↔ 30cm (12in) ❀ ❀ ❀ H7 ☼ ☼ ◊ ◐

Campanula glomerata 'Superba'
The erect stems of this bellflower bear clusters of deep purple, bell-shaped flowers throughout the summer. Prolong the flowering season by cutting plants back to the top of the leaves after the first flush of blooms. This variety is vigorous and can even be invasive.

↕ 60cm (24in)   ↔ indefinite ❀ ❀ ❀ H7 ☼ ☼ ◊ ◐

Centaurea dealbata 'Steenbergii'
Tolerant of dry conditions, knapweed is a magnet for bees and butterflies. The rich pink flowers with feathery petals can be cut for indoor displays when they appear in summer. The plant looks attractive in wild parts of the garden, or as part of a cottage garden scheme.

↕ 60cm (24in)   ↔ 60cm (24in) ❀ ❀ ❀ H7 ☼ ◊

Clematis integrifolia
This herbaceous perennial carries flowers on the current year's shoots, from midsummer to late autumn. The mid-blue flowers have slightly twisted 'petals' and cream anthers, and are followed by silvery seedheads which provide an extended season of interest. The plant may need supporting.

↕ 60cm (24in)   ↔ 60cm (24in) ❀ ❀ ❀ H6 ☼ ☼ ◐

Diascia 'Hopleys'
For sheer flower power, this perennial is hard to beat. It produces multitudes of charming soft pink flowers all summer and autumn until the first hard frosts. The little blooms are held in airy branched sprays atop rather brittle, leafy stems. It needs plenty of sun and some shelter, and the stems are best given support.

↕ 1m (3ft)   ↔ 80cm (32in) ❀ ❀ H4 ☼ ◊

Digitalis GOLDCREST ('Waldigone')
This delightful compact perennial foxglove forms a rosette of semi-evergreen foliage. It produces little red-spotted, peach-pink to yellow tubular flowers that dangle from slender spires through summer and into autumn. It is lovely towards the front of a lightly shaded border or by a wall.

↕ 50cm (20in)   ↔ 50cm (20in) ❀ ❀ ❀ H6 ☼ ☼ ◐

Digitalis × *mertonensis*
This cross between the yellow foxglove and common foxglove has resulted in a free-flowering perennial bearing large pink tubular flowers in late spring and early summer. An excellent plant for attracting bees. Self-sown seedlings will appear around the parent plant.

↕ to 90cm (36in)   ↔ 30cm (12in) ❀ ❀ ❀ H5 ☼ ◐

EARLY-FLOWERING PERENNIALS

- *Ajuga reptans* p.330
- *Crambe cordifolia* p.318
- *Dicentra* 'Bacchanal' p.331
- *Doronicum* 'Little Leo' p.332
- *Epimedium* × *perralchicum* p.332
- *Euphorbia myrsinites* p.332
- *Iris lazica* p.333
- *Helleborus argutifolius* p.325
- *Helleborus foetidus* p.325
- *Helleborus* WALBERTON'S ROSEMARY ('Walhero') p.333
- *Heuchera* 'Plum Pudding' p.333
- *Lamprocapnos spectabilis* 'Alba' p.326
- *Lathyrus vernus* p.334
- *Pulmonaria* 'Blue Ensign' p.334
- *Pulmonaria* 'Diana Clare' p.335
- *Rhodanthemum hosmariense* p.335

Medium-sized perennials

Dryopteris erythrosora
This slowly spreading fern from China and Japan emerges from the soil as coppery-red young fronds. These gradually turn pink and then silvery-green with age, forming a lacy network over the ground. Keep soil around the roots moist and site in a sheltered area. It makes a striking plant for a border.

↕60cm (24in) ↔40cm (16in) ❄ ❄ H4 ☼ ◊

Echinacea 'Glowing Dream'
With its large, unusual reddish coral-pink daisy flowerheads, each with a dark brown centre, this perennial coneflower bears its blooms from June well into autumn. It is clump forming and likes a good position without too much competition. It attracts pollinators well and the flowers are slightly scented.

↕50–60cm (20–24in) ↔50cm (20in) ❄ ❄ ❄ H5 ☼ ◐ ◊

Echinacea purpurea 'Virginia'
A beautiful coneflower with green-tinged, ivory-white daisy flowerheads, each with a dark green centre, held atop sturdy stems. The scented flowers are produced in succession through summer and into autumn, attracting pollinating insects. Choose a place that is not too crowded as it likes space around it.

↕60–70cm (24–28in) ↔50cm (20in) ❄ ❄ ❄ H5 ☼ ◐ ◊

Eremurus stenophyllus
The lovely tapering flower spikes of foxtail lilies emerge and bloom in summer. Staking may be required to prevent the tall stems blowing over. Provide a site with free-draining soil, and mulch around the crowns with garden compost in autumn. Suited to the back of a garden border.

↕1m (3ft) ↔60cm (24in) ❄ ❄ ❄ H6 ☼ ◊

Eryngium bourgatii 'Picos Blue'
An impressive sea holly that forms clumps of divided, prickly grey-green foliage with silver veins. In summer, blue-tinged stems support thistle-like flower heads, which are an arresting metallic blue and starry. The heads look bold as standing stems in autumn too. It is good for attracting pollinators.

↕60–70cm (24–28in) ↔40cm (16in) ❄ ❄ ❄ H5 ☼ ◊

Euphorbia griffithii 'Dixter'
This is a striking herbaceous perennial that contrasts well with other green-leaved euphorbias. Its copper-tinted, dark green leaves make an effective background to the orange bracts that surround the inconspicuous true flowers. The best colour comes from plants grown in dappled shade.

↕75cm (30in) ↔1m (3ft) ❄ ❄ ❄ H7 ☼ ◊

Euphorbia × martinii
With unusual flowers in a mixture of greens and reds, produced on the previous year's shoots, this euphorbia would be a welcome addition to any garden. It flowers over a long season from spring to midsummer and is a very adaptable plant, tolerating sun and shade.

↕1m (3ft) ↔1m (3ft) ❄ ❄ ❄ H5 ☼ ◐ ◊

Euphorbia schillingii
Pale yellow flowerheads perch above a mass of wiry, leafy stems on this strong-growing herbaceous perennial. Plant it with other border perennials, choosing colours carefully to bring out the subtleties of this late summer- to autumn-flowering plant. Provide rich soil in dappled shade.

↕1m (3ft) ↔30cm (12in) ❄ ❄ ❄ H5 ◐ ◊

Geranium 'Brookside'
This densely growing perennial is ideal for border edges; it is a vigorous, spreading plant and makes attractive ground cover, for sun or part-shade. Abundant violet-blue flowers with pale centres appear in summer, held above a mass of finely divided green leaves.

↕60cm (24in) ↔45cm (18in) ❄ ❄ ❄ H7 ☼ ◐ ◊

PLANT GUIDE

❊❊❊ H7–H5 fully hardy ❊❊ H4–H3 hardy in mild regions/sheltered sites ❊ H2 protect from frost over winter
❊ H1c–H1a no tolerance to frost ☼ full sun ☼ partial sun ☀ full shade ◊ well-drained soil ◐ moist soil ● wet soil

Geranium macrorrhizum
This plant has strongly aromatic, toothed, sticky leaves that turn an attractive red in autumn. Clusters of flat pink flowers with protruding stamens are borne in early summer from a mass of sprawling stems. This is a good plant for ground cover or underplanting in a shady site.

↕50cm (20in) ↔60cm (24in) ❊❊❊ H7 ☼ ◊

Geranium 'Nimbus'
A very vigorous and floriferous geranium that becomes a sea of blue when the lavender blue flowers appear in summer. This plant is very tolerant of shade and is a good choice for darker borders or corners that receive little direct sunlight. Clip to encourage repeat flowering.

↕to 1m (3ft) ↔45cm (18in) ❊❊❊ H7 ☼ ☼ ◊

Geranium phaeum
The dusky cranesbill is undemanding in its garden requirements. It will tolerate sun but is also a useful plant for deep shade. Dark maroon flowers with white eyes are produced in early summer. For a brighter-flowered geranium, try *G. psilostemon*, with its black-centred magenta flowers.

↕80cm (32in) ↔45cm (18in) ❊❊❊ H7 ☼ ☼ ☀ ◊

Geum 'Totally Tangerine'
From late spring and all through the season, this exceptional summer-flowering perennial bears a succession of soft orange single or semi-double flowers. These are held in clusters atop tall, hairy stems above clumps of soft green foliage. It is marvellously profuse throughout the season.

↕80cm–1m (32in–3ft) ↔60cm (24in) ❊❊❊ H7 ☼ ◐

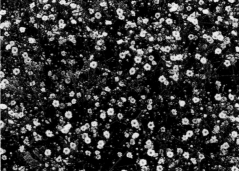

Gypsophila paniculata 'Bristol Fairy'
Also known as baby's breath, branching gypsophila creates a cloud of blossom as a profusion of tiny double-white flowers emerge in midsummer. It looks marvellous in a mixed border and also provides good cut flowers. 'Bristol Fairy' has double white flowers but may be shorter-lived than *G. paniculata*.

↕1.2m (4ft) ↔1.2m (4ft) ❊❊❊ H7 ☼ ◊

Helenium 'Moerheim Beauty'
Copper-red daisy flowers, each with a distinctive round central disc, are borne in early to late summer, filling the garden with warm colour. Deadhead through the season to encourage repeat flowering. The plant's striking colour and form mix well with either hot colours or pastel shades.

↕90cm (36in) ↔60cm (24in) ❊❊❊ H7 ☼ ◊ ◐

Helleborus argutifolius
The Corsican hellebore is a valuable plant for the designer in search of winter interest. A glossy-leaved perennial, it comes into flower in late winter and early spring, and the blooms are an unexpected pale green. It grows well in most conditions but will not thrive in acid soils.

↕to 1.2m (4ft) ↔90cm (36in) ❊❊❊ H5 ☼ ☼ ◐

Helleborus foetidus
The stinking hellebore is named for the unpleasant smell its leaves give off when crushed. However, the greenish-white flowers the plant bears in mid-winter and early spring make up for this downside. Other good varieties to choose from include the Wester Flisk Group, with red-tinted main stems.

↕to 80cm (32in) ↔45cm (18in) ❊❊❊ H7 ☼ ☼ ◐

LATE-FLOWERING PERENNIALS

- *Agastache* 'Blackadder' p.322
- *Anemone* x *hybrida* p.318
- *Aster amellus* 'Veilchenkönigin' p.330
- *Diascia* 'Hopleys' p.323
- *Helianthus* 'Lemon Queen' p.319
- *Kniphofia* 'Percy's Pride' p.326
- *Liriope muscari* p.334
- *Rudbeckia fulgida* var. *sullivantii* 'Goldsturm' p.329
- *Rudbeckia laciniata* 'Goldquelle' p.329
- *Salvia* 'Hot Lips' p.329
- *Salvia nemorosa* p.335
- *Salvia uliginosa* p.320
- *Selinum wallichianum* p.320
- *Symphyotrichum ericoides* 'White Heather' p.329
- *Symphyotrichum novae-angliae* 'Andenken an Alma Pötschke' p.329
- *Symphyotrichum* 'Ochtendgloren' p.321
- *Verbena bonariensis* p.321
- *Veronicastrum virginicum* 'Fascination' p.321

Medium-sized perennials

Hemerocallis 'Marion Vaughn'
A late afternoon-flowering daylily, 'Marion Vaughn' is a dependable evergreen with clear lemon-yellow flowers and bright green strap-like foliage, making a crisp addition to a mixed border. It looks good growing in a drift with other daylilies. Full sun will promote best flowering.

↕85cm (34in) ↔75cm (30in) ❋ ❋ ❋ H6 ☼ ◐ ◊

Hemerocallis 'Selma Longlegs'
A flamboyant daylily, it has unusual, rather spidery flowers in soft shades of peach and orange, held in a cluster atop tall, slender stems. Although individual flowers only last a day, this is a repeat bloomer, with more flowers produced on separate stems through summer. It has attractive green arching foliage.

↕80cm (32in) ↔60cm (24in) ❋ ❋ ❋ H6 ☼ ◐ ◊

Hosta sieboldiana var. elegans
With its heavily puckered, blue-green leaves, this large hosta makes a dramatic border plant. It tolerates shade although a very dark position will subdue the production of lilac-coloured flowers in early summer. Place a group of hostas together for a stunning foliage effect.

↕1m (3ft) ↔1.2m (4ft) ❋ ❋ ❋ H7 ☼ ◊ ◐

Hosta 'Sum and Substance'
A superbly bold and architectural hosta rising each year from purple-tinged shoots. Its leaves are glowing yellow-green and rather leathery, each at first cupped, then expanding to 50cm (20in) or more across. It is fairly resistant to slug and snail damage. Tall spires of lavender flowers open in summer.

↕90cm (36in) ↔1.2m (4ft) ❋ ❋ ❋ H7 ☼ ◊

Hylotelephium 'Matrona'
Fleshy leaves, initially green and later flushed purple, and dark red stems form the backdrop to the flattened heads of tiny pink star-like flowers in late summer. The dried flower heads add structure and interest to the winter garden.

↕60–75cm (24–30in) ↔to 30–45cm (12–18in) ❋ ❋ ❋ H6 ☼ ◊

Knautia macedonica
Similar to a scabious, this knautia carries purple-red pincushion flowerheads, held above the foliage on branching stems, from mid- to late summer. It is attractive to bees and butterflies and ideally suited to a wildflower or cottage garden. It is fairly drought-tolerant.

↕60–80cm (24–32in) ↔45cm (18in) ❋ ❋ ❋ H7 ☼ ◊

Kniphofia 'Bees' Sunset'
This is a yellow-orange variety of the deciduous plant familiarly known as the red hot poker. Upright, fleshy stems support a bottlebrush-like array of the downward-pointing, tubular flowers from early to late summer. Grow in the herbaceous border in groups for a dramatic display.

↕90cm (36in) ↔60cm (24in) ❋ ❋ ❋ H5 ☼ ◐ ◊

Kniphofia 'Percy's Pride'
This cultivar of the red hot poker produces long spikes of greenish-yellow flowers, maturing to cream, which emerge in late summer and early autumn on long, fleshy stems. The unusual flower colour makes it suitable for a colour-themed border using white, green and pale yellow.

↕to 1.2m (4ft) ↔60cm (24in) ❋ ❋ ❋ H6 ☼ ◐ ◊

Lamprocapnos spectabilis 'Alba'
When in flower, the graceful, arching stems of the bleeding heart (or Dutchman's breeches) look like a miniature washing line. New shoots appear in spring with rose pink or white flowers. 'Alba' is a less vigorous selection with pure white blooms. It will tolerate some sun if the roots are kept moist.

↕to 1.2m (4ft) ↔45cm (18in) ❋ ❋ ❋ H6 ☼ ◊

PLANT GUIDE

✿✿✿ H7–H5 fully hardy   ✿✿ H4–H3 hardy in mild regions/sheltered sites   ✿ H2 protect from frost over winter
✿ H1c–H1a no tolerance to frost   ☼ full sun   ☼ partial sun   ☼ full shade   ◊ well-drained soil   ◊ moist soil   ◆ wet soil

Liatris spicata 'Kobold'
The spikes of deep purple flowerheads on this plant are unusual in that the flowers open from the top downwards. 'Kobold' flowers from late summer to early autumn and suits a mixed border, but needs regular moisture to thrive. Stems can be cut for a cheerful indoor display.

↕70cm (30in) ↔45cm (18in) ✿✿✿ H7 ☼ ◊ ◊

Lupinus 'Chandelier'
If space allows, grow lupins in drifts, allowing complementary colours to sit close to one another. The pale yellow, pea-like blooms of clump-forming 'Chandelier' (part of the Band of Nobles series) appear in early and midsummer and are ideal for a mixed or herbaceous border in a cottage-style or informal design.

↕90cm (36in) ↔75cm (30in) ✿✿✿ H5 ☼ ☼ ◊

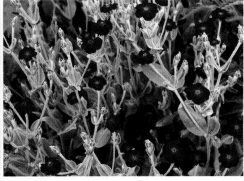

Lychnis coronaria
Known by the common names of dusty miller and rose campion, this short-lived perennial has soft silvery-grey stems and leaves. Late summer sees a long succession of rounded vermilion blooms. It self-seeds freely. For a pure white form, choose 'Alba'.

↕80cm (32in) ↔45cm (18in) ✿✿✿ H7 ☼ ☼ ◊

Lysimachia ephemerum
Woodland or streamside plants in the wild, these pretty herbaceous perennials are best suited to a damp border, bog garden or pond margin. In early and midsummer, erect spikes of saucer-shaped white flowers rise above mid-green tapered leaves. Plants may need protection in winter.

↕1m (3ft) ↔30cm (12in) ✿✿✿ H6 ☼ ☼ ◊ ◆

Lythrum salicaria 'Feuerkerze'
Masses of tiny star-shaped, intensely red-purple flowers cover the slender upright spikes of this purple loosestrife cultivar, making a beautiful display from midsummer to early autumn. The leaves are downy. The plant needs moisture and suits a damp border or bog garden.

↕to 90cm (36in) ↔45cm (18in) ✿✿✿ H7 ☼ ◊

Monarda didyma x fistulosa 'Oneida'
Bergamots are grown for their long-lasting, colourful flowers which appear from midsummer to early autumn. 'Oneida' is particularly striking, with its bright scarlet flowers held above dark bracts. Bergamots attract bees and butterflies and are ideal for a wildflower garden.

↕to 1.2m (4ft) ↔45cm (18in) ✿✿ H4 ☼ ☼ ◊ ◊

Nepeta grandiflora 'Dawn to Dusk'
As the name grandiflora suggests, the pale mauve-pink flowers on this catmint are larger than usual. Grow this cultivar near a path or garden seat to make the most of the distinctive fragrance released as the leaves are crushed. The plants are attractive to bees – and cats.

↕65cm (26in) ↔30cm (12in) ✿✿✿ H6 ☼ ☼ ◊

Nepeta 'Six Hills Giant'
This is a vigorous perennial bearing masses of lavender-blue flowers throughout the summer months. Be prepared for it to take up some space in the border. The leaves are light grey, and noticeably aromatic when touched. Clumps can be divided in spring or autumn to rejuvenate plants.

↕90cm (36in) ↔60cm (24in) ✿✿✿ H7 ☼ ☼ ◊

PERENNIALS FOR FOLIAGE INTEREST

- Adiantum venustum p.330
- Alchemilla mollis p.330
- Artemisia ludoviciana 'Silver Queen' p.322
- Arum italicum subsp. italicum 'Marmoratum' p.330
- Astelia chathamica p.322
- Athyrium niponicum var. pictum p.330
- Beesia calthifolia p.331
- Euphorbia epithymoides p.332
- Euphorbia x martinii p.324
- Foeniculum vulgare 'Purpureum' p.319
- Heuchera 'Plum Pudding' p.333
- Hosta sieboldiana var. elegans p.326
- Hosta 'Sum and Substance' p.326
- Hosta Tardiana Group 'June' p.333
- Hylotelephium 'Matrona' p.326
- Lychnis coronaria p.327
- Phlox paniculata 'Norah Leigh' p.328
- Sedum selections p.335
- Sempervivum tectorum p.335

Medium-sized perennials

Origanum laevigatum 'Herrenhausen'
Strongly aromatic leaves and bright clusters of pink flowers characterize this woody perennial, an ornamental cultivar of the culinary herb marjoram. The leaves are purple-flushed when young and in winter, and the flowers appear from late spring to autumn. Suited to a herb garden or border margin.
↕60cm (24in) ↔ 45cm (18in) ❀ ❀ ❀ H6 ☼ ◊

Paeonia 'Bartzella'
In summer, this spectacular intersectional peony produces large, blousy, double soft yellow flowers marked with dark red in the centre and with a lemony scent. These appear above foliage that forms a neat mound of growth, the leaves leathery and quite shiny. Occasionally, it repeat flowers in early autumn.
↕70–90cm (28–36in) ↔ 90cm (36in) ❀ ❀ ❀ H6 ☼ ☼ ◊

Papaver Oriental Group 'Black and White'
The bold, beautiful flowers of the Oriental poppy make an immediate impact. There are many cultivars; the large, ruffled petals of 'Black and White', each with a black blotch at the base, are papery white and surround a boss of dark stamens.
↕45–90cm (18–36in) ↔ 60–90cm (24–36in) ❀ ❀ ❀ H7 ☼ ◊

Penstemon 'Alice Hindley'
A favourite with many gardeners, foxglove-like penstemons are reliable and rewarding to grow. Large, tubular bell-like flowers open in succession along upright stems from midsummer to autumn. There are many cultivars; the flowers of 'Alice Hindley' are pale lilac-blue. Feed well.
↕90cm (36in) ↔ 45cm (18in) ❀ ❀ H4 ☼ ☼ ◊

Penstemon 'Andenken an Friedrich Hahn'
This hardy, vigorous, bushy penstemon carries elegant spikes of bright garnet-red flowers in profusion from midsummer through to mid-autumn, above masses of narrow green leaves. Deadheading will significantly prolong the flowering display.
↕75cm (30in) ↔ 60cm (24in) ❀ ❀ ❀ H5 ☼ ☼ ◊

Persicaria amplexicaulis 'Firetail'
This semi-evergreen perennial is a robust, undemanding garden plant. From midsummer to early autumn, the lush green foliage is joined by tall, rigid stems bearing small, bright red bottlebrush flowers. Grow as border plants, as ground cover, or naturalize in a woodland garden.
↕ to 1.2m (4ft) ↔ to 1.2m (4ft) ❀ ❀ ❀ H7 ☼ ☼ ◊

Persicaria bistorta 'Superba'
A long-flowering, semi-evergreen plant with rounded spikes of soft pink, miniature blooms, which present a good show all summer and well into autumn. Grow behind 'Firetail' (*left*) for interesting contrast. Divide particularly vigorous clumps in spring or summer to control their size and spread.
↕ to 90cm (36in) ↔ 90cm (36in) ❀ ❀ ❀ H7 ☼ ☼ ◊

Phlomis russeliana
This sage-like plant looks very effective grown in a large group in a border. The pale yellow, hooded flowers begin to appear in late spring and continue until autumn, with the best show of colour in early summer. The cut stems are good for dried arrangements.
↕ to 90cm (36in) ↔ 75cm (30in) ❀ ❀ ❀ H6 ☼ ☼ ◊

Phlox paniculata 'Norah Leigh'
Variegated forms of phlox are a relatively new phenomenon. The tapering leaves of 'Norah Leigh' have green mid-ribs but are mainly creamy-white with splashes of green. Clusters of pale lilac flowers with deeper pink centres are borne from summer to autumn over a long season.
↕ to 90cm (36in) ↔ 60–100cm (24–39in) ❀ ❀ ❀ H7 ☼ ◊

PLANT GUIDE

❊❊❊ H7–H5 fully hardy ❊❊ H4–H3 hardy in mild regions/sheltered sites ❊ H2 protect from frost over winter
❊ H1c–H1a no tolerance to frost ☼ full sun ☼ partial sun ☼ full shade ◊ well-drained soil ◐ moist soil ● wet soil

Phormium cookianum subsp. *hookeri* 'Tricolor'
The mountain flax from New Zealand comes in a number of forms. Here, the narrow, arching, strap-like leaves are green with cream and red margins. Yellow-green flowers emerge in summer on long, stiff stems, although it is for the foliage that the plant is grown. Ideal for a coastal garden.

↕1.2m (4ft) ↔3m (10ft) ❊❊ H4 ☼ ☼ ◊

Rudbeckia fulgida var. *sullivantii* 'Goldsturm'
Coneflowers are popular late-season plants, producing quantities of yellow flowerheads with dark eyes, held on bristly stems, from late summer to mid-autumn. The rich green leaves are tapering and toothed. Pair 'Goldsturm' with *Verbena bonariensis* and grasses for a dramatic display.

↕to 60cm (24in) ↔45cm (18in) ❊❊❊ H6 ☼ ☼ ◊

Rudbeckia laciniata 'Goldquelle'
The deeply cut green leaves make an effective background for this double-flowered, lemon-yellow coneflower. 'Goldquelle' makes a fine addition to the late summer border and will continue flowering until the middle of autumn. Rudbeckias and grasses make a happy combination in a large border.

↕to 90cm (36in) ↔45cm (18in) ❊❊❊ H6 ☼ ☼ ◊

Salvia 'Hot Lips'
An easy shrubby perennial with evergreen, blackcurrant-scented oval foliage, topped all summer and autumn by upright sprays of small white and red flowers. The blooms have red tips during the hottest part of summer. It is seldom out of flower through the growing season, and needs an occasional trim.

↕1m (3ft) ↔1m (3ft) ❊❊❊ H5 ☼ ☼ ◊

Sanguisorba 'Tanna'
This unusual-looking, easy-to-grow, clump-forming perennial produces masses of dark red, bobble-like flowerheads all through summer. These are held on tall, rather wiry stems above handsome blue-green foliage. It is a fairly compact plant and a great mixer with grasses and other perennials.

↕50cm (20in) ↔50cm (20in) ❊❊❊ H7 ☼ ☼ ◊

Sisyrinchium striatum 'Aunt May'
Excellent front-of-the-border plants, sisyrinchiums also suit a gravel garden. Less vigorous than the green-leaved species, the cultivar 'Aunt May' has cream-edged, grey-green, narrow leaves. In summer, the stiff flower stems are studded with small pale yellow flowers.

↕50cm (20in) ↔25cm (10in) ❊❊❊ (borderline) H4 ☼ ◊

Symphyotrichum ericoides 'White Heather'
A reliable and easy-to-grow perennial, 'White Heather' produces sprays of small daisy blooms at the end of summer, prolonging the season of interest in the garden. A sunny site will ensure an extended spell of flowering. To increase the stock, divide larger plants in spring.

↕1m (3ft) ↔30cm (12in) ❊❊❊ H7 ☼ ◊

Symphyotrichum novae-angliae 'Andenken an Alma Pötschke'
These Michaelmas daisies bear rich cerise-pink blooms in profusion from late summer to mid-autumn. Mix varieties together to create your own aster display or plant among other perennials for late summer colour.

↕1.2m (4ft) ↔60cm (24in) ❊❊❊ H7 ☼ ☼ ◐

PERENNIALS FOR DAMP SOIL CONDITIONS

- *Aconitum* 'Spark's Variety' p.318
- *Adiantum venustum* p.330
- *Alchemilla mollis* p.330
- *Astelia chathamica* p.322
- *Athyrium filix-femina* p.323
- *Beesia calthifolia* p.331
- *Dicentra* 'Bacchanal' p.331
- *Dryopteris wallichiana* p.319
- *Helleborus argutifolius* p.325
- *Inula magnifica* p.319
- *Pachysandra terminalis* p.334
- *Persicaria bistorta* 'Superba' p.328
- *Pulmonaria* 'Diana Clare' p.335
- *Salvia uliginosa* p.320
- *Symphyotrichum novae-angliae* 'Andenken an Alma Pötschke' p.329
- *Valeriana phu* 'Aurea' p.321
- *Veronicastrum virginicum* p.321

Small perennials

Adiantum venustum
The evergreen Himalayan maidenhair fern is a decorative plant for a shady wall crevice or a damp, shady corner. It looks delicate but is in fact surprisingly robust. Old growth should be removed in late winter before new pink croziers unfurl in spring, developing into fresh green fronds.

↕15cm (6in) ↔ indefinite ❈ ❈ ❈ H7 ☼ ◐

Ajuga reptans
Spikes of deep blue flowers emerge from the low-growing, dark green leaves of this evergreen perennial from late spring to early summer. The plant spreads rapidly and makes excellent ground cover. For a less invasive form try 'Catlin's Giant', which has large bronze-purple leaves.

↕15cm (6in) ↔ 60–90cm (24–36in) ❈ ❈ ❈ H7 ☼ ◐

Alchemilla mollis
Dependable and drought-tolerant, lady's mantle is grown for its pretty foliage and frothy sprays of tiny greenish-yellow flowers, which appear from early summer to autumn and are good for cutting. Deadhead after flowering to prevent self-seeding. Plant it at the front of a border or in a gravel garden.

↕to 60cm (24in) ↔ 75cm (30in) ❈ ❈ ❈ H7 ☼ ☼ ◐

Agapanthus 'Silver Baby'
This compact, clump-forming perennial is best in late summer, when stems bear showy heads of white, trumpet-shaped flowers tinged with soft blue, providing a shimmering silvery effect. Stems rise over arching, strappy, evergreen foliage. It is ideal for a sheltered, sunny spot and is great in a container.

↕50cm (20in) ↔ 50cm (20in) ❈ H3 ☼ ◐

Anemone sylvestris
This spring-flowering perennial is ideal for woodland conditions or below shrubs in moist soil, forming clumps of lobed foliage. In spring, it bears charming white flowers with golden stamens. The initially nodding then upward-facing blooms are 6cm (2½in) across. It is lovely beside ferns and fresh spring growth.

↕50cm (20in) ↔ 50cm (20in) ❈ ❈ ❈ H6 ◐

Anthemis punctata subsp. *cupaniana*
This Sicilian chamomile naturally prefers a sunny site, such as an open, well-drained rock garden. Flowers are long-lasting and bloom over a long season from late spring to late summer. The plant forms a tight mat at ground level and in winter the silvery-grey leaves turn grey-green.

↕30cm (12in) ↔ 90cm (36in) ❈ ❈ H4 ☼ ◐

Arum italicum subsp. *italicum* 'Marmoratum'
A truly exotic-looking plant whether in leaf, flower, or fruit, Italian arum 'Marmoratum' is excellent for filling in gaps in border displays. The glossy green leaves are veined with white, while the pale cream spathes give way to stalks of bright orange berries. It's at its best in a sheltered site.

↕30cm (12in) ↔ 15cm (6in) ❈ ❈ ❈ H6 ☼ ◐

Aster amellus 'Veilchenkönigin'
A clump-forming perennial, this aster produces a mass of tiny, violet-purple, daisy-like flowers in late summer, which are attractive to butterflies. The mid-green leaves are narrow and slightly hairy. Divide plants in spring and replant the strongest sections for most vigorous regrowth.

↕30–60cm (12–24in) ↔ 45cm (18in) ❈ ❈ ❈ H7 ☼ ◐

Athyrium niponicum var. *pictum*
These graceful, deciduous ferns (also known as lady ferns) are easy to grow and will thrive in a shady, sheltered border or woodland garden, as long as there is sufficient moisture. The arching fronds are light green or greyish, sometimes flushed purple, with a purple midrib.

↕to 30cm (12in) ↔ indefinite ❈ ❈ ❈ H6 ☼ ☼ ◐

❄❄❄ H7–H5 fully hardy ❄❄ H4–H3 hardy in mild regions/sheltered sites ❄ H2 protect from frost over winter
❄ H1c–H1a no tolerance to frost ☼ full sun ◐ partial sun ● full shade ◊ well-drained soil ◐ moist soil ● wet soil

Beesia calthifolia
A splendid woodland perennial, it is grown for its shiny, evergreen, slightly puckered heart-shaped foliage held on dark stems. In spring, wands of white starry flowers arise above the leaves. It forms elegant clumps and looks perfect next to ferns, enjoying the same cool, moist conditions.

↕50cm (20in) ↔50cm (20in) ❄❄❄ H6 ☼ ◐

Bergenia 'Pink Dragonfly'
In spring, this compact elephant's ear is impressive in flower, when red-tinged stems rise above the leathery, evergreen foliage bearing heads of pink flowers, each with a red throat. The leaves are also attractive, especially in winter, when they become red and purple tinged with frost, forming a bold clump.

↕45cm (18in) ↔50cm (20in) ❄❄❄ H6 ☼ ◐ ◊

Betonica officinalis 'Hummelo'
This mat-forming perennial was selected by garden designer Piet Oudolf and named after his home. It is semi-evergreen, retaining most leaves in a mild winter. From the rosettes of foliage arise stems of small, bright pink flowers during early summer – it is a great choice for the front of a border.

↕50cm (20in) ↔80cm (32in) ❄❄❄ H7 ☼ ◐ ◊

Calamintha grandiflora 'Variegata'
A plant for the woodland garden or a cool, sheltered position, this calamint has toothed, pale green leaves, speckled creamy-white, which are aromatic when crushed. From summer to autumn, pink-mauve, two-lipped flowers emerge above and level with the topmost leaves.

↕30cm (12in) ↔45cm (18in) ❄❄❄ H5 ☼ ◊ ◐

Campanula 'Pink Octopus'
This striking herbaceous perennial bears curious spidery flowers in summer. The long, slender, often curling pale pink petals with red spots are held above serrated green foliage. It is vigorous and clump forming, and spreads freely at the root. Grow it in sun or part shade and fairly moist soil.

↕50cm (20in) ↔1m (3ft) ❄❄❄ H7 ☼ ◐ ◊

Coreopsis verticillata 'Moonbeam'
A row of this brightly coloured plant will make a fine edging for a border. Finely cut leaves mingle together with a profusion of yellow, star-like flowers in early summer. A sunny position will promote the best show of blooms. Deadhead to encourage flowering.

↕to 50cm (20in) ↔45cm (18in) ❄❄ H4 ☼ ◐ ◊

Dianthus Allwoodii Group 'Bovey Belle'
This hardy pink bears clove-scented, bright pink double blooms on long stems above silver-grey strappy foliage in summer, making an impact in a mixed border or raised bed. Deadhead regularly to promote further flowering. Pinks make long-lasting cut flowers.

↕25–45cm (10–18in) ↔40cm (16in) ❄❄❄ H6 ☼ ◊

Dicentra 'Bacchanal'
Layer upon layer of deeply divided, grey-green leaves make an effective foil for the delicate, crimson, heart-shaped flowers, dangling from arching stems, which appear from mid- to late spring. 'Bacchanal' is one of the darkest cultivars. These are shade-loving plants and suit a moist, shady border.

↕45cm (18in) ↔60cm (24in) ❄❄❄ H5 ◐ ● ◐

SHADE-TOLERANT PERENNIALS

- *Aconitum* 'Spark's Variety' p.318 (damp shade)
- *Adiantum venustum* p.330 (damp shade)
- *Asplenium scolopendrium* Crispum Group p.322 (damp shade)
- *Athyrium niponicum* var. *pictum* p.330 (damp shade)
- *Beesia calthifolia* p.331
- *Dicentra* 'Bacchanal' p.331 (damp shade)
- *Dryopteris erythrosora* p.324 (damp shade)
- *Epimedium* selections p.332
- *Galium odoratum* p.332
- *Geranium macrorrhizum* p.325 (dry shade)
- *Geranium* 'Nimbus' p.325 (dry shade)
- *Geranium nodosum* 'Silverwood' p.333
- *Lamprocapnos spectabilis* 'Alba' p.326 (damp shade)
- *Liriope muscari* p.334
- *Pachysandra terminalis* p.334 (dry shade)

Small perennials

Doronicum 'Little Leo'
A spring-flowering perennial, this bright daisy is as attractive to wildlife as it is to gardeners. Bees, butterflies, and hoverflies are regular visitors to the large yellow blooms. Plant in small groups at the front of borders, or in containers. The flowers are also good for cutting.

↕25cm (10in) ↔30–60cm (12–24in) ❀ ❀ ❀ H5 ☼ ◊ ◊

Epimedium 'Amber Queen'
This first-rate, mid-spring-flowering perennial bears clouds of orange, spidery flowers held on slender stems above semi-evergreen foliage, which forms useful ground cover. Delicate young leaves are marked with purple splashes. It is great below trees or large shrubs in soil that is not too dry.

↕50cm (20in) ↔60cm (24in) ❀ ❀ ❀ H6 ☼ ◊

Epimedium × perralchicum
Strong-growing woodland plants, epimediums make excellent ground cover under trees or shrubs. This hybrid has both interesting foliage – bronze when young, deep green when mature – and pretty, bright yellow flowers, borne on leafless stems in spring. It is also drought-tolerant.

↕40cm (16in) ↔60cm (24in) ❀ ❀ ❀ H6 ☼ ◊ ◊

Eryngium bourgatii 'Oxford Blue'
This is one of the smaller sea hollies. It is a herbaceous, clump-forming plant with dark green lower leaves and, in summer, spiny, silvery flower stems bearing silver-blue, thistle-like flowers, surrounded by blue-tinged bracts. The flower stems can be dried for indoor arrangements.

↕15–45cm (6–18in) ↔30cm (12in) ❀ ❀ ❀ H5 ☼ ◊

Euphorbia epithymoides
This euphorbia forms a loose, dome-shaped mound of lime-green stems that carry canary-yellow flowerheads from mid-spring to midsummer. The blooms are at their brightest when young. The plant dies down in winter, re-emerging the next year with a batch of fresh young shoots.

↕40cm (16in) ↔60cm (24in) ❀ ❀ ❀ H6 ☼ ☼ ◊ ◊

Euphorbia myrsinites
Sprouting from a central crown, this distinctive evergreen perennial has serpentine ground-covering stems clad with whorls of blue leaves. Heads of green flowers appear in spring. It likes a sunny site in well-drained soil. Sprawling over rocks, atop a wall, or down a bank, it looks great year-round.

↕15cm (6in) ↔60cm (24in) ❀ ❀ ❀ H5 ☼ ◊

Galium odoratum
A UK native, this mat-forming perennial, also known as sweet woodruff, provides dense ground cover even in shaded sites. Dark green, pointed, hay-scented foliage is held in whorls around slender, brittle stems; then heads of tiny white flowers bloom in spring. It spreads by stolons.

↕20cm (8in) ↔60cm (24in) ❀ ❀ ❀ H7 ☼ ◊

Geranium 'Ann Folkard'
Bearing black-eyed, magenta flowers amid a filigree of golden green foliage, this perennial is great all summer. It rises from a central crown and develops a mass of lengthening stems that, as the season progresses, trail through neighbouring plants. It dies back in winter, reappearing in spring.

↕50cm (20in) ↔1m (3ft) ❀ ❀ ❀ H7 ☼ ☼ ◊

Geranium × cantabrigiense 'Biokovo'
This low-growing perennial forms a dense carpet of shining, green, rounded, aromatic foliage, enlivened in late spring with masses of charming pink-tinged white flowers held above the leaves. The leaves turn attractive tints in autumn, although many are retained in a mild year. It is tough and vigorous.

↕15cm (6in) ↔80cm (32in) ❀ ❀ ❀ H6 ☼ ◊

PLANT GUIDE 333

❄❄❄ H7–H5 fully hardy   ❄❄ H4–H3 hardy in mild regions/sheltered sites   ❄ H2 protect from frost over winter
☙ H1c–H1a no tolerance to frost   ☼ full sun   ☼ partial sun   ☀ full shade   ◊ well-drained soil   ◐ moist soil   ● wet soil

Geranium clarkei 'Kashmir White'
Cranesbills make versatile, undemanding garden plants. The 'Kashmir' cultivars are spreading, herbaceous perennials with dissected green foliage; they come in blue, pink, purple, and white. This cultivar produces large, whitish summer flowers with pale lilac-pink veining. Divide vigorous plants in spring.

↕ to 45cm (18in)   ↔ indefinite   ❄❄❄ H6 ☼ ☼ ◊

Geranium nodosum 'Silverwood'
A perfect plant for dry shade, this geranium produces masses of white flowers non-stop through summer and autumn, shining in the shade amid apple-green lobed foliage. It forms a neat, low mound of growth and survives quite poor conditions, looking great next to ferns and epimediums.

↕ 30cm (12in)   ↔ 50cm (20in)   ❄❄❄ H7 ☼ ☼ ☀ ◊

Geranium ROZANNE ('Gerwat')
This outstanding flowering perennial bears masses of blue, white-centred flowers in succession from spring to autumn. Flowers are carried on stems that rise from a central crown. It forms a mound but is useful to thread through and over other planting. Stems die back to the crown in winter.

↕ 50cm (20in)   ↔ 1m (3ft)   ❄❄❄ H7 ☼ ☼ ◊

Helleborus WALBERTON'S ROSEMARY ('Walhero')
This early-flowering perennial is perfect for winter and spring, bearing starry, five-petalled, rose-pink flowers atop tall stems. It is clump forming with evergreen foliage that is best trimmed away before flowers arise; new leaves soon follow. Plants in flower withstand frost well, but shelter improves performance.

↕ 50cm (20in)   ↔ 40cm (16in)   ❄❄❄ H7 ☼ ◊

Heuchera 'Pink Pearls'
A useful evergreen perennial, it forms colourful, leafy ground cover, enlivened in summer by masses of tiny pale pink flowers held in wands above the plant. The leaves are pink-orange and are useful for providing contrast with other plants. Divide every few years to keep plants thriving.

↕ 40cm (16in)   ↔ 45cm (18in)   ❄❄ H4 ☼ ☼ ◊

Heuchera 'Plum Pudding'
Compact, evergreen perennials, heucheras are useful for all-year-round interest. This cultivar has purple ruffled leaves with deeper purple veins. Small white flowers are held aloft on thin wiry stems in late spring. Grow alongside silvery-leaved 'Pewter Moon' to show both off to good effect.

↕ 65cm (26in)   ↔ 50cm (20in)   ❄❄❄ H6 ☼ ☼ ◊ ◐

Hosta Tardiana Group 'June'
This beautiful and reliable perennial has impressive blue-green leaves, each variegated with yellow green in the centre. It is clump forming, forming a mound through summer, when spires of lavender flowers arise as a bonus. It is relatively resistant to slug and snail damage.

↕ 40cm (16in)   ↔ 80cm (32in)   ❄❄❄ H7 ☼ ☼ ◊

Iris lazica
Evergreen and late-winter-flowering, this compact perennial is a useful choice. Its erect, lance-shaped foliage forms clumps that look fresh year round, and it produces showy lavender-blue flowers in late winter and early spring. It is not fussy, thriving at the front of a sunny or partly shaded border.

↕ 30cm (12in)   ↔ 60cm (24in)   ❄❄❄ H5 ☼ ☼ ◊

PERENNIALS FOR CONTAINERS

- *Agapanthus* 'Silver Baby' p.330
- *Astelia chathamica* p.322
- *Beesia calthifolia* p.331
- *Geranium* ROZANNE ('Gerwat') p.333
- *Gypsophila paniculata* 'Bristol Fairy' p.325
- *Helleborus foetidus* p.325
- *Heuchera* 'Pink Pearls' p.333
- *Heuchera* 'Plum Pudding' p.333
- *Hosta* 'Sum and Substance' p.326
- *Hosta* Tardiana Group 'June' p.333
- *Melianthus major* p.320
- *Musa basjoo* p.320
- *Origanum laevigatum* 'Herrenhausen' p.328
- *Phlox paniculata* 'Norah Leigh' p.328
- *Phormium cookianum* subsp. *hookeri* 'Tricolor' p.329
- *Phormium tenax* Purpureum Group p.320
- *Rhodanthemum hosmariense* p.335
- *Sedum* 'Vera Jameson' p.335
- *Sempervivum tectorum* p.335

Small perennials

Lamium maculatum 'White Nancy'
Spreading, low-growing plant, excellent for ground cover. The toothed leaves are silver with a green edge, while the summer flowers are pure white. Grow to cover bare soil and to suppress weeds. 'Red Nancy' has silver leaves with purplish-red flowers.

↕ to 15cm (6in) ↔ to 1m (3ft) or more
❄ ❄ ❄ ❄ H7 ☼ ☼ ◊ ◊

Lathyrus vernus
A great herbaceous plant for early spring interest, this pea-flowered perennial forms a clump of low, divided foliage. Masses of pink, white, or purple flowers, held in small sprays, appear in early spring. It looks great with crocus and early primroses at the front of a border, and is tough and easy to grow.

↕ 45cm (18in) ↔ 50cm (20in) ❄ ❄ ❄ ❄ H6 ☼ ☼ ◊

Liriope muscari
This handsome, grassy-leaved, evergreen perennial is one of those helpful plants that survives dry shade under shrubs, although it looks better given some care. It produces striking spikes of purple flowers in autumn, just as much in the garden is petering out. It will form clumps but spreads slowly.

↕ 50cm (20in) ↔ 60cm (24in) ❄ ❄ ❄ ❄ H5 ☼ ☼ ◊

Oenothera fruticosa 'Fyrverkeri'
From late spring to late summer, the large, bright yellow flowers of this evening primrose appear on upright stems above the purple-brown-flushed leaves below. The flowers bloom during the day and are short-lived but are borne over a long season. The plant will perform best in a sunny site.

↕ 30–90cm (12–36in) ↔ 30cm (12in) ❄ ❄ ❄ H5 ☼ ◊

Omphalodes cappadocica 'Cherry Ingram'
In spring, this perennial has masses of forget-me-not blue flowers held in dainty sprays above oval leaves. It is the perfect companion for small daffodils that bloom at the same time and is marvellous below shrubs or at the front of a cool border. Avoid over-tidying as stems are brittle and easily broken.

↕ 30cm (12in) ↔ 40cm (16in) ❄ ❄ ❄ H5 ☼ ◊

Pachysandra terminalis
Good for ground cover, this tough evergreen perennial is grown for its foliage and will spread freely given enough moisture. It has coarsely toothed, glossy, dark green leaves and tiny white flowers, which are carried in spikes in early summer. A useful plant for shady sites.

↕ 20cm (8in) ↔ indefinite ❄ ❄ ❄ H5 ☼ ☼ ◊

Phuopsis stylosa
This unusual summer-flowering perennial deserves to be better known. It forms a mat of bright green foliage, the fine, almost evergreen aromatic leaves arranged around stems in whorls, topped by rounded heads of bright pink flowers. It is eye-popping at its peak and ideal for the front of a border or gravel garden.

↕ 30cm (12in) ↔ 50cm (20in) ❄ ❄ ❄ H5 ☼ ◊

Polypodium × *mantoniae* 'Cornubiense'
The finely dissected fronds of this ground cover fern easily cover the soil and break up the hard lines of path edges. New growth starts in spring with the fronds taking several weeks to unfurl. This is a handsome, resilient plant for a damp and shady spot in the garden.

↕ 30cm (12in) ↔ indefinite ❄ ❄ ❄ H7 ☼ ☼ ◊ ◊

Pulmonaria 'Blue Ensign'
This clump-forming perennial bears heads of violet-blue, bell-shaped flowers from early spring. These appear in profusion on short stems above foliage. It is broad, bristly, and spreading, but foliage can get scruffy by winter so is best removed before flowering. New leaves form good ground cover.

↕ 30cm (12in) ↔ 50cm (20in) ❄ ❄ ❄ H6 ☼ ☼ ◊

PLANT GUIDE   **335**

❋ ❋ ❋ H7–H5 fully hardy   ❋ ❋ H4–H3 hardy in mild regions/sheltered sites   ❋ H2 protect from frost over winter
❋ H1c–H1a no tolerance to frost   ☼ full sun   ☼ partial sun   ☀ full shade   ◌ well-drained soil   ◐ moist soil   ● wet soil

Pulmonaria 'Diana Clare'
An early spring-flowering perennial, 'Diana Clare' is easy to grow and needs very little attention once established. In late winter and spring, clusters of violet-blue flowers, striped red, open above green leaves marked with silver. It makes good ground cover where the soil is not too dry.

↕30cm (12in) ↔45cm (18in) ❋ ❋ ❋ H7 ☼ ☼ ◐

Rhodanthemum hosmariense
Plants that flower from spring until autumn are much prized in the garden and this daisy-flowered, shrubby perennial amply fulfils this role. The leaves are silver and deeply lobed while the flowers are white-petalled with a yellow eye. A plant for a sunny border or rock garden with very free-draining soil.

↕10–30cm (4–12in) ↔30cm (12in) ❋ ❋ H4 ☼ ◌

Salvia nemorosa
Wrinkly green leaves form a neutral backdrop to the main attraction of purple, white, or pink flowers during the summer and autumn months. The flower stems stand stiff and upright and, when seen from a low viewpoint, create a sea of colour. Grow in sun or dappled shade in well-drained soil.

↕to 1m (3ft) ↔60cm (24in) ❋ ❋ ❋ H7 ☼ ☼ ◌

Sedum rupestre 'Angelina'
A carpet-forming succulent with golden foliage, it is great for the front of a border, low cover in a gravel garden, on a green roof, or cascading over a bank. Its small cylindrical leaves are at their brightest in sun. Heads of yellow flowers appear in summer. It is easy to grow, with stems rooting as they travel.

↕15cm (6in) ↔1m (3ft) ❋ ❋ ❋ H7 ☼ ☼ ◌

Sedum takesimense ATLANTIS ('Nonsitnal')
This low-growing, ground-covering succulent is good for a sunny site, or a green roof or gravel garden. It is grown for its bright variegated foliage, serrated leaves variably edged with cream, and showy heads of small yellow flowers in summer. It will die back in winter, but pinkish shoots appear in spring.

↕15cm (6in) ↔40cm (16in) ❋ ❋ H4 ☼ ◌

Sedum 'Vera Jameson'
A striking stonecrop to grow for colour impact. Purplish, fleshy leaves and stems sprawl sideways while rounded heads of rose-pink flowers are held aloft in late summer and early autumn. Mix with silvers and greys to accentuate the bold colouring; grow in a rock garden or at a border edge.

↕20–30cm (8–12in) ↔45cm (18in) ❋ ❋ ❋ H7 ☼ ◌

Sempervivum tectorum
The common houseleek creates starry patterns over the ground as its tight red rosettes hug the soil. Grow in old sinks, troughs, or terracotta pots to show off the architectural shapes. Reddish-purple flowers are borne in summer. A gritty, well-drained compost and full sun are desirable.

↕15cm (6in) ↔50cm (20in) ❋ ❋ ❋ H7 ☼ ◌

Veronica gentianoides
This pretty veronica is grown for its spikes of pale blue, early-summer flowers held on erect stems above a mound of glossy, bright green foliage. In hot-hued borders it makes a contrast with reds and oranges, and is also effective when planted in drifts on its own. It performs best in moist soil.

↕45cm (18in) ↔45cm (18in) ❋ ❋ ❋ H7 ☼ ☼ ◌ ◐

EVERGREEN PERENNIALS

- *Adiantum venustum* p.330
- *Asplenium scolopendrium* Crispum Group p.322
- *Astelia chathamica* p.322
- *Beesia calthifolia* p.331
- *Dryopteris erythrosora* p.324
- *Euphorbia myrsinites* p.332
- *Helleborus argutifolius* p.325
- *Helleborus foetidus* p.325
- *Heuchera* 'Plum Pudding' p.333
- *Liriope muscari* p.334
- *Pachysandra terminalis* p.334
- *Phormium cookianum* subsp. *hookeri* 'Tricolor' p.329
- *Phormium tenax* Purpureum Group p.320
- *Polypodium* x *mantoniae* 'Cornubiense' p.334
- *Sedum rupestre* 'Angelina' p.335
- *Sisyrinchium striatum* 'Aunt May' p.329

Bulbs, corms, and tubers

Allium cristophii
Huge, rounded flowerheads made up of many star-like, pinkish-purple blooms ensure this plant's place as a designers' favourite. A scattering of these bulbs among low-growing plants adds unexpected interest in early summer. The dried seedheads are spectacular in indoor arrangements.

↕30–60cm (12–24in) ↔15cm (6in) ❀ ❀ ❀ H5 ☀ ◊

Allium hollandicum 'Purple Sensation'
The deep purple, spherical flowerheads of 'Purple Sensation' look stunning when planted with silver-leaved, shorter plants. This is a summer-flowering bulb that will self-sow around the garden, although the resulting seedlings may not be so richly coloured. The blooms make decorative dried flowers.

↕1m (3ft) ↔7cm (3in) ❀ ❀ ❀ H6 ☀ ◊

Allium stipitatum 'Mount Everest'
Bearing bold, spherical heads of starry, ivory-white flowers, this ornamental onion makes an impressive statement in late spring and early summer. Flowers are held aloft on sturdy green stems, allowing them to be used between other perennials. The strappy foliage is grey-green and begins to wither as flowers peak.

↕1–1.2m (3–4ft) ↔40cm (16in) ❀ ❀ ❀ H5 ☀ ◊

Anemone blanda 'White Splendour'
Quick to establish and form a carpet, this white anemone brings a gleam of light to gardens in spring. For a different colour, try 'Radar', which has magenta flowers with a white eye, or 'Pink Star', with bright pink blooms. All look delightful in large drifts below spring-flowering trees.

↕15cm (6in) ↔15cm (6in) ❀ ❀ ❀ H6 ☀ ☼ ◊

Canna 'Durban'
Vividly coloured foliage and bright, "hot" flowers, which appear from late summer to autumn, make cannas an exotic addition to mixed borders. The deep purple, paddle-shaped leaves sometimes have contrasting midribs. Cannas look very attractive in containers, adding a tropical element to a patio.

↕1.2m (4ft) ↔60cm (24in) ❀ ❀ H3 ☀ ◊ ◊

Canna 'Striata'
A statement plant for a bed or border, 'Striata' has broad, rich green leaves striped with yellow, and showy, bright orange flowers, carried on dark red-purple stems, from midsummer to early autumn. As with most cannas, in cold areas rhizomes should be lifted to overwinter in a frost-free place.

↕1.5m (5ft) ↔50cm (20in) ❀ ❀ H3 ☀ ◊

Convallaria majalis
Lily-of-the-valley is a creeping perennial loved for its sweetly fragrant, white, bell-shaped flowers. Dark green leaves are upward-pointing, with leafless flowerstalks rising among them in late spring. The plant relishes moist, fertile soil in either full or partial shade. All parts are toxic.

↕23cm (9in) ↔30cm (12in) ❀ ❀ ❀ H7 ☀ ☼ ◊

Crinum × powellii
A very decorative plant, this lily produces flared trumpet blooms, up to ten at a time, at the top of rigid stems from late summer until mid-autumn. It suits a position at the base of a sheltered, sunny wall. In cooler areas, provide a deep winter mulch. For a pure white form, choose the cultivar 'Album'.

↕1.5m (5ft) ↔30cm (12in) ❀ ❀ ❀ H5 ☀ ◊ ◊

Crocosmia × crocosmiiflora 'Coleton Fishacre'
In a sunny border, the lemon-yellow trumpets of this South African plant will shine brightly against a background of bronze-tinted, mid-green foliage. Split the clumps every few years for a good supply of flowers. Crocosmias make excellent cut flowers and can be grown solely for this purpose.

↕75–90cm (30–36in) ↔45cm (18in) ❀ ❀ ❀ H5 ☀ ◊ ◊

PLANT GUIDE

❄❄❄ H7–H5 fully hardy ❄❄ H4–H3 hardy in mild regions/sheltered sites ❄ H2 protect from frost over winter
❀ H1c–H1a no tolerance to frost ☼ full sun ☼ partial sun ☀ full shade ◊ well-drained soil ◉ moist soil ● wet soil

Crocosmia x *crocosmiiflora* 'Venus'
The dense green, strappy foliage of this crocosmia is attractive even before the red blooms appear in summer. As each flower opens, a distinctive deep yellow throat is revealed. Overgrown clumps can be split and divided in spring and used to expand your border display.

↕70cm (28in) ↔45cm (18in) ❄❄ H4 ☼ ◊ ◉

Crocosmia 'Firebird'
A strong-growing crocosmia, 'Firebird' has tapering, strap-like dark green foliage, joined in summer by arching stems of showy, bright orange-red tubular flowers with speckled throats. It tolerates drier conditions than many crocosmias, and flowers freely.

↕80cm (32in) ↔30–45cm (12–18in) ❄❄ H4 ☼ ◊ ◉

Crocus goulimyi
This is one of the autumn-flowering crocuses, producing scented, long-tubed, lilac flowers at the same time as the leaves. It can be naturalized in a lawn in drifts, grown around the edges of mixed borders, or planted in containers on a patio (use a gritty potting mix to ensure free drainage).

↕10cm (4in) ↔5cm (2in) ❄❄❄ H6 ☼ ◊

Crocus tommasinianus
Silvery-lilac to purple petals are the distinguishing features of this late winter- to early spring-flowering crocus. Grow in naturalized drifts in grassy areas or in small clumps in terracotta pots on a windowsill. For a white-flowered selection, try *Crocus tommasinianus* f. *albus*.

↕8–10cm (3–4in) ↔2.5cm (1in) ❄❄❄ H6 ☼ ◊

Cyclamen hederifolium
These fluted pink flowers are carried above the soil surface in mid- to late autumn before the appearance of any foliage. The triangular or heart-shaped leaves are dark green with intricate silver patterning. The plant self-seeds freely and suits a site under trees or shrubs in partial shade. Mulch annually.

↕10–13cm (4–5in) ↔15cm (6in) ❄❄❄ H5 ☼ ◊

Dahlia 'Bishop of Llandaff'
The vivid red, semi-double flowers of this dahlia look dramatic against the black-red foliage, making it a striking addition to a mixed border from summer to autumn. 'Bishop of Llandaff' also suits containers. In frost-prone areas, tubers should be lifted after the first frost and stored in a cool, dry place.

↕1.1m (3½ft) ↔45cm (18in) ❄❄ H3 ☼ ◊

Dahlia 'David Howard'
The dark green-purple leaves and stems make an excellent foil for the large, double, burnt orange flowers of this dahlia. Stems can be used for indoor arrangements, and regular cutting will encourage further flowering. Site in a sunny border. See *D*. 'Bishop of Llandaff' for overwintering advice.

↕75cm (30in) ↔60cm (24in) ❄❄ H3 ☼ ◊

Dahlia 'Gay Princess'
Waterlily dahlias are so-called because of the flowerhead form, which is double and resembles a waterlily. This cultivar has lilac-pink blooms in summer and autumn, above rich green foliage. At 1.5m (5ft) tall, it can be planted behind shorter perennials in a border or grown for cut flowers.

↕1.5m (5ft) ↔75cm (30in) ❄❄ H3 ☼ ◊

BULBOUS PLANTS FOR SPRING COLOUR

A range of bulbous plants will provide spring colour, including tulips, daffodils (*Narcissus*), crocuses, snowdrops (*Galanthus*), winter aconites (*Eranthis*), and hellebores.

- *Crocus tommasinianus* p.337
- *Erythronium dens-canis* p.338
- *Fritillaria meleagris* p.338
- *Hyacinthoides non-scripta* p.338
- *Hyacinthus orientalis* 'Blue Jacket' p.338
- *Muscari armeniacum* 'Blue Spike' p.340
- *Muscari latifolium* p.340
- *Narcissus* 'Tête-à-tête' p.340
- *Scilla siberica* p.341
- *Tulipa* 'Flaming Parrot' p.341
- *Tulipa* 'Prinses Irene' p.341
- *Tulipa* 'Queen of Night' p.341

Bulbs, corms, and tubers

Eranthis hyemalis
Buttercup-yellow cup-shaped flowers, surrounded by a collar of deeply cut green leaves, are a welcome sight in the depths of winter. Relatives of buttercups, winter aconites rapidly spread by way of their underground tubers. Plant where the soil does not dry out in summer.

↕5–8cm (2–3in) ↔8cm (3in) ❄ ❄ ❄ H6 ☼ ◊ ♦

Erythronium dens-canis
The European dog's-tooth violet produces heavily marked green leaves and dainty nodding flowers from winter to early spring, in colours ranging from white through to pink. The plant likes well-drained soil in dappled shade, and looks attractive grown underneath deciduous trees or shrubs.

↕10–15cm (4–6in) ↔10cm (4in) ❄ ❄ ❄ H5 ☼ ◊

Eucomis bicolor
The pineapple lily from South Africa needs full sun and rich soil in order to flourish. Maroon-spotted stems appear among the leaves in late summer, bearing pale green flowers with purple markings. It will grow best in a sheltered bed against a warm wall. Mulch dormant bulbs in very hard winters.

↕30–60cm (12–24in) ↔20cm (8in) ❄ ❄ H4 ☼ ◊

Fritillaria imperialis
Tall, stately, and strong-growing, the crown imperial stands regally in the centre of an island bed or within a mixed border or rock garden. Clusters of orange flowers, yellow if you choose the cultivar 'Maxima Lutea', radiate from the top of tall stems in early summer.

↕to 1.5m (5ft) ↔25–30cm (10–12in) ❄ ❄ ❄ H7 ☼ ◊

Fritillaria meleagris
A native of English grasslands, the snake's head fritillary looks stunning when planted en masse in grassy areas, each petal featuring a distinctive chequered pattern. These spring-flowering bulbs in pinkish-purple or white can be mixed to create a patchwork effect.

↕to 30cm (12in) ↔5–8cm (2–3in) ❄ ❄ ❄ H5 ☼ ☼ ◊

Galanthus 'Atkinsii'
The cold season would not be the same without snowdrops, and there are plenty of cultivars to choose from. They flower from late winter and can be planted in grass or in small pots on their own. Lift and divide clumps when the leaves die back. 'Atkinsii' is vigorous, with slender green-marked flowers.

↕20cm (8in) ↔8cm (3in) ❄ ❄ ❄ H5 ☼ ◊ ♦

Galtonia viridiflora
A hyacinth relative from South Africa, galtonia has funnel-shaped, pale green flowers which add glistening highlights to a border. The flowers appear in late summer, suspended from tall arching stems. In very cold areas, lift the bulbs over winter and store in a cool spot indoors.

↕to 1m (3ft) ↔10cm (4in) ❄ H3 ☼ ◊ ♦

Hyacinthoides non-scripta
This is the English bluebell rather than the more upright-growing Spanish species. Plant the bulbs in broad drifts under trees in dappled shade for maximum impact in spring. Flowers are traditionally blue, although pink or white forms can be found. It can become invasive if planted in the border.

↕20–40cm (8–16in) ↔8cm (3in) ❄ ❄ ❄ H6 ☼ ◊ ♦

Hyacinthus orientalis 'Blue Jacket'
Famed for their exquisitely perfumed flowers, hyacinths are very easy to grow. They are available in a range of colours and the bulbs can be planted as spring bedding, singly in pots, or even rooted in water on a windowsill indoors. 'Blue Jacket' has navy-blue, waxy flowers with purple veins.

↕20–30cm (8–12in) ↔8cm (3in) ❄ ❄ ❄ H4 ☼ ◊

✿ ✿ ✿ H7–H5 fully hardy ✿ ✿ H4–H3 hardy in mild regions/sheltered sites ✿ H2 protect from frost over winter
❀ H1c–H1a no tolerance to frost ☼ full sun ☼ partial sun ☀ full shade ◊ well-drained soil ◗ moist soil ◆ wet soil

Iris 'Golden Alps'
This cream and yellow, tall bearded iris should be planted with its lower stem and rhizome just above soil level. Sword-shaped green leaves form a fan, while summer flowers are held high on sturdy stems. Bearded irises come in a range of colours, and all are ideal for a sunny, mixed border.

↕90cm (36in) ↔60cm (24in) ✿ ✿ ✿ H4 ☼ ◊

Iris pallida 'Variegata'
The long, tapering, yellow-striped leaves of this iris surround a succession of showy, scented blue flowers in late spring and early summer. This is a perfect plant for a hot border or exposed site where the sun can bake the soil surface. Lift clumps, divide, and replant in early autumn.

↕to 1.2m (4ft) ↔45–60cm (18–24in) ✿ ✿ ✿ H6 ☼ ◊

Iris 'Superstition'
Purple-brown and blue-black combine here with dramatic effect in this deeply coloured, tall bearded iris. Plant with pale-coloured selections such as 'White Knight' to create a contrasting combination. The dark flowers are also fragrant, and appear almost black in fading light.

↕90cm (36in) ↔60cm (24in) ✿ ✿ ✿ H7 ☼ ◊

Leucojum aestivum 'Gravetye Giant'
Similar to a large snowdrop, the summer snowflake is an attractive plant for damp areas of the garden. Nodding white flowers with green petal tips emerge in spring; the narrow green leaves providing a subtle backdrop. 'Gravetye Giant' is robust and will grow quite tall next to water.

↕90cm (36in) ↔8cm (3in) ✿ ✿ ✿ H7 ☼ ◗ ◆

Lilium 'African Queen' (African Queen Group)
Place some pots of these by your back door and you will be greeted by deliciously fragrant, bright orange trumpet flowers every time you step outside from mid- to late summer. This lily can also be grown in a border, if the flowers are in the sun while the roots are kept shaded.

↕1.5–2m (5–6ft) ↔25cm (10in) ✿ ✿ ✿ H6 ☼ ◊

Lilium 'Black Beauty'
Lilies with this flower form are known as turk's caps because of the way the petals curve back on themselves, revealing pollen-laden anthers. 'Black Beauty' is a vigorous type and can be positioned among herbaceous plants in the border, or grown in containers for a movable midsummer display.

↕1.4–2m (4½–6ft) ↔25cm (10in) ✿ ✿ ✿ H6 ☼ ◊

Lilium henryi
Perhaps the finest late-summer-flowering lily for garden use, this handsome bulb develops tall but sturdy stems with lance-shaped foliage, topped late in the season by clusters of large orange flowers, each marked with small red spots. The petals are recurved in a show that lasts about two weeks.

↕2m (6ft) or more ↔50cm (20in) ✿ ✿ ✿ H6 ☼ ☼ ◊

Lilium martagon
Scatter bulbs of the common turk's-cap lily around a mixed border and plant them where they land. The pretty flowers, which have recurved purple petals with dark markings, appear from early to midsummer. The flowers of *Lilium martagon* var. *album* are pure white.

↕0.9–2m (3–6ft) ↔20cm (8in) ✿ ✿ ✿ H6 ☼ ☼ ◊

BULBOUS PLANTS FOR SUMMER COLOUR

- *Allium cristophii* p.336
- *Allium hollandicum* 'Purple Sensation' p.336
- *Allium stipitatum* 'Mount Everest' p.336
- *Canna* 'Durban' p.336
- *Canna* 'Striata' p.336
- *Crinum* x *powellii* p.336
- *Crocosmia* x *crocosmiiflora* 'Coleton Fishacre' p.336
- *Crocosmia* x *crocosmiiflora* 'Venus' p.337
- *Crocosmia* 'Firebird' p.337
- *Dahlia* selections p.337
- *Fritillaria imperialis* p.338
- *Iris* 'Golden Alps' p.339
- *Iris pallida* 'Variegata' p.339
- *Lilium* African Queen Group 'African Queen' p.339
- *Lilium* 'Black Beauty' p.339
- *Lilium henryi* p.339
- *Lilium martagon* p.339
- *Lilium* Pink Perfection Group p.340
- *Lilium* 'Star Gazer' p.340

Bulbs, corms, and tubers

Lilium Pink Perfection Group
First introduced in 1950, the large, pinkish-red trumpets of this lily hybrid soon caught the attention of keen gardeners. In midsummer, short flower stems are laden with lightly-scented blooms with protruding orange anthers. Choose a sunny site with some shade for the roots for best results.

↕1.5–2m (5–6ft) ↔25cm (10in) ❋ ❋ ❋ H6 ☼ ◊

Lilium regale
The large, white, trumpet-shaped flowers of the regal lily are purple on the outside and held in clusters on tall stems, creating an eye-catching display in midsummer. The lilies are very fragrant and are ideal for use in mixed borders or as cut flowers. The stems may need staking.

↕0.6–2m (2–6ft) ↔25cm (10in) ❋ ❋ ❋ H6 ☼ ◊

Lilium 'Star Gazer'
Both the colour and the perfume of 'Star Gazer' attract attention and make this Asian lily one of the most popular cut flowers ever developed. The pink and white flowers with speckled petals are upward-facing and robust, and appear in midsummer. Plant in a border or in a stylish container.

↕1–1.5m (3–5ft) ↔25cm (10in) ❋ ❋ ❋ H6 ☼ ◊

Muscari armeniacum 'Blue Spike'
This is a double-flowered form of the common grape hyacinth. Fleshy green narrow leaves form a carpet as small fat spikes of blue flowers push their way through in spring. The plant can become invasive, so restrict its spread by growing it in a container. Choose a site in full sun.

↕20cm (8in) ↔5cm (2in) ❋ ❋ ❋ H6 ☼ ◊ ◊

Muscari latifolium
The flowers of this grape hyacinth seem to be wearing little hats. Blue flowerspikes are topped by small, paler-coloured flowers, while the leaves are mid-green and more flattened than those of *Muscari armeniacum* (*left*). Attractive in drifts at the front of a border, it is also good for a rock garden.

↕20cm (8in) ↔5cm (2in) ❋ ❋ ❋ H6 ☼ ◊ ◊

Narcissus 'Bridal Crown'
'Bridal Crown' has sweetly-scented double white blooms with pale orange centres. The flowers cluster together at the top of the stems and appear in early spring. Plant bulbs during autumn in well-drained soil in a sunny border, or in a container. 'Bridal Crown' makes a pretty cut flower.

↕40cm (16in) ↔15cm (6in) ❋ ❋ ❋ H6 ☼ ◊

Narcissus poeticus var. recurvus
Known as the old pheasant's eye, this late spring-flowering daffodil differs from *Narcissus poeticus* in having backward-curving petals. Pure white petals surround a yellow eye, which has a dainty orange frilled edge. It can be naturalized in a lawn, and is also good for cut flowers for the house.

↕35cm (14in) ↔5–8cm (2–3in) ❋ ❋ ❋ H6 ☼ ◊

Narcissus 'Tête-à-Tête'
Tiny flowers on short stems make this a favourite spring bulb for planting at the front of borders, in rock gardens, and in containers of all shapes and sizes. Plant en masse for the best effect, as small clumps can look insignificant. Container-grown plants can be grown on a windowsill indoors.

↕15cm (6in) ↔5cm (2in) ❋ ❋ ❋ H6 ☼ ◊

Narcissus 'Thalia'
This delicately beautiful daffodil carries two milky-white flowers per stem. Mid-spring sees these emerge from papery buds to lighten border plantings or provide early interest in a "white" border. Grow in a tall container and place against a painted wall to make a bold statement.

↕35cm (14in) ↔8cm (3in) ❋ ❋ ❋ H6 ☼ ◊

PLANT GUIDE 341

✽✽✽ H7–H5 fully hardy ✽✽ H4–H3 hardy in mild regions/sheltered sites ✽ H2 protect from frost over winter
❀ H1c–H1a no tolerance to frost ☼ full sun ☼ partial sun ☀ full shade ◊ well-drained soil ◊ moist soil ● wet soil

Nectaroscordum siculum subsp. *bulgaricum*
The flowers on this onion relative are green, white, and burgundy. Grouped in sprays of 10–30 on top of tall stems, they make an attractive display in early summer. Grow in a wild garden or herbaceous border where the flowers will catch the eye. Deadhead to prevent it spreading.

↕ to 1.2m (4ft) ↔ 30–45cm (12–18in) ✽✽✽ H5 ☼ ◊

Nerine bowdenii
South Africa has given gardeners worldwide many wonderful plants and this spectacular bulb is no exception. Stems of vivid pink, spidery flowers appear from bare soil in autumn. Nerines look good in groups at the foot of a sunny, light-coloured wall. Provide a deep mulch in winter in very cold areas.

↕ 45cm (18in) ↔ 8–12cm (3–5in) ✽✽✽ H5 ☼ ◊

Scilla siberica
The Siberian squill produces bright blue, pendent flowers in spring, giving the garden a dash of colour. The bulbs can be grown in groups in a rock garden, between paving stones or at the front of herbaceous and mixed borders. Plant in full sun or part shade, and water well when in growth.

↕ 10–20cm (4–8in) ↔ 5cm (2in) ✽✽✽ H6 ☼ ☼ ◊

Trillium grandiflorum
A vigorous plant for a shady woodland area, wake robin forms clumps of dark green, rounded leaves with distinctive, three-petalled white flowers in spring and summer. The cultivar 'Flore Pleno' is slower-growing and has double flowers.

↕ to 40cm (16in) ↔ 30cm (12in) or more
✽✽✽ H5 ☼ ◊

Tulipa 'Flaming Parrot'
This late spring-flowering tulip has fringed yellow petals, each with a distinctive red blaze. Inside is a cluster of black anthers. Grow as a single variety in formal beds or in drifts, merging with other colours. Alternatively, plant a number of the bulbs in a tall pot or container in a sunny position.

↕ 55cm (22in) ↔ 15cm (6in) ✽✽✽ H6 ☼ ◊

Tulipa 'Prinses Irene'
The orange petals of this striking tulip look like they have been painted with delicate brush strokes of purple. Flowering in mid-spring, 'Prinses Irene' is effective when grouped in swathes in a border or planted as part of a container display with decorative grasses. It can also be cut for indoor arrangements.

↕ 35cm (14in) ↔ 15cm (6in) ✽✽✽ H6 ☼ ◊

Tulipa 'Queen of Night'
Popular because it is so deeply coloured and satiny, this late spring-flowering tulip looks striking if planted among purple and black-leaved perennials and low shrubs, or with grey or silver-leaved plants. Alternatively, use it in front of a pale-painted fence or wall for contrast.

↕ 60cm (24in) ↔ 15cm (6in) ✽✽✽ H6 ☼ ◊

Tulipa 'Spring Green'
This Viridiflora Group tulip sports a green feathery flash on each of its ivory-white petals and adds an elegant touch to a mixed or colour-themed border. Plant where it can be appreciated at close quarters, as it is only 40cm (16in) high when flowering in late spring.

↕ 40cm (16in) ↔ 10cm (4in) ✽✽✽ H6 ☼ ◊

BULBOUS PLANTS FOR SCENT

Plant groups with a range of scented cultivars include many daffodils (*Narcissus*), crocuses, lilies, some snowdrops (*Galanthus*), *Leucojum* (snowflake), hyacinths, cyclamen, and freesias.

- *Convallaria majalis* p.336
- *Crocus goulimyi* p.337
- *Hyacinthus orientalis* 'Blue Jacket' p.338
- *Leucojum aestivum* 'Gravetye Giant' p.339 (light scent)
- *Lilium* African Queen Group 'African Queen' p.339
- *Lilium* 'Black Beauty' p.339
- *Lilium* Pink Perfection Group p.340
- *Lilium regale* p.340
- *Lilium* 'Star Gazer' p.340
- *Narcissus* 'Bridal Crown' p.340
- *Narcissus poeticus* var. *recurvus* p.340

Grow bulbs in pots by the house or in drifts for maximum appreciation.

Grasses, sedges, and bamboos

Acorus calamus 'Argenteostriatus'
An undemanding evergreen, the sweet rush, or sweet flag, thrives in damp or boggy soils, making it the perfect plant for the shallows of a pond edge. Like all acorus, it is non-invasive, and its strong cream variegation will remain vivid, even in deep shade.

↕45cm (18in) ↔45cm (18in) ❀ ❀ ❀ H7 ☼ ☀ ◊ ●

Anemanthele lessoniana
Fine-leaved pheasant's tail grass has a pleasing arching habit. In summer, it produces purplish flower spikes; in winter, the evergreen leaves turn an eye-catching orange-brown. Leave the seedheads – hungry birds will quickly tidy them up during winter. The plant may need protection in cold areas.

↕1m (3ft) ↔1.2m (4ft) ❀ ❀ H4 ☼ ☀ ◊ ●

Arundo donax var. versicolor
The striking variegation of the evergreen giant reed (the white stripes turn a creamy yellow in summer) makes it a popular choice, although it is less vigorous than the green form and not as hardy. In cold areas, enjoy it outdoors in summer, then bring it under cover for the winter; grow it in a pot for flexibility.

↕2.2m (7ft) ↔2m (6ft) ❀ ❀ H4 ☼ ☀ ◊ ●

Borinda papyrifera
This is an impressive clump-forming, evergreen bamboo with remarkable powder-blue stems, at their brightest on younger growth. The leaves are soft and rich green, the growth upright then arching, forming in time an elegant clump. It must have moisture as new growth develops.

↕6m (20ft) ↔2–3m (6–10ft) ❀ ❀ ❀ H5 ☼ ☀ ◊

Briza maxima
One of the most attractive of the annual grasses, quaking grass is easy to grow from seed (sow into individual modules for the best results). The nodding flowerheads rattle in the lightest breeze, making it clear how the common name arose. The stems dry well for flower arranging.

↕30cm (1ft) ↔23cm (9in) ❀ ❀ ❀ H6 ☼ ◊ ●

Calamagrostis x acutiflora 'Overdam'
Use the striped feather reed to make a strong vertical accent in prairie-style planting. As the leaves emerge in spring, there is a pink tinge to the green and white variegation; cutting the foliage back in late summer will encourage a second flush of new growth. Unfussy, the plant tolerates most soils.

↕1m (3ft) ↔1.2m (4ft) ❀ ❀ ❀ H6 ☼ ◊

Carex buchananii
This striking evergreen sedge from New Zealand has slender, coppery-brown leaves with a hint of a curl. It is stiffly upright when young, becoming more arching with age, and it contrasts well with golden sedges and blue grasses. In early spring, comb out any dead leaves with a fork, or cut them back.

↕60cm (2ft) ↔60cm (2ft) ❀ ❀ ❀ H5 ☼ ☀ ◊ ●

Carex elata 'Aurea'
Deservedly one of the most widely grown sedges, Bowles' golden sedge produces a broad spray of vibrant yellow leaves, edged in green. In summer there is the added bonus of feathery brown flower spikes. A compact, deciduous plant, it produces its best colour in partial shade.

↕75cm (30in) ↔1m (3ft) ❀ ❀ ❀ H6 ☼ ☀ ◊ ●

Carex oshimensis 'Evergold'
The low-arching habit of this neat evergreen sedge makes it a useful plant for containers or as ground cover in shade, where its long golden yellow and thinly striped green leaves add a touch of light colour. Like many sedges, it is happy in boggy soil and makes a decorative addition to poolside plantings.

↕50cm (20in) ↔45cm (18in) ❀ ❀ ❀ H7 ☼ ☀ ◊ ●

PLANT GUIDE

❀❀❀ H7–H5 fully hardy ❀❀ H4–H3 hardy in mild regions/sheltered sites ❀ H2 protect from frost over winter
❀ H1c–H1a no tolerance to frost ☼ full sun ☼ partial sun ☀ full shade ◊ well-drained soil ◊ moist soil ◆ wet soil

Cortaderia selloana 'Aureolineata'
Ideal for small gardens, this dwarf pampas is half the size of the parent species, and has broad leaves with golden edges that become more richly coloured as the season progresses. The colourful leaves and silky plume-like flowerheads add a dramatic highlight to late summer borders and gravel gardens.

↕1.5m (5ft) ↔1.5m (5ft) ❀❀❀ H6 ☼ ◊ ◊

Cortaderia selloana 'Pumila'
Hardier and more free-flowering than the taller species, this dwarf pampas grass mixes surprisingly well in a border. Long-lasting golden-brown plumes are produced in summer on stout stems. Combing through the leaves with a hand fork in winter will keep the clump looking tidy.

↕2m (6ft) ↔2m (6ft) ❀❀❀ H6 ☼ ◊ ◊

Deschampsia flexuosa 'Tatra Gold'
Wavy hair grass forms slowly spreading tufts of fine evergreen leaves. 'Tatra Gold' grows well in moist shade, where its acid-green leaves look almost luminous. In summer, it produces a shimmering haze of red-brown flowers. Plant it in large drifts among bright leaved sedges for a dramatic effect.

↕15cm (6in) ↔15cm (6in) ❀❀❀ H6 ☼ ☀ ◊

Elymus magellanicus
Blue wheatgrass is so-named because of its wonderful blue colour – it looks stunning against a gravel mulch – and the herringbone flowerheads that look like ears of wheat. It forms slow-spreading, rather sprawling clumps of evergreen leaves that need winter protection in cold areas.

↕45cm (18in) ↔45cm (18in) ❀❀❀ H6 ☼ ◊ ◊

Fargesia murielae
A tough plant for tough situations, this evergreen bamboo copes well with dry soils and exposed sites, and makes an effective windbreak or screen. The closely spaced, arching canes are slow-spreading, and it won't engulf its neighbours. Use it at the back of a border or in a container.

↕4m (12ft) ↔4m (12ft) ❀❀❀ H5 ☼ ☀ ◊ ◊

Festuca glauca 'Elijah Blue'
One of those useful plants that look good all year round, the silvery-blue, needle-like leaves of this fescue form neat, round mounds. In summer, the plant produces spikes of small blue flowers that age to brown. It is particularly effective grown as a container plant, contrasting well with terracotta and metal.

↕30cm (12in) ↔60cm (24in) ❀❀❀ H5 ☼ ☀ ◊ ◊

Hakonechloa macra 'Aureola'
A beautiful slow-growing, deciduous grass from Japan that deserves to be the centrepiece in a container or a dry gravel border. The low-arching, golden yellow leaves, which are thinly striped with lime green, develop a warm reddish tinge in autumn. Cut back in early spring to encourage new growth.

↕25cm (10in) ↔1m (3ft) ❀❀❀ H7 ☼ ☀ ◊

Imperata cylindrica 'Rubra'
Japanese blood grass is undisputedly one of the finest foliage plants – fluffy white flowerspikes are a bonus in summer. Position it carefully, so the crimson-tipped, upright leaves are backlit by the sun. In cold areas, grow it in a container and bring under cover during winter.

↕45cm (18in) ↔1.8m (6ft) ❀❀ H4 ☼ ☀ ◊

GRASSES, SEDGES, AND BAMBOOS FOR CONTAINERS

- *Acorus calamus* 'Argenteostriatus' p.342
- *Arundo donax* var. *versicolor* p.342
- *Carex buchananii* p.342
- *Carex oshimensis* 'Evergold' p.342
- *Deschampsia flexuosa* 'Tatra Gold' p.343
- *Elymus magellanicus* p.343
- *Fargesia murielae* p.343
- *Festuca glauca* 'Elijah Blue' p.343
- *Hakonechloa macra* 'Aureola' p.343
- *Imperata cylindrica* 'Rubra' p.343
- *Miscanthus sinensis* cultivars p.344
- *Ophiopogon planiscapus* 'Nigrescens' p.344
- *Phyllostachys nigra* p.345
- *Phyllostachys vivax* f. *aureocaulis* p.345
- *Uncinia rubra* p.345

Grasses, sedges, and bamboos

Lagurus ovatus

A popular garden plant because of its fluffy flowerheads, the hare's-tail grass is a tufted annual that can be grown easily from seed sown *in situ* in spring. The soft, hairy spikelets, pale green at first, maturing to pale cream, form in summer and can be cut for indoor displays.

↕to 50cm (20in) ↔30cm (12in) ❋ ❋ H4 ☼ ◊

Miscanthus sinensis 'Gracillimus'

A dainty-looking subject for a grass garden or mixed border, 'Gracillimus' produces a shock of narrow green leaves with white midribs. After the late summer flush, the curved leaves take on a bronzy hue as temperatures cool. Leave in place as a structural element through the winter.

↕1.3m (4½ft) ↔1.2m (4ft) ❋ ❋ ❋ H6 ☼ ◊ ♦

Miscanthus sinensis 'Kleine Silberspinne'

An attractive ornamental grass with colourful, curving plumes, this miscanthus does not grow as tall as the species. In late summer and early autumn, silky white and red flower spikes appear, turning to silver as they age and lasting all winter. Cut down to ground level in spring before new growth emerges.

↕1.2m (4ft) ↔1.2m (4ft) ❋ ❋ ❋ H6 ☼ ◊ ♦

Miscanthus sinensis 'Malepartus'

One of the easiest of the miscanthus to establish, 'Malepartus' looks good spilling onto a lawn or path edge where it can be seen at close quarters. Feathery reddish-brown flowerheads, maturing to cream, appear from late summer to autumn among the cascading green foliage.

↕2m (6ft) ↔2m (6ft) ❋ ❋ ❋ H6 ☼ ◊ ♦

Miscanthus sinensis 'Silberfeder'

This cultivar is grown mainly for its autumn show of red-tinged, creamy flowers that last well and are held above narrow, green foliage. 'Silberfeder' needs space to be seen at its best and a site that doesn't get waterlogged. Plant in front of a dark-leaved hedge for a perfect backdrop.

↕2.5m (8ft) ↔1.2m (4ft) ❋ ❋ ❋ H6 ☼ ◊ ♦

Miscanthus sinensis 'Zebrinus'

Easily confused with the more upright-growing *M. sinensis* 'Strictus', 'Zebrinus' has a more lax habit and spreads more readily. The unusual horizontal bands of pale cream variegation make it an interesting subject for a grass garden or large zinc planter. The brown deciduous foliage offers winter interest.

↕to 1.2m (4ft) ↔1.2m (4ft) ❋ ❋ ❋ H6 ☼ ◊ ♦

Molinia caerulea subsp. *caerulea* 'Variegata'

This is a densely tufted perennial with boldly variegated green and cream leaves. From spring through to autumn, purple-tinged flowers are borne on yellow flower stems. The whole plant matures to a pale bronzy-brown in autumn, an effect that looks striking in a gravel garden.

↕45–60cm (18–24in) ↔40cm (16in) ❋ ❋ ❋ H7 ☼ ☼ ◊

Ophiopogon planiscapus 'Nigrescens'

Few plants are as deeply coloured as this clump-forming, tufted perennial. Although not strictly a grass, its appearance and habit make it a useful plant in schemes where grasses predominate. It also looks dramatic in pale-coloured containers. Small, pale purplish-white flowers appear in summer.

↕20cm (8in) ↔30cm (12in) ❋ ❋ ❋ H5 ☼ ☼ ◊ ♦

Panicum virgatum 'Heavy Metal'

A deciduous perennial grass with stiff, upright, steely grey-green leaves. In favourable conditions, the foliage will turn yellow in autumn, gradually fading to pale brown in winter. Wispy flowerheads bearing purple-green flowers emerge during summer. Plant in clumps of threes or fives for impact.

↕1m (3ft) ↔75cm (30in) ❋ ❋ ❋ H5 ☼ ◊

PLANT GUIDE

❄❄❄ H7–H5 fully hardy ❄❄ H4–H3 hardy in mild regions/sheltered sites ❄ H2 protect from frost over winter
❄ H1c–H1a no tolerance to frost ☼ full sun ◐ partial sun ● full shade ◊ well-drained soil ◊ moist soil ● wet soil

Pennisetum alopecuroides
Also known appropriately as the fountain grass, this evergreen perennial has narrow, mid-green leaves that tumble from the centre of the plant, joined in summer and autumn by flowing, bristly, decorative flowerheads. It needs a warm, sheltered site since it is not fully hardy.

↕ 0.6–1.5m (2–5ft) ↔ 0.6–1.2m (2–4ft) ❄❄ H3 ☼ ◊

Phyllostachys nigra
The black bamboo is grown for its distinctive stems, which are initially green and then turn glossy black, contrasting well with the fresh green leaves. It has a tall, upright habit, so grow for impact in a border, or in blocks in a modernist scheme.

↕ 3–5m (10–15ft) ↔ 2–3m (6–10ft)
❄❄❄ H5 ☼ ◐ ◊ ●

Phyllostachys vivax f. aureocaulis
Like many bamboos, this is a vigorous, fast-growing plant. The bright yellow canes are flecked with green and it has slim, arching foliage. Plant it in a large container, or surround the plant's roots below soil level with an impenetrable barrier to control its spread.

↕ to 8m (25ft) ↔ 4m (12ft) ❄❄❄ H5 ☼ ◐ ◊ ●

Sesleria autumnalis
This mound-forming evergreen grass is most impressive in mid- to late summer, when appealing slender spikes of off-white flowers appear. The foliage is a bright, fresh green, and the colour holds well over winter. Useful as underplanting for small trees or open-growing deciduous shrubs.

↕ 50cm (20in) ↔ 50cm (20in) ❄❄❄ H7 ☼ ◐ ●

Stipa gigantea
Giant feather grass is a fabulous plant for the garden, commanding a prime position in an island bed or mixed border in full sun. Tall, fluttering plumes of flowers emerge above the evergreen foliage in summer; the stems create a transparent screen, allowing shorter plants to be seen behind them.

↕ to 2.5m (8ft) ↔ 1.2m (4ft) ❄❄❄ (borderline) H4 ☼ ◊

Stipa ichu
A super-elegant grass at its best in late summer and early autumn, when it produces upright then arching stems topped by long, slender, feathery silver flowerheads above clumps of fine, bright green foliage. Flowers are followed by buff seedheads of equal beauty. A statuesque stunner, ideal in a gravel garden.

↕ 1m (3ft) ↔ 60cm (24in) ❄❄ H4 ☼ ◊

Stipa tenuissima
In summer, this neat, compact, deciduous perennial produces soft feathery stems with green flowerheads that fade to buff. The fine leaves gently wave in the slightest breeze, and contrast well with dark green foliage plants. The autumn seedheads are very attractive to birds.

↕ 60cm (24in) ↔ 30cm (12in) ❄❄ (borderline) H4 ☼ ◊

Uncinia rubra
The tough ochre-red leaves of this evergreen perennial are three-angled and upright, joined in mid- and late summer by dark brown flowers. It makes an unusual specimen for a gravel or scree garden where the soil is free-draining but not too dry. Protect from the elements in very cold winters.

↕ 30cm (12in) ↔ 35cm (14in) ❄❄ H3 ☼ ◐ ◊ ●

EVERGREEN GRASSES, SEDGES, AND BAMBOOS

- *Acorus calamus* 'Argenteostriatus' p.342
- *Borinda papyrifera* p.342
- *Carex buchananii* p.342
- *Carex oshimensis* 'Evergold' p.342
- *Deschampsia flexuosa* 'Tatra Gold' p.343
- *Fargesia murielae* p.343
- *Festuca glauca* 'Elijah Blue' p.343
- *Ophiopogon planiscapus* 'Nigrescens' p.344
- *Phyllostachys nigra* p.345
- *Phyllostachys vivax f. aureocaulis* p.345
- *Sesleria autumnalis* p.345
- *Stipa gigantea* p.345
- *Uncinia rubra* p.345

Water and bog plants

Actaea simplex Atropurpurea Group 'Brunette'
A herbaceous perennial for a damp, shady area in the garden, 'Brunette' has bronze, deeply cut foliage and slender spires of fluffy, fragrant white flowers in late summer, which show up well against a dark background. Plant in moisture-retentive soil in a woodland or shady bog garden.

↕1.2m (4ft) ↔60cm (24in) ❄ ❄ ❄ H7 ☼ ☀ ~

Aruncus dioicus 'Kneiffii'
Fern-like foliage and tumbling flowerheads resembling small white caterpillars combine to create this striking plant. The flowers appear in summer and make a bright focal point in a bog garden or at a pond edge. It looks delicate, but is in fact robust and will tolerate full sun or part shade.

↕75cm (30in) ↔45cm (18in) ❄ ❄ ❄ H6 ☼ ☀ ~

Astilbe 'Fanal'
Producing feathery plumes of long-lasting, crimson flowers in early summer, 'Fanal' adds fiery interest to a garden with boggy soil. Finely cut, dark green leaves provide a suitable backdrop for the intense flower colour. Plant in groups of threes or fives to make a bold statement.

↕60–100cm (2–3ft) ↔60cm (24in) ❄ ❄ ❄ H7 ☼ ☀ ~

Astilbe 'Professor van der Wielen'
A plant that needs space to show off its full potential, this astilbe produces large, arching sprays of delicate creamy-white flowers in midsummer above fern-like foliage. Place at the back of a wet border or pond-edge planting scheme, and divide clumps every three to four years.

↕1.2m (4ft) ↔to 1m (3ft) ❄ ❄ ❄ H7 ☼ ~

Astilbe 'Willie Buchanan'
This astilbe cultivar produces a haze of pink when its tiny white flowers with red stamens, borne on fine, branching flower stems, open from mid- to late summer. Ideal for a pond or path edge, plant en masse for a wonderful floral display. The flowers attract beneficial insects.

↕23–30cm (9–12in) ↔20cm (8in) ❄ ❄ ❄ H7 ☼ ☀ ~

Butomus umbellatus
The flowering rush is a deservedly popular plant for pond margins, where it can immerse its feet in wet soil. The leaves are narrow and angled, bronze-purple when young, turning to mid-green. In late summer, delicate, pale pink, fragrant flowers are borne on slender stems.

↕1m (3ft) ↔unlimited ❄ ❄ ❄ H5 ☼ ☀ ≈⊥ 5–15cm (2–6in)

Caltha palustris
Marsh marigolds bring colour to pond margins as their intense yellow, cup-shaped blooms appear in late spring. Grow in planting baskets to control their spread. Try *C. palustris* var. *alba* for white flowers.

↕60cm (24in) ↔45cm (18in) ❄ ❄ ❄ H7 ☼ ☀
≈⊥ at water level

Darmera peltata
The umbrella plant is a slow-spreading perennial that looks good alongside streams and pond margins. Heads of white to pink flowers appear in late spring on long stems before the large, rounded green leaves appear. The foliage gradually turns red in autumn before dying down.

↕1.2m (4ft) ↔unlimited ❄ ❄ ❄ H6 ☼ ☀ ~

Eupatorium maculatum 'Atropurpureum Group'
A great plant for late summer and early autumn colour, this stately perennial bears clusters of small pink flowers on tall, purple stems. Toothed, purple-green leaves circle the stems right up to the flowerheads. It attracts bees and butterflies, and makes a superb addition to a wildlife bog garden.

↕2m (6ft) ↔1m (3ft) ❄ ❄ ❄ H7 ☼ ☀ ~

PLANT GUIDE **347**

❄❄❄ H7–H5 fully hardy ❄❄ H4–H3 hardy in mild regions/sheltered sites ❄ H2 protect from frost over winter
❄ H1c–H1a no tolerance to frost   ☼ full sun   ☼ partial sun   ☀ full shade   ∼ bog plant   ≈ marginal plant   ≋ aquatic plant   ⊥ planting depth

Filipendula rubra 'Venusta'
Meadowsweet needs space to spread, so choose a planting position for this perennial carefully. Green jagged leaves sit below wiry stems bearing a frothy display of deep rose-pink flowers in early and midsummer. Use its height to form a screen at the back of a bog garden display.

↕ 2m (6ft) ↔ unlimited ❄❄❄ H5 ☼ ☼ ∼

Gunnera manicata
A real giant of the bog garden with huge, rhubarb-like leaves, gunnera demands plenty of room, even for just one plant. A herbaceous perennial, it makes a dramatic statement at the waterside. Plant in permanently moist soil and cover the crowns with a dry mulch in hard winters.

↕ 4.5m (15ft) ↔ 3m (10ft) ❄❄❄ H5 ☼ ☼ ∼

Iris 'Butter and Sugar'
Bred from the Siberian iris, 'Butter and Sugar' bears shapely flowers with white upper petals and butter-yellow lower petals from mid- to late spring. Each stem is surrounded by green strappy foliage and can hold up to five blooms. Divide the tight clumps in spring or once flowers have faded.

↕ 50cm (20in) ↔ 25cm (10in) ❄❄❄ H7 ☼ ∼

Iris laevigata
This iris flourishes reliably in the wet soil in the shallows of ponds and streams. Blue-purple flowers crown green stems in early and midsummer, and sit among broad, sword-shaped, mid-green leaves. Clumps will spread steadily.

↕ 75cm (30in) ↔ 1m (3ft) ❄❄❄ H6 ☼
≈ ⊥ 10–15cm (4–6in)

Iris pseudacorus 'Variegata'
This is the variegated-leaved version of the well-known yellow flag iris. Pale yellow stripes decorate the green, upright leaves when young; the yellow blooms appear in summer. A spreading iris, it needs restricting if it is not to become invasive. Plant in a basket at the margins of a pond.

↕ 1m (3ft) ↔ 75cm (30in) ❄❄❄ H7 ☼ ≈⊥ 15cm (6in)

Iris sibirica 'Perry's Blue'
This is a traditional cultivar producing closely spaced flower stems that carry mid-blue flowers with rusty-coloured veins. It flowers in early summer and will bring colour to the edges of small ponds and borders with boggy soil. Plant with lighter-flowered irises for duo-tone effect.

↕ 1m (3ft) ↔ 60cm (24in) ❄❄❄ H7 ☼ ∼

Iris versicolor 'Kermesina'
From eastern North America, the blue flag is a small iris for small ponds. In summer, the species has lavender-blue flowers with white markings, while 'Kermesina' bears red-purple blooms. The long, strappy leaves add architectural interest to a pond margin from spring until autumn when they die down.

↕ 75cm (30in) ↔ 60cm (24in) ❄❄❄ H7 ☼ ≈⊥ 5cm (2in)

Kirengeshoma palmata
An unusual plant for the bog garden, this clump-forming perennial has jagged green leaves with reddish-purple stems. Pale yellow, bell-shaped flowers hang from the slim stems above the foliage in late summer and early autumn. Plant in moist acid soil in a part-shaded sheltered site.

↕ 1.2m (4ft) ↔ 75cm (30in) ❄❄❄ H7 ☼ ∼

PLANTS FOR YOUR POND

AQUATIC ≋
- *Nymphaea* 'Darwin' p.348
- *Nymphaea* 'Froebelii' p.348
- *Nymphaea* 'Gonnère' p.348
- *Nymphaea* 'Marliacea Chromatella' p.348

MARGINAL ≈
- *Butomus umbellatus* p.346
- *Caltha palustris* p.346
- *Iris laevigata* p.347
- *Iris pseudacorus* 'Variegata' p.347
- *Iris versicolor* 'Kermesina' p.347
- *Myosotis scorpioides* p.348
- *Orontium* species p.214
- *Sagittaria sagittifolia* p.214
- *Saururus* species p.214
- *Thalia dealbata* p.349
- *Typha minima* p.349
- *Zantedeschia aethiopica* p.349

OXYGENATING PLANT
- *Ranunculus aquatilis* p.214

Water and bog plants

Ligularia 'The Rocket'
A plant of contrasts with jet black flower stems and bright yellow flowers, this bog lover is a must for larger gardens. The leaves form a carpet through which the flower spikes emerge from early to late summer. Choose a bright site but one that is shaded from the midday sun.

↕2m (6ft) ↔1.1m (3½ft) ❋ ❋ ❋ H6 ☼ ◐ ～

Matteuccia struthiopteris
The common names of shuttlecock fern and ostrich fern can be easily understood when the enormous finely dissected fronds emerge from the ground in spring. During late summer, fertile, narrow brown fronds cluster at the centre of the plant and last through winter. Grow in moist shade.

↕1.7m (5½ft) ↔to 1m (3ft) ❋ ❋ ❋ H5 ◐ ～

Myosotis scorpioides
Plant the water forget-me-not close to a pond edge, where its flowers can be seen clearly. The tiny blue blooms have white, pink, or yellow eyes and appear in early summer. The cultivar 'Mermaid' has a more compact habit.

↕45cm (18in) ↔unlimited ❋ ❋ ❋ H6 ☼ ◐
≈ ⊥ at water level

Nymphaea 'Darwin'
The almost peony-like, fragrant flowers of this waterlily are pale pink in the centre while the outermost petals are white with a tinge of pink. With its large, flat, dark green leaves and vigorous growth, 'Darwin' (also sold as Hollandia) is best suited to medium-sized to large ponds.

↔1.5m (5ft) ❋ ❋ ❋ H3 ☼ ≈⊥ 60–100cm (2–3ft)

Nymphaea 'Froebelii'
Tiny burgundy-red flowers with golden stamens open between the dark green leaves (bronze when young) Of 'Froebelii' to make a perfect miniature water lily. Ideal for small ponds, tubs, or half-barrels, it will put on a beautiful flower display from midsummer to autumn.

↔75cm (30in) ❋ ❋ ❋ H5 ☼ ≈⊥ 30–45cm (12–18in)

Nymphaea 'Gonnère'
A stunning water lily for medium-sized ponds, 'Gonnère' sends up pure white fragrant flowers with yellow stamens from mid- to late summer. The circular lily pads are bronze when young but soon turn a light pea-green. Grow in full sun for the best results.

↔1.5m (5ft) ❋ ❋ ❋ H5 ☼ ≈⊥ 60–75cm (24–30in)

Nymphaea 'Marliacea Chromatella'
This is a very old cultivar that has stood the test of time. Lemon-yellow flowers, with broad incurved petals and deep yellow centres, are produced from mid- to late summer and appear between floating olive-green leaves with bronze markings. Plant in a medium-sized pond or pool in full sun.

↔1.5m (5ft) ❋ ❋ ❋ H5 ☼ ≈⊥ 60–100cm (2–3ft)

Osmunda regalis
The royal fern makes an arresting sight at the edge of a pond with its toes just in the water. It is deciduous, producing a crop of fresh, mid-green sterile fronds that gracefully unfurl each spring. In summer, upright, fertile, tassel-like fronds form in the centre of the plant. This fern needs space to spread.

↕2m (6ft) ↔4m (12ft) ❋ ❋ ❋ H6 ☼ ◐ ～

Primula alpicola
Originally from Tibet, this moisture-loving primula flowers in midsummer with fragrant, white, yellow or violet, tubular blooms on whitish stems. The deciduous leaves are mid-green and have toothed or scalloped margins. Plant in a bog garden or in soil that stays reliably damp.

↕50cm (20in) ↔30cm (12in) ❋ ❋ ❋ H7 ☼ ～

PLANT GUIDE

❄❄❄ H7–H5 fully hardy   ❄❄ H4–H3 hardy in mild regions/sheltered sites   ❄ H2 protect from frost over winter
❄ H1c–H1a no tolerance to frost   ☼ full sun   ☼ partial sun   ☀ full shade   ~ bog plant   ≈ marginal plant   ≋ aquatic plant   ⊥ planting depth

Primula beesiana
A semi-evergreen candelabra primula, *P. beesiana* has vivid magenta flowers in summer. The spherical flowerheads appear at intervals up greenish-white stems, giving rise to the plant's common name. Plant in a boggy border, or at a pond edge, in large groups with ferns to create a colourful, textured display.

↕60cm (24in) ↔60cm (24in) ❄❄❄ H6 ☼ ☼ ~

Primula 'Inverewe'
In summer, up to 15 bright red flowers appear on each white stem on this semi-evergreen candelabra primula. The mid-green leaves are oval with toothed margins. The plant is a vigorous grower that prefers partial shade, but will tolerate full sun as long as the roots are kept moist.

↕75cm (30in) ↔60cm (24in) ❄❄❄ H5 ☼ ☼ ~

Rheum palmatum 'Atrosanguineum'
This ornamental rhubarb needs a large garden to accommodate its metre-long, toothed leaves and huge plumes of cerise-pink summer flowers. The young leaves are purple, but fade to green as they age. The soil has to be deep, moist, and very fertile to sustain healthy growth.

↕to 2.5m (8ft) ↔to 1.8m (6ft) ❄❄❄ H6 ☼ ☼ ~

Rodgersia pinnata 'Superba'
Grown for its foliage, the young, purplish-bronze leaves of this plant mature to dark green with distinctive veins, giving a puckered appearance. From mid- to late summer, clusters of tiny bright pink flowers reach above the leaves, followed by brown seedheads. Protect from cold winds.

↕to 1.2m (4ft) ↔75cm (30in) ❄❄❄ H6 ☼ ☼ ~

Sanguisorba canadensis
This is a tall plant that needs to be placed at the back of a bog garden or moist border. It produces lush green foliage on branching stems, and long, bottlebrush-like spikes of small white flowers, which open from the bottom upwards, in late summer and early autumn. Divide clumps in spring or autumn.

↕to 2m (6ft) ↔1m (3ft) ❄❄❄ H7 ☼ ☼ ~

Thalia dealbata
This distinctive-looking plant is best used as a pond marginal. It forms a clump of upright reed-like stems that bear bold, upward-pointing, blue-green lance-shaped leaves. In late summer, clusters of dark violet flowers sway above the leaves. It often proves frost hardy if roots are below water.

↕2m (6ft) or more ↔1m (3ft) ❄ H2 borderline ☼ ≈

Typha minima
An ideal plant for small ponds or tubs, this perennial has clusters of narrow vertical leaves, which are joined in late summer by cylindrical flower spikes. The flower stalks can be cut and used in indoor arrangements.

↕to 75cm (30in) ↔30–45cm (12–18in) ❄❄❄ H7
≈⊥ 30cm (12in)

Zantedeschia aethiopica
One of the most exotic-looking marginal plants, the arum lily brings grace and style to ponds and bog gardens. Large pure white flowers, which gleam against the bright green foliage, open from late spring through to midsummer. Grow in shallow water, dividing the rootstock if necessary in spring.

↕90cm (36in) ↔90cm (36in) ❄❄ H4 ☼ ≈⊥ 15cm (6in)

PLANTS FOR BOGGY SOIL

- *Actaea simplex* Atropurpurea Group 'Brunette' p.346
- *Aruncus dioicus* 'Kneiffii' p.346
- *Astilbe* 'Professor van der Wielen' p.346
- *Astilbe* 'Willie Buchanan' p.346
- *Eupatorium maculatum* 'Atropurpureum Group' p.346
- *Filipendula rubra* 'Venusta' p.347
- *Gunnera manicata* p.347
- *Iris* 'Butter and Sugar' p.347
- *Iris sibirica* 'Perry's Blue' p.347
- *Kirengeshoma palmata* p.347
- *Ligularia* 'The Rocket' p.348
- *Matteuccia struthiopteris* p.348
- *Osmunda regalis* p.348
- *Primula alpicola* p.348
- *Primula* 'Inverewe' p.349
- *Rheum palmatum* 'Atrosanguineum' p.349
- *Rodgersia pinnata* 'Superba' p.349
- *Sanguisorba canadensis* p.349

Materials guide

Hard landscaping materials provide the essential structure that every garden needs to create a usable space. As well as their practical functions, walls, paving, fences, and structures also help to shape the overall design, forming a permanent framework for the more ephemeral planting. Factors to consider when choosing materials include their cost, colour range, ease of installation, durability, and their environmental impact – look online for options and check readers' reviews of those you select. With paving, for example, always consider if the material or grouting is porous, which helps reduce the risk of localized flooding. This at-a-glance directory shows you what materials are available and their essential properties.

Surfaces

Bricks
Clay bricks are timeless and can be laid in a variety of patterns. The colour range is determined by the clay and the firing; also the higher the temperature (and the cost), the more durable the brick. For paths and patios, bricks must be frostproof and hardwearing; house bricks are not suitable.

£–££ ♦♦ reds, buffs, browns, blue/greys

Concrete blocks
In place of bricks you can use less costly concrete blocks, which come in a wide range of sizes, shapes, colours, and textures. You can also buy blocks set on a fabric backing ("carpet stones") or moulded into a slab for easy laying. Concrete blocks can easily take the weight of a car and are ideal for driveways.

£ ♦♦ concrete can be dyed almost any colour

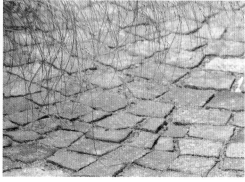

Granite setts
Fast disappearing from our city streets, granite setts have great charm and are increasingly available from reclamation yards for use in the garden – where they make a hardwearing surface for paths and drives. Individual setts vary in size and depth, which can make levelling and fitting them together a challenge.

££ ♦♦ blue/greys, pink, black

Terracotta tiles
These offer the warmth and colour of the Mediterranean, but most are not frostproof. Their porosity creates a safe, nonslip surface, but makes them vulnerable to staining, so apply a sealant. Available in a huge range of sizes and shapes, the colours are determined by the kiln firing of natural clays.

£–£££ ♦ orange, red, mellow yellow

Stone and tiles
You can have some fun with mixed coloured materials – here, granite setts, terracotta and glazed tiles. If you have a handful of expensive tiles, this is a great way to eke them out. Laying the blocks and tiles on a dry mortar mix will help you to adjust the different levels and avoid an uneven surface.

£–£££ ♦♦ various

Crazy paving
A 1970s favourite, crazy paving is brought up-to-date by using just one type of stone – here, reclaimed Yorkstone. It makes a hardwearing surface for paths, patios, and drives, although laying a random pattern isn't as easy as it appears and you may need professional help to achieve a decorative mosaic effect.

£–££ ♦♦ large range

MATERIALS GUIDE

£££ high cost   ££ medium cost   £ low cost   ♦♦ high durability   ♦ low durability   🎨 colour options

Granite
A popular stainproof surface for kitchens, polished granite is diamond-hard and tough enough for use in the garden. It comes in a huge range of colours; some also include speckled and streaked detailing. Affordable composite and terrazzo (granite chips bonded with cement and polished) are available.

££–£££ ♦♦ 🎨 black and greens to pinks, reds, cream

Limestone
A sedimentary rock, limestone often has shells and fossils embedded in it. Riven stone (*shown here*) is popular in gardens because it is split in a way that leaves a roughened, nonslip surface. Limestone darkens when it is wet and it can stain, so consider sealing it. Available as composite.

££–£££ ♦♦ 🎨 grey, white, pale red, yellow or black

Marble
More familiar in sunnier climes, marble is increasing in popularity as a sophisticated landscaping material. When polished, it has a lustrous quality that will smarten up any patio. The characteristic veining is caused by mineral impurities. Consider sealing. Available as composite.

££–£££ ♦♦ 🎨 white, black, grey, green, pink, red, brown

Sandstone
Made up of small mineral grains, sandstone is easy to cut and lay. The import market has made available a wide range of colours and patterns, including streaking and stripes. The colour darkens when wet. Reclaimed sandstone paving is a less expensive option. Sealing is advisable. Available as composite.

££–£££ ♦♦ 🎨 gold, jade, rose, brown, grey, white, black

Slate
Stylish and modern, slate is a hardwearing fine-grained stone. Unless polished, it's nonslip, even when wet, making it ideal for pathways. Note the colour darkens when wet. Various surface textures are available, including rough cut (visible saw marks), sandblasted, and polished (called "honed"). Consider sealing.

££–£££ ♦♦ 🎨 black, blue-grey, green, purple

Travertine
Popular as a building material since Roman times, travertine is a dense form of calcium carbonate. Pure travertine is white, but impurities add colour. The characteristic pitting is caused by gases trapped in the molten rock. The best quality travertine has smaller holes that are infilled and polished.

££–£££ ♦♦ 🎨 white, pink, yellow, brown

Yorkstone
Most of Britain's cities are paved with this hardwearing fine-grained sandstone. The colour, which darkens when wet, depends on where it was quarried in Yorkshire. Reclaimed and composite paving slabs with a nonslip, riven surface (*as shown*) are available. Consider sealing.

££–£££ ♦♦ 🎨 grey, black, brown, green or red tinged

Porcelain tiles
Porcelain tiles have become increasingly popular, as a range of colours, textures, finishes, and styles has come to market. Needing a little extra preparation for laying (a slurry needs to be painted on the back to help adhesion) and pin-point level accuracy, they make a great statement in any garden.

£–£££ ♦♦ 🎨 various

ENVIRONMENTAL ISSUES
Our purchasing power as consumers can have a huge impact on the environment, especially when choosing materials for the garden.

- Wood and stone that's been transported halfway around the world has a large carbon footprint, so first check what's available from local quarries. If you do decide to use imported stone, check that it isn't produced by child labourers.
- Soft- and hardwoods should be from a sustainably managed source. Look for accreditation from a recognized authority, such as the Forest Stewardship Council (FSC), or try to use recycled wood.
- Always consider porous materials for paving, so rainwater can seep into the ground and reduce the risk of flooding.
- Low-solvent or water-based paints and wood preservatives are a responsible choice.

Surfaces

Patio kit
Used as a centrepiece for a patio or path, this stylized sun comes in kit form ready to fit together like a jigsaw. Other popular designs include fish, butterflies, and geometrical patterns. Usually made from hardwearing moulded composite stone, it can add a decorative note to a patio.

££ ♦♦ 🎨 various stone colours

Wood-effect tiles
There are many benefits of porcelain or ceramic wood-effect tiles compared to wood. Long-term durability is considerably higher with wood-effect tiles, as there will be no rotting or degrading of the material over time. In addition, there are more stable colours and textures available than with natural wood.

££–£££ ♦♦ 🎨 various

Metal grille
Parallel steel tracks (*one shown here*) follow the route of car tyres on a driveway, creating a modern, strong, safe surface for parking; when the car is not there, the ground cover beneath is revealed. Commission a specialist blacksmith or metalworker to make a similar grille to suit your needs.

£££ ♦♦ 🎨 shiny metallic

Wooden decking tiles
Choose decking tiles with battens attached on the underside and lay them straight on to a level concrete or asphalt surface. Made from softwood, they are lightweight and ideal for roof terraces, balconies, and patios. When they start to wear, just lift the damaged squares and replace like carpet tiles.

£ ♦ 🎨 oil or stain tiles

Wooden decking
Hardwoods, such as balau (shown) and oak, are a popular choice for decks. They are warp- and weather-resistant and more durable than softwoods. Most decks, however, are made from pressure-treated softwoods, which are less costly and also available as kits. If well maintained, they should last 20 years.

£–£££ ♦♦ 🎨 oil or stain

Composite decking
Made from recycled waste, composite decking is weatherproof, UV stable, rot-proof, and low maintenance. Construction is the same as when using wood, the difference is in the aftercare. It needs no oiling or retreating, just an occasional hose-down. There is a good range of colour and texture options.

£–££ ♦♦ 🎨 "natural" wood, green, black, blue

Wooden sleepers
Old railway sleepers are no longer available; saturated in creosote and bitumen, they are now considered a health risk. You can buy untreated timber lookalikes (often oak) that are just as heavy to lift and as hard to cut – you will need a chain saw. Good for stepping stones, but slippery when wet.

£–££ ♦♦ 🎨 natural wood, could be stained

Concrete sleepers
Made from cast concrete, these composite sleepers are amazingly realistic and very hardwearing. They come in varying lengths (minimizing cutting) but, like paving slabs, the depth is consistent, making them easy to lay on a bed of mortar. The wood-grain pattern provides a sure grip in the wet.

££ ♦♦ 🎨 "natural" wood

Bark
Bark provides a springy surface for paths and play areas. Fine shredded is kinder on children's knees, but will break down and need replacing more frequently than coarse chipped bark. You can lay it directly on soil (it acts as a soil improver), but for best results, spread it over a weed-suppressing membrane.

£ ♦ 🎨 usually brown; dyed chips are also available

MATERIALS GUIDE   353

£££ high cost   ££ medium cost   £ low cost   ♦♦ high durability   ♦ low durability   ◊ colour options

Gravel

Gravel comes in a wide range of colours and sizes and is a tough, quick-to-lay surface for paths and drives. Spread in a thick layer over a weed-suppressing membrane, or, to stop it spilling everywhere, use a honeycomb gravel containment mat. Guests – welcome or not – are announced by loud crunching.

£ ♦♦ ◊ wide range of stone colours

Cobblestones

Laying a cobblestone path – whether patterned or plain – is a painstaking exercise, but, if you have the patience, the result is worth the effort. Set the cobbles on a bed of mortar, then brush a dry mortar mix into the joints for a hardwearing surface. Use only smooth rounded stones; others are hard to walk on.

£ ♦♦ ◊ white, creams, greys, blacks, browns

Slate chips

If you use slate chips on a well-trodden path, they will crack and slowly break down. Renewing them every few years, however, is a small price to pay for the beautiful colour that provides a foil for edging plants. Lay over a weed-suppressing membrane. Sharp pieces of slate are not child- or pet-friendly.

£ ♦ ◊ grey with green, blue, purple, or plum tones

Paddlestones

Usually large pieces of slate, paddlestones are tumbled to round off the edges. In Japanese-style gardens they are used as decorative paths designed to resemble a winding river bed. Smooth and flat, they are fairly easy to walk on, but they are best reserved for areas of light traffic.

££ ♦♦ ◊ grey with green, blue, purple, or plum tones

Self-binding gravel

Soil and small stone particles are usually washed off gravel, but in this form they are retained and help bind the gravel together to form a more solid surface. Tamp down a thick layer over a solid bed of hardcore to form a hardwearing surface that is easy to walk on.

£–££ ♦♦ ◊ grey, gold, plum, red, green

Decorative shell

Shells are much too fragile to walk on, and should only be used as decorative surfaces. They are a waste product from the shellfish industry, and have a lovely light-reflective quality. Lay them over a weed-suppressing membrane and use them in Mediterranean-style or seaside gardens as a foil for plants.

£ ♦ ◊ cream, grey, pink, soft brown

Shredded rubber

As a decorative mulch, shredded rubber can look quite chic. Its spongy quality also makes it ideal for play surfaces, but it does have quite a distinctive odour (that deters cats) and is therefore unsuitable for areas close to seating and dining tables. It does not rot, so won't need replacing.

£ ♦♦ ◊ grey-black

Resin-bound gravel

Gravel is a popular choice for drives and paths, but sometimes its movement and noise can be offputting, as can treading gravel into the house. Resin-bound gravel (compacted gravel with a clear resin poured on top) is permeable and gives a clean, smooth look. Resin bonded is similar but not permeable.

£–££ ♦♦ ◊ various

Coloured aggregates

Usually made from glass fragments that have been tumbled to remove the razor-sharp edges, aggregates can be used between plants, or for secondary paths – they are not suitable for play areas. Lay the aggregate over a weed-suppressing membrane and hose down occasionally to refresh the colours.

£ ♦♦ ◊ various

Walls and railings

Brick
Acting like a storage heater, brick walls absorb the sun's heat during the day and release it at night to create a mild microclimate. While walls make a garden feel protected, permeable screens are actually better at filtering winds (see p.223). Brick is cheaper than stone and just as durable.

£ ♦♦ ⁄⁄ yellow, red, blue-grey, mottled

Weathered stone
Structures made from aged and weathered natural stone look particularly effective in the gardens of period homes, especially when they match the house walls. Stone that has to be worked or shaped for a wall will add to the cost. Reconstituted (or composite) stone made from concrete is a more affordable option.

£–££ ♦♦ ⁄⁄ various natural stone colours

Mortared stone
Rough-hewn stone forms a structure that is as much a work of art as it is a wall. "Gluing" it together with mortar makes it easier to build than a dry stone wall, where each stone has to fit neatly within a specific space. Top with coping stones and point between the joints to prevent water and frost damage.

££ ♦♦ ⁄⁄ various natural stone colours

Dry stone wall
The materials (a tonne of stone per cubic metre), skill, and time required to build a dry stone wall make it an expensive, though beautiful, option. Two parallel walls, built on foundation stones, are bound together with an infill of rubble; the meticulous placement of the stones negates the need for mortar.

££–£££ ♦♦ ⁄⁄ various stone colours

Gabion
Rocks, cobbles, bricks, or tiles crammed into metal gabions, which are then wired together, create an instant, fairly inexpensive "dry stone" wall. The weight and strength of the filled cages make them ideal for retaining as well as decorative walls. Gabions come in various sizes.

£ ♦♦ ⁄⁄ grey metal; depends on the filling

Knapped flint
Popular as a building material, flint is a tough silica that forms as "nodules" in chalk beds. Here, the flints have been "knapped", i.e. split in half, and set in lime putty (which retains a degree of flexibility and is resistant to cracking) to form a decorative facing on a brick or block wall.

££–£££ ♦♦ ⁄⁄ black and white

Mosaic wall
A mixture of terracotta and glazed tiles, cobbles, setts, and bricks, this wall is both colourful and tactile. In practical terms, the materials are set into a layer of rendering (a mix of cement and sand) covering a brick or block wall. For a neat finish, smooth out the pointing in between each piece.

£–££ ♦♦ ⁄⁄ as colourful as you wish to make it

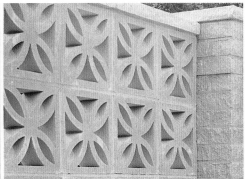

Screen wall
Concrete blocks offer the strength of brick without cutting out the light. Prices are similar, too, but walls made from blocks are quicker to build. Use them for low patio walls, or to top an existing wall, adding extra height and privacy. Their open structure makes them effective windbreaks.

£ ♦♦ ⁄⁄ cement grey unless you paint them

Shell mosaic
Mosaics are a weatherproof decoration for the garden. Here, a low retaining wall has been brightened up with a collection of shells, fossils, and stones. The pieces are set into a thin skim of still-damp render (cement and sand). Once dried, a coat of water-based varnish helps protect the mosaic.

£ ♦ ⁄⁄ various, depending on the materials used

MATERIALS GUIDE 355

£££ high cost   ££ medium cost   £ low cost   ♦♦ high durability   ♦ low durability   🎨 colour options

Shuttered concrete
For a textured finish, concrete is poured into moulds made from timber shuttering. Walls taller than knee height need foundations and steel reinforcement rods for strength. Red sand in the concrete mix gives a buff colour; yellow sand the usual grey; for stronger colours, use concrete dyes or paint.

£ ♦♦ 🎨 buff or grey; various if using dyes or paint

Rendered walls
Applying a skim of render (a mix of cement and sand) is a relatively quick – and inexpensive – way to tidy up rough block walls or crumbling brick. Once dry, you have a smooth blank canvas for applying exterior masonry paints. These come in a range of colours, from subtle to shocking – like this pink.

£ ♦♦ 🎨 various

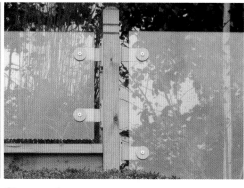

Glass panels
Surrounding a patio, balcony, or raised deck with glass panels provides a degree of shelter without blocking the view. For safety and strength, use toughened glass fixed to sturdy posts. Treat the glass with a silicone-based rain-repellent coating to make it easier to clean and to prevent smears.

££ ♦ 🎨 clear

Aluminium panels
Hide an ugly fence or view and provide an unfussy backdrop for planting with powder-coated aluminium panels. The coating is fade- and flake-resistant. At night, treat them like a projectionist's screen, creating shadow play with spotlights. For a cheaper option, paint sheets of marine ply.

£–££ ♦–♦♦ 🎨 various

Wooden block wall
Building a wall using random materials is a skilful job; like a 3-D jigsaw puzzle, each piece must fit neatly with its neighbour. Here, cedarwood offcuts and squares of rusted steel have been glued and screwed together and mounted on a sheet of marine ply, which, in turn, is fixed to a solid wall.

£–££ ♦–♦♦ 🎨 various

Wooden pallets
Use pallets to make a "wildlife wall", wiring them together and packing the gaps with moss, wool, and grass (nesting material for birds), and crocks, rotting wood, and hollow canes (homes for insects and amphibians). Usually made from pine, better quality pallets are available from specialist suppliers.

£ ♦ 🎨 natural wood shades

Corrugated iron
A maintenance-free fencing option, corrugated iron has one drawback – sharp edges. To cover these, use protective metal edging strips, and fix panels to sturdy posts to hold them steady in gusting winds. Galvanized metal (shown here) has a matt finish, while metal paints can add a splash of colour.

£–££ ♦♦ 🎨 metallic grey or, if painted, various

Iron railings
Off-the-peg cast-iron railings make an attractive divider in a garden. After a few years, however, they will need repainting. While "no-paint", plastic-coated metal seems a good idea, the coating eventually becomes brittle and chips off, allowing rust to get a hold.

£–££ ♦♦ 🎨 usually black or dark green

Bespoke ironwork
Many blacksmiths specialize in decorative metal work – this whimsical fence made from steel horseshoes is a bespoke commission. The shoes, which are mounted on horizontal metal bars, are painted to protect against rust and make an eye-catching feature, as well as a functional boundary.

££–£££ ♦♦ 🎨 usually black, especially if wrought iron

Screens and gates

Shiplap
This is one of the cheapest and most popular ready-made fencing options, though not the most durable. Even though the panels are pre-treated, it is best to apply a preservative every few years. The larch strips often warp, leaving small gaps. Available in standard fence panel sizes.

£ ♦ 🌀 often pre-stained orange, but will tone down

Featheredge
Ready-made panels come in various sizes, but the design (vertical softwood timbers nailed on to horizontal rails at the top and bottom) makes it easy to construct. If fixed to strong post supports, the sturdy panels are good for boundaries. Best given a coat of preservative every few years, even if pre-treated.

£ ♦♦ 🌀 often pre-stained orange, but will tone down

Hit and miss
While offering privacy, the alternating panels of hit and miss fencing are wind permeable, making it ideal for exposed sites. Attached to sturdy posts, it is unlikely to blow down, and the wood strips (fixed vertically or horizontally) are easy to replace. Buy ready-made or construct panels yourself.

£ ♦♦ 🌀 often pre-stained orange, but will tone down

Chevron panel
Decorative panels are not usually strong enough for use as a boundary fence, but this chevron design, a variation on the sturdy hit and miss (*see above right*), is suitable. It is also ideal for dividing up a garden into rooms, or screening an ugly view, perhaps where the compost bins are stored.

££ ♦♦ 🌀 usually stained a subtle tan

Trellis panel
Another hit and miss variant, but this time with an inset of trellis down the centre. It would make a good windbreak, but the lack of privacy could be a problem for a boundary. One way to mask the gaps would be to train a climber through the trellis, thereby creating a colourful display of flowers.

££ ♦♦ 🌀 usually stained a subtle tan

Slatted wood
This fence creates a contemporary, durable screen that allows both light and wind to pass through. Use it to divide up the garden or to mask bins or a shed; it also doubles as a plant support for climbers. Paint or a wood stain will help protect the timber, and introduce colour into your design.

£ ♦ 🌀 natural wood or painted

Picket fence
This simple wood fence has rustic charm, yet it also works well with a modern property. Leave it natural, or paint it to match your house or planting scheme. Its open structure and low profile makes it more of a visual boundary than a barrier to keep out unwanted visitors. Available ready-made.

££ ♦ 🌀 natural wood or painted

Oak panel
This made-to-order fence is perfect for a country-style front garden, where you want the world to admire your planting design. The hardwood has a beautiful appearance and is best left unpainted, but a clear oil will preserve its colour; over time, if left untreated, oak develops lovely silver hues.

£–£££ ♦♦ 🌀 natural wood

Chestnut paling
Often seen on farms, this fencing is naturally rot-resistant and perfect for a subtle, rustic barrier between a country garden and the natural landscape beyond. The wood pieces come on a roll and are linked, at the top and bottom, by a double row of twisted wires. This fence is fixed to wood rails for extra strength.

£ ♦♦ 🌀 natural wood

£££ high cost   ££ medium cost   £ low cost   ♦♦ high durability   ♦ low durability   🍃 colour options

Willow hurdle
Surprisingly robust, willow hurdles make effective windbreaks. They can be woven to order, or are available in standard panel sizes. Willow makes a beautiful backdrop for naturalistic or cottage-style plantings, or fix it to the top of a wall to increase privacy. Protect with linseed oil.

££ ♦ 🍃 golden brown

Willow screen
If you love the look of willow but want a more contemporary look, choose a framed willow screen – it provides a neat yet natural backdrop for planting. Good for privacy around the patio, the screen is clamped into a timber frame for extra strength, but the size range is limited. Treat with linseed oil.

£ ♦ 🍃 golden brown

Bamboo/reed screen
Ideal for when you want an instant screen to block out an ugly view. You could also use it to make a roof for a pergola. For extra strength, attach it to an existing fence – it works especially well on chain-link. It's not suitable for exposed sites, and it will start to deteriorate after a few seasons.

£ ♦ 🍃 soft browns

Formal hedge
While slow growers, such as yew and beech, may take a few years to thicken up, quick-fix conifers require endless cutting. It's tempting to buy established plants for instant results; but young "whips" are cheaper and quickly catch up. Plant thorny *Berberis*, *Pyracantha*, and *Rosa rugosa* to keep out intruders.

£ ♦♦ 🍃 various

Fedge
The backbone of this lovely hedge is a chain-link fence with climbing plants grown through it. Results are not instant, but the low price makes this a good choice for a long boundary in an informal or wildlife garden. Plant a mixture of prickly plants for security, and flowering climbers for colour.

£ ♦♦ 🍃 various

Living willow
Is it an art installation or is it a screen? Both really, and that is the fun of woven willow structures. Plant the young willow "whips" in winter or early spring in a sunny spot, then come summer, you can start weaving. To stop your screen maturing into a forest, prune back to the framework in late winter.

£ ♦ 🍃 golden stems and lush green foliage

Gate in a fence
Choosing a gate that closely matches the fence panels gives a visually unbroken line for a crisp, clean design. If you can, position the gate in a gap between two whole fence panels – reducing the size of some panels, such as featheredge, involves complicated carpentry.

£ ♦ 🍃 natural wood or painted

Bespoke gate
This spiral metal gate was made to order, but there are plenty of lovely designs available off-the-peg. Set between two sturdy steel posts, it makes a beautiful focal point in a country-style hedge. Regularly trim the foliage away from the hinges and the catch.

£–£££ ♦ 🍃 wrought iron, rusting steel, painted metal

Wooden door
An arched wooden door set in a stone or brick wall is a design classic. Peeling paint and rusting fittings will only add to its charm. This door was custom-made to fit the space, but, if you are building a wall from scratch, it is worth checking out the sizes of off-the-peg doors before you start.

££ ♦ 🍃 natural wood or painted

Structures and storage

Contemporary garden room
These garden rooms range from compact, relatively inexpensive structures to luxury state-of-the-art buildings that include the latest technology and equipment. Most are constructed from timber and glass, with heating and cooling systems, and an electricity supply connected to the house mains.

£–£££ ♦–♦♦ natural wood, steel, glass

Garden office/studio
Usually made from wood, you can work in peace away from the hubbub of family life in these buildings. Ideal as an art studio, workshop, or home office. For comfort and to protect books, etc, opt for insulation and a heater fitted with a thermostat. Fit blinds, and a lock for security.

££–£££ ♦♦ natural wood, painted, or stained

Traditional garden room
Built straight on to the house but surrounded by greenery, garden rooms allow you to enjoy the outdoors whatever the weather. A timber and brick construction with an insulated sheet metal roof makes the room more usable year round than the average glass conservatory, though not so light and airy.

£££ ♦♦ brick, stone, rendered walls; stained wood

Colonial-style gazebo
Relatively small, this type of gazebo can be slotted in almost anywhere, such as next to a pool or surrounded by pots of subtropical plants on a deck. Offers shaded seating for drinks or afternoon tea. Usually wooden with a thatched roof; some designs are more weatherproof with removable slatted sides.

££–£££ ♦ wood shades and muted period colours

Garden shed
DIY or off-the-peg, a shed is a must for anyone needing extra storage or space for a hobby. Can be painted or stained a wide range of colours. Sheds made from shiplap (overlapping wood) panels may warp; tongue-and-groove models are more expensive but superior in quality. Usually with a felted roof.

£–££ ♦ natural wood, painted, or stained

Green roof
A shed roof may need shoring up with extra timbers for it to take the weight of a planted roof. Before laying the sedum matting and moisture-retentive growing medium you will need to protect the roof with butyl or polythene sheeting. Green roofs provide good insulation and increase biodiversity.

££ ♦♦ sedums and other succulents provide colour

Lean-to greenhouse
Space-saving design. Best for south- or west-facing walls, which act like storage heaters releasing warmth at night. Off-the-peg and bespoke models available in wood or aluminium, with glass or polycarbonate (the latter offers good insulation and safety). Cheap tubular steel frame models with plastic covers available.

£–£££ ♦–♦♦ white/dark green, cedar, or painted

Obelisk
A sturdy wooden obelisk (this traditional design is topped with a finial) is a feature in its own right, adding extra height to a border as well as providing support for climbers. DIY or buy off-the-peg; they are made from wood or metal (the latter available in more decorative designs).

£–££ ♦–♦♦ natural wood, painted, or stained

Willow arch
Easy to construct and adaptable for the smallest garden, use long "rods" of living willow (plant in winter) or buy dried and pre-soak to make them flexible and workable. Push into the ground, weave together, then tie the tops to form an arch. If the willow starts to sprout, prune it back in late winter.

£ ♦ natural willow

MATERIALS GUIDE

£££ high cost   ££ medium cost   £ low cost   ♦♦ high durability   ♦ low durability   colour options

Arbour seat
Self-assembly kits range in price and quality. Custom-built and corner models are available. In a sunny spot, the roof provides shade. Trellis sides and/or roof are ideal for scented climbers. Usually constructed in wood, but also available in wrought iron or a wood/metal mix.

£–£££ ♦–♦♦ natural wood, painted, or stained

Modern arbour with brazier
This designer piece with a Moorish flavour incorporates bench seating and a metal brazier – perfect for entertaining on summer evenings. Made from wood, the structure is a sculptural focus for a modern or period garden. A canvas awning would provide extra weather proofing.

£££ ♦♦ natural wood, painted, or stained

Traditional pergola
Easy to construct for a competent woodworker. Substantial uprights and horizontal supports can carry heavy climbers, such as grapevines, roses, and wisteria. Creates dappled shade for a pathway or seating area. Flat-pack timber kits, wrought iron, and bespoke models also available.

££–£££ ♦♦ natural wood, painted, or stained

Folly
A focal point, especially for period gardens. May be any design, but often hinting at a specific point in history. Examples include mock Gothic ruins, "ancient" stone circles, classical temples, rustic buildings, and grottoes. DIY construction is possible, such as with reclaimed masonry.

£–£££ ♦♦ depends on construction materials

Wendy house
From the simplest wooden box to a two-storey chalet with windowboxes, owning a Wendy house is every child's dream. Bespoke; mid-price, off-the-peg self-assembly; and cheaper click-together plastic are available. Ensure the base is stable. Paints and fixings must be child-safe.

£–£££ ♦–♦♦ natural wood; child-safe paints/stains

Children's play area
The best play structures are made to order and erected on site. When buying – especially self-assembly – look for the CE (European compliance) symbol and the European Safety Standard BS EN 71 for toys. Needs safe flooring material, ie at least 15cm (6in) depth of play bark or a bonded-rubber surface.

££–£££ ♦♦ natural wood; child-safe paints/stains

Storage/tool box
A spacious mini shed for tools and lawn mowers, garden furniture or bicycles, can be made from panels of larch lap fencing bolted together, or bought ready-made (usually with a felted roof). It only needs to be as high as your tallest tool. Tuck away in a corner and paint green to blend in.

£–££ ♦♦ natural wood, painted, or stained

Recycling cupboard
A great way to disguise unsightly wheelie bins and plastic recycling boxes in a front garden. Wide-opening doors give good access. Make yourself or buy ready-made in wood, plastic, trellis screening, or even woven willow. This one has a green roof, further increasing its eco credentials.

£–££ ♦–♦♦ paint/stain to blend in or match house

Garden furniture storage
This bench seat opens to reveal a weatherproof box for storing loose cushions, throws, and covers from garden furniture. Wood and plastic ready-made models available. Site next to the patio for convenience. Also useful as a toy box or compact tool storage for courtyard gardens.

£–££ ♦♦ natural wood, painted, or stained

Containers

Terracotta clay pots
Today's clay pots are mostly machine-moulded rather than hand-thrown, but you can still buy handmade pots from specialist potteries or antique shops. The higher the temperature of the firing, the greater the frost resistance – and cost. Clay is porous, and pots dry out quickly in hot sun.

£–££ ♦ 🌣 soft orange and sandy yellow clay

Terracotta-style trough
Versatile clay can be moulded to almost any shape; but take a good look, could this be plastic? These days it is hard to tell the two apart. While replicating the look of clay, plastic is lighter, frostproof, and usually cheaper. It's also better at keeping compost and plant roots moist during hot dry spells.

£–££ ♦ 🌣 clay colours or, if plastic, a huge colour range

Glazed ceramic
Glazing a clay pot transforms it. During the kiln firing, the glaze melts to coat the pot in a thin layer of glassy material. As a result, the pot becomes stronger, frost- and waterproof, if it is glazed inside and out, and, depending on the glaze, more colourful. Match your pots with planting for a unified display.

£–££ ♦♦ 🌣 huge colour range

Water feature
For water features, such as bubble fountains and patio ponds, choose pots that are glazed (or at least glazed inside) to minimize water loss. This urn is set on a cobble-covered metal grille over a reservoir; water is pumped up through the drainage hole in the base to overflow back into the tank.

£–££ ♦ 🌣 huge colour range if glazed

Strawberry pot
Hand-thrown or moulded (the cheaper option) clay strawberry pots, with their "balcony" planting shelves, are also ideal for herbs. With this type of pot, big is best as the increased volume of compost prevents the plants drying out too quickly. May not be frostproof. Also available in plastic.

£–££ ♦ 🌣 usually terracotta

Stone urn
Whether empty or planted up, stone urns have a classic, timeless quality. You can find originals in reclamation yards at a price; but composite stone (cast concrete) is a more affordable and widely available option. Stand an urn on a plinth and it instantly becomes a focal point.

££–£££ ♦♦ 🌣 natural stone colours

Cast concrete
Strong and cheap, concrete is a versatile material for making planters, like this rough-cast bowl. Containers made from concrete are available in both contemporary and classic designs, and, because they are very heavy, they make a good choice for top heavy plants, such as trees and shrubs.

£ ♦♦ 🌣 concrete can be dyed almost any colour

Terrazzo
Hardwearing, easy to clean, and very tactile, terrazzo is the ideal material for contemporary containers. Granite or marble chips are bonded with cement, then polished to create a smooth surface – a technique that has been around since Roman times. Lightweight polyester terrazzo planters are available.

£–££ ♦♦ 🌣 marble and granite greys, white, and black

Weathering steel
Never has rust looked so good. Weathering steel, of which Corten is the best-known brand, is a high-strength steel alloy. It is designed to develop a layer of rust that, ironically, helps to protect the metal underneath. Strong and durable, it is perfect for long-term plantings, and, as here, water features.

£££ ♦♦ 🌣 rusty orange

MATERIALS GUIDE   361

£££ high cost   ££ medium cost   £ low cost   ♦♦ high durability   ♦ low durability   ◉ colour options

Powder-coated metal
A much tougher, non-flaking finish than paint, powder coating (a mix of pigments and resin) is baked on to the surface of metal. Available in a huge range of colours and finishes, the coating inhibits rust. To protect the surface, clean with soapy water and a soft dry cloth, and avoid abrasive solvents.

£–££ ♦♦ ◉ huge colour range

Galvanized metal
The mottled patina of galvanized metal is created by "hot dipping" – a chemical process that coats steel and iron with rust-resistant zinc. Planters come in a range of styles and sizes; most are lightweight and single skinned. In winter, protect plant roots by wrapping the container with bubble plastic.

£ ♦♦ ◉ mottled matt grey

Lead planter
Lead is a soft, malleable metal that is easy to work. This planter is made from a sheet of lead hammered into shape; the raised pattern is formed by pressing it into a mould. Lead is toxic and shouldn't come into contact with food plants. Glass fibre lead-style planters are a "food-safe" option.

££–£££ ♦ ◉ grey

Wooden barrel
Traditionally made from oak, the wooden pieces (called staves) are shaped to fit tightly together and held in place with metal hoops. You may be lucky enough to find half wine or whisky barrels; cheaper replicas are also available. Best lined with plastic or butyl, especially if using as a patio pond.

£–££ ♦ ◉ wood with black metal bands

Wooden trough
Lightweight and insulating in winter, this rustic planter is made from woven hazel twigs set in a timber frame. For longevity, choose pressure-treated timber, and check that the planter is lined with plastic (with drainage holes at the bottom) to prevent compost and water leaking through the sides.

£ ♦ ◉ natural wood

Versailles planter
Relatively light for the volume of compost they contain, these planters were originally designed for the orange trees at Versailles so they could be brought indoors over winter. Lining them with plastic extends the life of both hard- and softwood planters. Good quality plastic imitations are also available.

£–££ ♦ ◉ natural wood or, if painted or plastic, various

Old boots
The more holes in the soles, the better the drainage! Fill the boots with compost, packing it firmly into the toe, and plant up. Be warned that even if you have enormous feet, boots still hold relatively little compost and plants are at risk of dehydrating in hot sun, so consider using water-retaining gel.

£ ♦ ◉ various fashion colours

Recycled kitchenware
Old colanders, chipped teapots, saucepans that have lost their handles – almost any old household vessel has planting potential for a sustainable garden design. Kitchen cupboards are an especially rich hunting ground. You may need to drill holes for drainage or go easy on the watering.

£ ♦ ◉ depends on your crockery and cookware

Car and lorry tyres
Get extra mileage out of old tyres by giving them a splash of paint and a new lease of life as a raised bed. Place the tyres straight on to the soil and fill with compost (line them first with plastic if you're growing food). The rubber absorbs the sun's heat and warms up the compost for early plantings.

£ ♦♦ ◉ black (brightened up with a splash of colour)

Water features

Wall fountain
Made from a variety of materials – steel, resin, stone, and so on – these can add a dashing and classically inspired focal point to any space. With recirculating water coming from a subterranean reservoir, the range of sizes and designs for these fountains is endless.

£–££ ♦–♦♦ various

Wall spout
Some fountains are simple designs, with water coming from a central outlet fixed on a wall or a free-standing feature. The water is then captured at ground level, either in a bowl, container, or onto pebbles, which hide a collecting sump beneath them.

£–££ ♦–♦♦ various

Blade
Sheer lines of water running over a smooth, hard material have become highly fashionable in garden design. The water power can be increased or decreased as required, and many blades are set within a wall, ensuring the water flows seamlessly into a collecting station at ground level. Lighting can be added.

£–£££ ♦–♦♦ mainly stainless steel

Free-standing blade
A feature in their own right, these water blades are more obvious than wall-mounted blades and add a sculptural element. They give height and scale to any shape, and often come with a fitting kit and water sump to contain the water, which then gets recirculated. Lighting can be added at the blade.

£–£££ ♦–♦♦ mainly stainless steel

Bowl
Simple but very effective, water bowls are often just a container filled with water – no pumps, lighting, or moving parts are needed. They are great for tranquillity and for wildlife to use. Water evaporates over time and will need refreshing; it can get slimy with algae if in direct sunlight or temperatures are high.

£–££ ♦–♦♦ various

Rill
Rills have a classical feel to them: they help divide a space, can give interest to the changing level of a space (some are stepped), and are made from a range of materials. There are multiple ways of making the rill, from liner to slabs or mosaics, and most have a pump to gently push down the water.

£–£££ ♦–♦♦ various

Fishpond
Welcoming fish into your garden can bring movement and interest year-round. Ensure you follow guidance on the volume and depth of water needed for fish to flourish, and the amount of greenery surrounding it. Aerating water may be necessary to ensure a healthy environment.

£–£££ ♦–♦♦ mainly black

Wildlife pond
Bringing water into any garden is a vital lifeline for wildlife. Whether you use a raised wooden tub or a large pond, ensure a balance is struck with making the water accessible for wildlife, providing areas for it to live and thrive, and creating a feature beautiful enough to become a vital part of your garden's appeal.

£–££ ♦–♦♦ various

Ornamental pond
Water is an intrinsic part of many garden designs and can be used in any style – whether traditional or modern – and in large and small gardens. When designed to be enjoyed as part of the whole scheme, ensure the materials and planting are consistent with what surrounds the pond.

£–£££ ♦–♦♦ various

MATERIALS GUIDE

£££ high cost   ££ medium cost   £ low cost   ♦♦ high durability   ♦ low durability   ⓒ colour options

Dome
Easy to buy and to install, domes can bring life and tranquillity to a small corner of any garden. They are often made from stainless steel, sometimes from stone or acrylic, with water recirculating. You can add lighting and planting to surround them to give added appeal to your garden.

£–££ ♦ ⓒ stainless steel, stone, or acrylic

Raised trough
Bringing a contemporary edge to any garden, these raised water bodies can sit on their own – as a focal point – or be softened by surrounding planting. They can be rectangular, square, or round. You can install fountains, jets, and lighting to bring movement and interest.

££–£££ ♦–♦♦ ⓒ various

Barrel
A barrel is possibly the easiest way to bring water into a garden. Filling an old whisky barrel with water and aquatic plants will swiftly bring a host of wildlife. To keep the water healthy, top it up regularly and monitor the plants-to-water ratio so there is a good balance.

£ ♦ ⓒ wooden

Dancing fountain
Dancing fountains can bring great fun and movement to any outside space. You can install timers to deploy the fountains, and use motion sensors to surprise people as they walk through the space. Ensure there is enough water for the pump to maintain an ongoing display.

£–££ ♦ ⓒ mainly stainless steel

Tiered fountain
Classical in design, these fountains work perfectly as a feature to a central island or larger planting scheme. They can be a range of heights and widths. Either standalone or sitting atop a larger reservoir of water, they often have more than two levels, adding movement and flow to the design.

£–£££ ♦–♦♦ ⓒ mainly stone

Natural swimming pond
Rapidly increasing in popularity, these ponds are a more natural way to enjoy swimming outside. They are more complicated to make than they look, needing pumps, filters, filtration zones, and planting to help establish the water environment. But what better way to enjoy outdoor swimming?

££–£££ ♦♦ ⓒ not applicable

Swimming pool
The design of swimming pools has evolved substantially over the years, with different materials used within the pool – from stone and tiles to a mosaic liner. Infinity edges, internal seating, and retractable covers all add to the overall attraction of swimming outdoors.

££–£££ ♦–♦♦ ⓒ not applicable

Rain chain
Interesting additions to an outside space, and replacing standard downpipes, rain chains can add ornament and interest when it rains. They don't cope well with heavy downpours and can get overwhelmed, but are an unusual and attractive feature.

£ ♦ ⓒ metal

Stream
Some of the most charming approaches to water in a garden is when there is an air of naturalistic serendipity to them – such as a natural stream weaving its way through the greenery. You can construct a stream in your garden, but they often look best when they are natural.

£–££ ♦–♦♦ ⓒ mainly stone

Furniture

Fire pit
Sitting around a fire pit at the end of a sunny summer's day, or during a cold winter's afternoon, can be a lovely experience. Myriad materials and designs are available, as are the range of fuels you can burn – mainly timber but also natural gas. Fire pits can be built in or moveable.

££–£££ ♦–♦♦ metal, variety of finishes

Eco fire pit
Fire pits that are less damaging to the environment are available in a range of designs. Instead of burning wood that causes smoke pollution, they use biodegradable vegetable wax. This still gives a feel of sitting snugly around a fire, but with less harm to the environment.

£–££ ♦–♦♦ various

Fire and water feature
To really add drama to a patio or outside space, some manufacturers are now able to combine a sculptural focal point that includes fire and water. Water circulates around the central burner, and both can be used independently. It's a feature in its own right and highly effective.

£££ ♦♦ metal, variety of finishes

Fire table
Available in a range of sizes and different heights (either coffee table or dining table height), these designs put fire and warmth at the heart of the outdoor eating experience. Using natural gas or more efficient propane, with either manual or electronic ignition, they can give off heat and light for many hours.

££–£££ ♦–♦♦ various

Bean bag
Perfect for slouching on a warm summer's day, bean bags offer a child-friendly approach to enjoying the garden. Often shower proof and with washable covers, the only downside is that they need quite a lot of storage space. Some versions float and can be taken into the pool.

£–££ ♦–♦♦ fabric

Hammock
Whether tied between two trees (just like on a desert island) or supported on its own metal or timber frame, a hammock can give a gentle, undemanding, and informal approach to relaxation outside. Most people find the rocking motion soothing and calming. They are mainly made from fabric.

£–££ ♦–♦♦ various

Swing seat
Made from timber, metal, or Accoya® (modified timber), swing seats add movement and drama to any outdoor seating environment. Available in a variety of sizes and finishes, they can be placed in discreet areas for intimate seating or as part of a large patio. Cushions and lighting can be added for effect.

£–£££ ♦–♦♦ mainly timber

Awning
Awnings can be a cost-efficient and simple way of providing shade and some privacy to a patio and house. Fully retractable and available in a range of colours and finishes, awnings fit well in a variety of locations and architectural styles. LED lighting is often available as an optional extra.

£–££ ♦–♦♦ various

Shade sail
Shade sails, which are often made from sailcloth, can be tied to buildings, posts, or walls, and are available in a variety of shapes and sizes. They are water resistant, and easy to take down and store away when necessary, but are not practical in heavy rain or wind.

£–££ ♦–♦♦ various

MATERIALS GUIDE

£££ high cost   ££ medium cost   £ low cost   ♦♦ high durability   ♦ low durability   ⁄⁄ colour options

Garden pavilion
Garden pavilions can be part of the furniture, and are a modern and easy way to include shade in the garden. Many models offer lighting, manual or electric side screens, louvered shower-proof roofs, and integral speakers. They also have a contemporary feeling to their design.

£–£££ ♦–♦♦ ⁄⁄ metal

Daybed
Can there be anything better than lying in the garden, looking up at the passing clouds and blue sky? Daybeds have been around for hundreds of years and continue to entice people to be outside, resting and relaxing. A mixture of materials and styles can be used.

£–£££ ♦–♦♦ ⁄⁄ various

Hybrid daybed
With the ever-expanding range of garden furniture available, some pieces can offer a range of functions. Hybrid daybeds have an adjustable backrest and can be used as a bench, daybed, or lounger. A variety of styles, colours, and materials makes them relevant to any design style.

£–££ ♦–♦♦ ⁄⁄ various

Lounger and covered daybed
There are some really amazing designs for ways to lie down and relax in a garden – the antidote to a busy day gardening. Covered loungers and daybeds are equipped with canopies for screening, shade, and privacy. Shapes, sizes, and materials can all vary, as can price points.

£–£££ ♦–♦♦ ⁄⁄ various

Outdoor sofa
Modular sofas that can be increased or decreased in size, and be large or small in scale, are popular additions to most gardens. Waterproof cushions are increasingly available, so the large foam pads can stay out in more than just the sunniest of days. Some cheaper furniture does suffer from fading fabrics.

£–£££ ♦–♦♦ ⁄⁄ various

Polyrattan furniture
One of the big developments in recent years has been the arrival of non-wicker rattan. There are myriad designs and layouts, from corner sofas to dining sets, and they can be very stylish additions. Durability is higher than real rattan, and polyrattan requires little maintenance.

£–£££ ♦–♦♦ ⁄⁄ polyrattan

Wooden dining set
Wood has been used for years to sit or perch on, and there are plenty of wooden dining set designs that will work in any outside space. Tables, chairs, and benches are made from a range of sustainable timbers and treated for extra longevity. Weathering will change the colour of the timber over time.

£–£££ ♦–♦♦ ⁄⁄ timber

Aluminium garden furniture
Lightweight, sleek, and contemporary, aluminium garden furniture offers a lot of positives. Fitting well into more modern settings, the fabric colour and finish can tonally match with paintwork or other garden features. It is usually treated to withstand the elements, so requires little maintenance.

£–££ ♦–♦♦ ⁄⁄ metal

Bench
Whether made from stone, timber, concrete, plastic, or polyrattan, benches offer function and form. Some include storage compartments under the seat base, while others can be more sculptural or organic in style. They may need to be stored away during the winter.

£–£££ ♦–♦♦ ⁄⁄ various

Outdoor living

Outdoor bar
What is better than an outdoor bar for refreshing drinks on a warm summer's day? Modular and flexible by design, outdoor bars provide everything from bar stools and wine coolers to space for a barbecue or drinks dispenser. High levels of personalization can be achieved too.

£–£££ ♦–♦♦ various

Pizza oven
There are so many pizza ovens available – from portable devices to more high-tech and substantial types. They can take charcoal, wood, or gas, and can heat up the internal stone very quickly. The pizzas may not be as standardized as they are in restaurants, but they'll have a highly authentic and delicious flavour.

£–££ ♦–♦♦ various

Barbecue
Standard barbecues have been around for many years, whether on wheels, built into a wall, or standalone. Some still use charcoal; others, wood or gas – or perhaps a hybrid combination of two. Either way, there are styles, sizes, and approaches to suit all tastes.

£–£££ ♦–♦♦ metal

Tabletop barbecue
Sometimes cooking on a standalone barbecue can feel rather onerous if you haven't got much time or it's a meal for one or two. Tabletop cooking is far easier, and a small, compact, and easy-to-reach barbecue is all that's needed. Some barbecues are electric for smaller spaces like rooftops or balconies.

£–££ ♦–♦♦ metal

Built-in barbecue
If you are keen on barbecuing frequently, it's worth considering building one into a larger, more considered design feature. Stainless steel grills and trays are used to cook the food on, and a variety of fire-resistant materials can be used for the actual construction.

£–££ ♦–♦♦ metal

Smoker
Adding to the dining experience, smokers are a step up from plain barbecuing and are increasingly popular. They have a side chamber for the smoking box, where charcoal and wood chips are burnt. The smoke is then drawn into the main cooking chamber to cook and flavour the food.

£–£££ ♦–♦♦ metal

Open-air oven
Combining style and substance, outdoor ovens are not only a focal point in their own right, but also they can cook delicious food. Whether wood- or gas-fired, they can reach high temperatures, but are generally manageable enough to control for a great cooking experience.

££–£££ ♦–♦♦ various

Dutch oven
Cooking above an open fire goes back to when humans first started cooking meat. A Dutch oven, a cast iron pot that can hang over flames, is a great way to produce simple, all-in-one, pot-type cooking. Ensure heat-proof gloves are always to hand, and use a hanging chain to vary the cooking height.

£–££ ♦–♦♦ iron

Fire pit and swing arm
Fire pits are great additions to the outdoor living experience, and the enjoyment of them can be enhanced with swing arms. These are moveable arms that sit on top of the pit so the heat can be managed and food cooked well. In effect, it's just another way of barbecuing.

£–££ ♦–♦♦ metal

MATERIALS GUIDE   367

£££ high cost   ££ medium cost   £ low cost   ♦♦ high durability   ♦ low durability   ◔ colour options

Concrete kitchen
Concrete is often overlooked in garden design. Of course, there is a sustainability issue (the high carbon footprint of concrete manufacturing), but by recycling or reusing existing concrete products, a hard-wearing and tough-looking cooking environment can be created.

£–££ ♦–♦♦ ◔ concrete

Modular kitchen
A range of tempting outdoor kitchens is available from an array of manufacturers. Modular by design, they offer personalization in layout and materials. Some have sinks, barbecues, and grills; others, wine coolers and in-built ovens. They all take outdoor dining to a new level of enjoyment and ease.

£–£££ ♦–♦♦ ◔ various

Wooden kitchen
Outdoor kitchens come in a variety of materials, but perhaps wood is the most eye-catching. Offering a variety of facilities and cooking appliances, wooden outdoor kitchens weather, stain, and change appearance over time, almost like a living wooden sculptural feature.

£££ ♦♦ ◔ wood

Sink
Whether it's preparing and cooking food, or a needing a place to wash tools or hands, outdoor sinks can be a really useful addition to any garden. Ensure plumbing is undertaken professionally: pipes need to be insulated in winter, and dirty water needs to be able to drain away.

£–££ ♦–♦♦ ◔ various

Integrated dining area
Some people want more than just an outdoor barbecue or sink – they want to experience cooking, eating, and socializing all in the same place. Bespoke kitchen companies can offer everything from a canopy to integrated cooking appliances. It's a smart (but often expensive) way of eating al fresco.

£££ ♦♦ ◔ various

Gazebo
Having a simple metal frame around you helps enhance any dining experience – whether a quick cup of tea or a full-blown meal. You can grow plants around and over the structure for extra shade, privacy, and interest, while the shape can be as ornamental or simple as your garden requires.

£–££ ♦–♦♦ ◔ metal

Thatched gazebo
There is something lovely about being out in the garden yet ensconced in a cosy seating area. Thatched gazebos can be placed in any garden, and have a range of finishes and materials to suit everything from traditional to jungle and modern urban. Lighting, furniture, and electrics can all be specified.

£££ ♦♦ ◔ wood

Pergola awning
Not everyone wants to cook outside – some are happy to prepare food indoors, but they want the al fresco eating experience in a cosy, relatively sheltered environment. Pergolas with integrated awnings give that feeling of protection and are available in a range of sizes.

£–££ ♦–♦♦ ◔ various

Slatted side pavilion
It is human nature to want to feel protected and enclosed when sitting outside. A slatted side pavilion creates shade and enclosure in a contemporary design, with clean lines and modern timber finishes helping to make a design statement. Slats can be used as trellising for climbers.

£–£££ ♦♦ ◔ timber

370 Understanding hardiness ratings
371 Designers' details
372 Acknowledgments

RESOURCES

Understanding hardiness ratings

All plants in the Plant Guide (pp.288–349) have been assigned RHS hardiness ratings, using one of nine categories – H1a to H7 – determined by the lowest temperature range the plant is likely to withstand, along with various other factors, such as the relative exposure of the planting location. These ratings serve as a general guide to growing conditions and should be interpreted according to the table below. Bear in mind, however, that they are guidelines only, and many other factors will affect a plant's overall hardiness.

RATING	TEMPERATURE RANGE	CATEGORY	DEFINITION
H1a	warmer than 15°C (59°F)	Heated greenhouse – tropical	Grow as a house plant or under glass all year round.
H1b	10–15°C (50–59°F)	Heated greenhouse – subtropical	Can be grown outside in summer in hotter, sunny, and sheltered locations, but generally performs better as a house plant or under glass all year round.
H1c	5–10°C (41–50°F)	Heated greenhouse – warm temperate	Can be grown outside in summer throughout most of the UK while daytime temperatures are high enough to promote growth.
H2	1–5°C (34–41°F)	Cool or frost-free greenhouse	Tolerant of low temperatures, but will not survive being frozen. Except in frost-free inner-city areas or coastal extremities, requires greenhouse conditions in winter. Can be grown outside once risk of frost is over.
H3	-5–1°C (23–34°F)	Unheated greenhouse/ mild winter	Hardy in coastal and relatively mild parts of the UK, except in hard winters and at risk from sudden, early frosts. May be hardy elsewhere with wall shelter or a good microclimate. Can often survive with some artificial protection in winter.
H4	-10–-5°C (14–23°F)	Average winter	Hardy throughout most of the UK apart from inland valleys, at altitude, and central/northerly locations. May suffer foliage damage and stem dieback in harsh winters in cold gardens. Plants in pots are more vulnerable.
H5	-15–-10°C (5–14°F)	Cold winter	Hardy in most places throughout the UK, even in severe winters. May not withstand open or exposed sites or central/northerly locations. Many evergreens will suffer foliage damage and plants in pots will be at increased risk.
H6	-20–-15°C (-4–5°F)	Very cold winter	Hardy in all of the UK and northern Europe. Many plants grown in containers will be damaged unless given some artificial protection in winter.
H7	colder than -20°C (-4°F)	Very hardy	Hardy in the severest European continental climates, including exposed upland locations in the UK.

Designers' details

The publisher would like to thank the following garden designers for their contributions:

Charlie Albone
charliealbone.com

Marcus Barnett
marcusbarnett.com

Chris Beardshaw
chrisbeardshaw.com

Jinny Blom
jinnyblom.com

Jane Brockbank Gardens
janebrockbank.com

Declan Buckley
buckleydesignassociates.com

Alasdair Cameron
camerongardens.co.uk

Vladimir Djurovic
vladimirdjurovic.com

Prof. Nigel Dunnett
nigeldunnett.com

Sarah Eberle
sarah-eberle.com

Helen Elks-Smith
elks-smith.co.uk

Naomi Ferrett-Cohen
naomiferrettcohen.com

Adam Frost
adamfrost.co.uk

Bunny Guinness
bunnyguinness.com

Stephen Hall
(Giles Landscapes)
gileslandscapes.co.uk

Harry Holding
harryholding.co.uk

Colm Joseph Gardens
colmjoseph.co.uk

Maggie Judycki
(Green Themes, Inc.)
greenthemes.com

Matt Keightley
(Rosebank Landscaping)
rosebanklandscaping.co.uk

Carly Kershaw
(Hyland Edgar Driver)
heduk.com

Marcio Kogan
(Studio MK27)
studiomk27.com.br

Catherine MacDonald
(Landform Consultants)
landformconsultants.co.uk

Steve Martino
stevemartino.net

Tom Massey
tommassey.co.uk

Claire Mee
clairemee.co.uk

Robert Myers
robertmyers-associates.co.uk

Philip Nixon
philipnixondesign.com

Gabriella Pape and
Isabelle van Groeningen
koenigliche-gartenakademie.de

Sara Jane Rothwell
(London Garden Designer)
londongardendesigner.com

Charlotte Rowe
charlotterowe.com

Alan Rudden
alanrudden.ie

Nicola Stocken
nicolastocken.com

Andy Sturgeon
andysturgeon.com

Angus Thompson
angusthompsondesign.com

Jo Thompson
jothompson-garden-design.co.uk

Stuart Charles Towner
stuartcharlestowner.co.uk

Cleve West
clevewest.com

Will Williams
willwilliamsgardendesign.com

Nick Williams-Ellis
nickwilliamsellis.co.uk

Andrew Wilson
(McWilliam Studio)
mcwilliamstudio.com

Acknowledgments

The publisher would like to thank the following for their kind permission to reproduce their photographs:

(Key: a-above; b-below/bottom; c-centre; f-far; l-left; r-right; t-top)

2 GAP Photos: Pernilla Bergdahl/Design: Sarah Eberle.

4 GAP Photos: Heather Edwards/Design: Will Williams (br); Nicola Stocken, NHS Tribute Garden/Design: Naomi Ferrett-Cohen (tl); **Clive Nichols:** Clive Nichols/Design: Matt Keightley (bl).

4–5 GAP Photos: Anna Omiotek-Tott/Design: Richard Miers.

6–7 GAP Photos: Joanna Kossak/Design: Alan Rudden (tr); Robert Mabic/Design: Ekaterina Zasukhina & Carly Kershaw (tl); **MMGI:** Marianne Majerus/Design: Stuart Charles Towner (br); **Clive Nichols:** Clive Nichols/Design: Alasdair Cameron (bl).

8 The RHS Images Collection: RHS/Neil Hepworth, design: Adam Frost, RHS Chelsea 2015.

11 The Garden Collection: Jonathan Buckley/Design: Judy Pearce (cr); **The RHS Images Collection:** RHS/Neil Hepworth (tr); **The Garden Collection:** Derek Harris (bl); Torie Chugg/RHS Chelsea 2008 (tl).

12–13 GAP Photos: Joanna Kossak/Design: Andy Sturgeon.

14 The Garden Collection: Andrew Lawson/Design: Jinny Blom (tl); **MMGI:** Marianne Majerus/Design: Sara Jane Rothwell (tr); **Photolibrary:** David Cavagnaro (bl).

15 Harpur Garden Library: Jerry Harpur/Design: Shunmyo Masuno (tl); **MMGI:** Marianne Majerus/Palazzo Cappello, Venice (bl); **Photolibrary:** Michael Howes (br); **Richard Felber:** Design: Raymond Jungles Landscape Architect (tr).

16 Charles Hawes: "Artificial Paradise"/Design: Catherine Baas & Jean-Francis Delhay (France), Chaumont International Gardens Festival 2003 (tl); **MMGI:** Marianne Majerus/Claire Mee Designs (br); Marianne Majerus/Design: Andy Sturgeon, RHS Chelsea 2006 (tr); Marianne Majerus/Design: Charlotte Rowe (bl).

17 The Garden Collection: Liz Eddison (tr); **DK Images:** Peter Anderson/RHS Chelsea 2009 (tl); **Photolibrary:** Michael Howes/Design: Dean Herald, Fleming's Nurseries, RHS Chelsea 2006 (br).

18 The Interior Archive: Simon Upton (tr); **MMGI:** Bennet Smith/Design: Mary Nuttall (tl); Marianne Majerus/Henstead Exotic Garden/Andrew Brogan, Jason Payne (tc); **Photolibrary:** John Ferro Sims (br); **Richard Felber:** Design: Raymond Jungles Landscape Architect (bc).

19 Helen Fickling: Design: Williams, Asselin, Ackaqui et Associés/International Flora, Montreal (br); **Charles Hawes:** Design: Laureline Salisch & Seun-Young Song, Ecole Supérieure d'Art et de Design (ESAD) Reims, Chaumont International Festival 2007 (tr); **MMGI:** Marianne Majerus/Design: Arabella Lennox-Boyd, RHS Chelsea 2008 (tl); Marianne Majerus/Design: Charlotte Rowe (tc); **Clive Nichols:** Data Nature Associates (bl); Design: Stephen Woodhams (bc).

20 MMGI: Marianne Majerus/Design: Will Giles, The Exotic Garden, Norwich (tr); **Photolibrary:** Linda Burgess (tl).

21 MMGI: Bennet Smith/Design: Denise Preston, Leeds City Council, RHS Chelsea 2008 (tl); **Undine Prohl:** Dry Design (tr).

22 GAP Photos: Rachel Warne/Design: Angus Thompson.

23 GAP Photos: Paul Debois (br); Rob Whitworth/Design: Tony Wagstaff (tr).

24 GAP Photos: Christina Bollen (tr); Gary Smith (tl); Jenny Lilly/Design: Richard Wanless (bl); Robert Mabic (br); Elke Borkowski (cr).

25 GAP Photos: Heather Edwards/Design: Charles Dowding (tr); Heather Edwards/Charles Dowding (cr); Annie Green-Armytage (bl) (cl).

26 GAP Photos: GAP Photos/Nova Photo Graphik (bl); Jenny Lilly (tl). **Clive Nichols:** Clive Nichols/Design: Martha Krempel, London (tr) (br).

27 GAP Photos: Christa Brand (t); Heather Edwards/Design: Richard Miers (b).

29 Alamy Images: Holmes Garden Photos (tl); **The Garden Collection:** Derek St Romaine/Design: Woodford West, RHS Chelsea 2001 (tr); **MMGI:** Marianne Majerus/Gainsborough Road, Alastair Howe Architects (t). **Roger Foley:** (br); **Harpur Garden Library:** Jerry Harpur/Design: Philip Nixon, RHS Chelsea 2008 (bl); **MMGI:** Marianne Majerus/Design: Jonathan Baille (bc).

30 MMGI: Bennet Smith/Design: Mary Nuttall (tl); Marianne Majerus/Design: Charlotte Rowe (tr). **GAP Photos:** Lynne Keddie (bl); **Steve Gunther/**Design: Steve Martino (bc); **MMGI:** Marianne Majerus/Gunnebo House, Gardens of Gothenburg Festival, Sweden 2008, Joakim Seiler (br).

32 MMGI: Marianne Majerus/Design: Tom Stuart-Smith, RHS Chelsea 2000.

33 GAP Photos: Brian North (r).

34–35 The RHS Images Collection: RHS/Neil Hepworth, design: Charlie Albone, RHS Chelsea 2016.

35 The Garden Collection/Design: Tom Stuart-Smith, RHS Chelsea 2005 (4); **Harpur Garden Library:** Jerry Harpur (tl); **Clive Nichols/**Design: Dominique Lafourcade, Provence (l); www.stonemarket.co.uk (5).

36 GAP Photos: Joanna Kossak/Design: Charlie Albone, Sponsor: Husqvarna (tl).

36–37 James Silverman: www.jamessilverman.co.uk/Architect: Marcio Kogan, Brazil.

37 Alamy Images: Andrea Jones/Design: Buro Landrast, Floriade (4); Matthew Noble Horticultural/Design: Lizzie Taylor & Dawn Isaac, RHS Chelsea 2005 (2); **DK Images:** Design: Marcus Barnett & Philip Nixon, RHS Chelsea 2007 (1); Design: Denise Preston, RHS Chelsea 2008 (3); Design: Philip Nixon, RHS Chelsea 2008 (5); **Peter Anderson:** (tl).

38 GAP Photos: Jerry Harpur/Design: L Giubbilei (clb); Jo Whitworth (cla); **MMGI:** Marianne Majerus/Design: Del Buono Gazerwitz (tr); **Photolibrary:** Marijke Heuff (br).

39 Andrew Lawson: Design: Christopher Bradley-Hole (b); **Charles Mann:** Sally Shoemaker, Phoenix AZ (tl); **B & P Perdereau:** Design: Yves Gosse de Gorre (c).

40 MMGI: Marianne Majerus/Design: Declan Buckley (br) (2)(5); **Photolibrary:** John Glover (6).

41 Clive Nichols: Clive Nichols/Design: Matt Keightley.

42 MMGI: Marianne Majerus/Design: Charlotte Rowe (br) (1).

43 GAP Photos: Clive Nichols (1); **Harpur Garden Library:** Jerry Harpur/Design: Andy Sturgeon, London (br) (2) (4); **Photolibrary:** John Glover (3).

44–45 The RHS Images Collection: RHS/Neil Hepworth, design: Marcus Barnett, RHS Chelsea 2015.

46 GAP Photos: Nicola Stocken.

47 GAP Photos: FhF Greenmedia (r).

48–49 The Garden Collection: Nicola Stocken.

49 The Garden Collection: Nicola Stocken (3); **Harpur Garden Library:** Marcus Harpur/Design: Gertrude Jekyll, Owners: Sir Robert and Lady Clark, Munstead Wood, Surrey (b); **MMGI:**

Marianne Majerus/Bryan's Ground, Herefordshire (2).

50 GAP Photos: Nicola Stocken (c). **Clive Nichols:** Clive Nichols/Design: Alasdair Cameron (tr).

51 GAP Photos: Joanna Kossak/Design: Tony Woods/Garden Club London (tl); Caroline Mardon/Design: Karen Rogers www.krgardendesign.com (br).

52 GAP Photos: John Glover/Design: Penelope Hobhouse (tr); Jerry Harpur/Design: Britte Schoenaic (br); **Harpur Garden Library:** Jerry Harpur/Design: Christopher Lloyd, Great Dixter (cla).

52–53 Andrew Lawson: Design: Arabella Lennox-Boyd.

53 The Garden Collection: Andrew Lawson/Design: Oehme van Sweden (tr); **Harpur Garden Library:** Jerry Harpur/Design: Piet Oudolf (r).

54 The Garden Collection: Liz Eddison/Design: Gabriella Pape & Isabelle Van Groeningen, RHS Chelsea 2007 (br); **Clive Nichols:** (4); **Photolibrary:** Kit Young (1); Tracey Rich (6).

55 Nicola Browne: Design: Jinny Blom (t).

56 GAP Photos: Jo Whitworth (6); **The Garden Collection:** Jane Sebire/Design: Nigel Dunnett (br) (4).

57 The Garden Collection: Gary Rogers/Design: Rendel & Dr James Bartons (t) (6); **MMGI:** Marianne Majerus (l).

58–59 Clive Nichols: Clive Nichols/Design: Matt Keightley.

60 GAP Photos: Hanneke Reijbroek.

61 MMGI: Marianne Majerus/Claire Mee Designs (cra); Marianne Majerus/Design: Lynne Marcus (crb).

62 DK Images: Design: Franziska Harman, RHS Chelsea 2008 (3); Design: Paul Stone Gardens, RHS Hampton Court 2007 (6); **MMGI:** Marianne Majerus/Claire Mee Designs (1); **TopFoto.co.uk:** (fcl).

63 Steve Gunther: Design and installation: Chuck Stopherd of Hidden Garden Inc. of CA.

64 DK Images: www.jcgardens.com (t); **Harpur Garden Library:** Jerry Harpur/Design: Ryl Nowell (bl); **MMGI:** Marianne Majerus/Design: Lucy Sommers (tl); Marianne Majerus/Design: David Rosewarne (br).

64–65 Steve Gunther: Design: Sandy Koepke, LA (c).

65 Harpur Garden Library: Jerry Harpur/Design: Bunny Guinness (b); **Ian Smith:** Design: Acres Wild (t); **Steve Gunther:** Design: Mia Lehrer, Malibu CA (cr).

66 GAP Photos: (tl).

66–67 GAP Photos: GAP Photos/Garden Design: Cube 1994 (c); Rachel Warne (tc).

67 GAP Photos: Annie Green-Armytage (tr).

68 DK Images: Brian North/Design: Catherine MacDonald, RHS Hampton Court 2012

69 Roger Foley: Design: Maggie Judycki for Green Themes, Inc. (br) (3) (6).

70 Richard Bloom: Richard Bloom/Design: Jane Brockbank.

71 MMGI: Marianne Majerus/Claire Mee Designs (tl) (tr).

72–73 The RHS Images Collection: RHS/Tim Sandall, design: Nick Buss & Clare Olof, RHS Hampton Court 2012.

74 The RHS Images Collection: RHS/Neil Hepworth, design: Cleve West, RHS Chelsea 2016.

75 GAP Photos: Jonathan Buckley/Design: Christopher Lloyd, Great Dixter (a); **The Garden Collection:** Liz Eddison/Design: Daniel Lloyd Morgan, RHS Hampton Court 2001 (b).

76 DK Images: Design: Teresa Davies, Steve Putnam, Samantha Hawkins, RHS Chelsea 2007 (l); **Harpur Garden Library:** Jerry Harpur/Design: Rosemary Weisse, West Park, Munich, Germany (l).

76–77 DK Images: Design: Stephen Hall, RHS Chelsea 2005.

77 DK Images: Design: Kate Frey, RHS Chelsea 2007 (3); Design: English Heritage Gardens (4).

78 GAP Photos: Jerry Harpur (t).

78–79 Helen Fickling: Design: Andy Sturgeon.

79 DK Images: Steven Wooster (2) (4); **GAP Photos:** Jerry Harpur/Pashley Manor (3); S & O (6).

80 Clive Nichols: Clive Nichols/Design: Harry Holding (t, bl).

80–81 GAP Photos: Annie Green-Armytage/Designers: Tom Massey and Sarah Mead (c).

81 Clive Nichols (br).

82 Richard Bloom: Richard Bloom/Design: Helen Elks-Smith (c); **GAP Photos:** Andrea Jones (t).

83 GAP Photos: Jenny Lilly (t, bl); Howard Rice, 26 St Barnabas Road, Cambridge (cr).

84 DK Images: Peter Anderson/Design: Jo Thompson, RHS Chelsea 2009.

85 Richard Bloom: Richard Bloom/Design: Jane Brockbank.

86–87 DK Images: Peter Anderson/Design: Cleve West, RHS Chelsea 2011.

88 GAP Photos: Richard Bloom.

89 MMGI: Marianne Majerus/Design: Paul Cooper (cra); **GAP Photos:** Richard Bloom/Design: Katharina Nikl Landscapes (crb).

90–91 Harpur Garden Library: Jerry Harpur/Design: Philip Nixon.

91 GAP Photos: Clive Nichols/Design: Amir Schlezinger My Landscapes (3); Jerry Harpur/Design: Fiona Lawrenson & Chris Moss (4); Jerry Harpur/Design: Luciano Giubbilei (1); **MMGI:** Marianne Majerus www.finnstone.com (2); Marianne Majerus/Design: Lucy Sommers (5).

92 GAP Photos: Friedrich Strauss (tl).

92–93 GAP Photos: Richard Bloom/Design: John Davies Landscape (tc); Joanna Kossak/Design: Tony Woods (cb).

93 GAP Photos: Nicola Stocken (tr); **MMGI:** Marianne Majerus/Design: Jane Brockbank (b); **Clive Nichols:** Clive Nichols/Design: Matt Keightley (tc).

94 GAP Photos: Matteo Carassale/Design: Cristina Mazzucchelli (tl); John Glover (bl); **Clive Nichols:** Clive Nichols/Design: Alasdair Cameron (t).

94–95 MMGI: Marianne Majerus/Design: Charlotte Rowe.

95 MMGI: Marianne Majerus/Design: Sara Jane Rothwell (t); **DK Images:** Design: Mark Gregory, RHS Chelsea 2008 (b).

96 GAP Photos: Suzie Gibbons (tl); **MMGI:** Marianne Majerus/Design: Sara Jane Rothwell (bl).

96–97 Marion Brenner: Design: Joseph Bellomo Architects, Palo Alto CA.

97 Henk Dijkman: www.puurgroen.nl (bc); **Harpur Garden Library:** Jerry Harpur/Design: Christopher Bradley-Hole (c) (r).

98 Harpur Garden Library: Jerry Harpur/Design: Vladimir Djurovic, Lebanon.

99 Nicola Browne: Design: Pocket Wilson (t) (1); **GAP Photos:** Richard Bloom (3/c); **Charles Hawes** (5/c); **GAP Photos:** Clive Nichols/Design: Nigel Dunnett & The Landscape Agency (b).

100–101 Clive Nichols: Clive Nichols/Design: Alasdair Cameron.

101 Clive Nichols: Clive Nichols/Design: Alasdair Cameron (br).

102 GAP Photos: John Glover/Design: Rosemary Verey.

103 GAP Photos: Mark Bolton (tr); Joanna Kossak/Design: Tracy Foster, Sponsor: Just Retirement Ltd (br).

104–105 GAP Photos: Elke Borkowski.

105 GAP Photos: Elke Borkowski (tl); **DK Images:** Peter Anderson/RHS Hampton Court 2014 (5).

106 GAP Photos: Maddie Thornhill (cl); **Clive Nichols** (tr, cr, b).

107 Richard Bloom: Richard Bloom/Design: Jane Brockbank (tl); **GAP Photos:** (tr); Jenny Lilly (br); **Clive Nichols** (cl).

108 GAP Photos: Friedrich Strauss (l).

108–109 DK Images: Peter Anderson/Design: Heather Culpan and Nicola Reed, RHS Hampton Court 2011.

109 GAP Photos: Elke Borkowski (c); **Red Cover:** Ron Evans (t); **DK Images:** Peter Anderson/Design: Bunny Guinness, RHS Hampton Court 2011 (b).

110 Clive Nichols.

111 The Garden Collection: Jonathan Buckley/Design: Bunny Guinness (t) (4); **Photolibrary:** Mark Winwood (3/c); **GAP Photos:** J. S. Sira/Design: Ron Carter (cb).

112 GAP Photos: Nicola Stocken.

113 GAP Photos: Stephen Studd/Design: Andy Bending.

114–115 GAP Photos: Brian North/Design: Nick Williams-Ellis.

116 The RHS Images Collection: RHS/Sarah Cuttle, design: Ruth Willmott, RHS Chelsea 2015.

117 GAP Photos: Richard Bloom (ar); **MMGI:** Andrew Lawson/Design: Philip Nash, RHS Chelsea 2008 (br).

118 Michael Schultz Landscape Design (br).

118–119 Harpur Garden Library: Jerry Harpur/Design: Steve Martino.

119 DK Images: Design: Matthew Rideout, RHS Hampton Court 2008 (1); Design: Paul Cooper, RHS Chelsea 2008 (3); **GAP Photos:** Fiona McLeod/Design: Cleve West, RHS Chelsea 2006 (5); **The Garden Collection:** Liz Eddison/Design: Reaseheath College, RHS Tatton Park 2007 (6); **Harpur Garden Library:** Jerry Harpur/Design: Sonny Garcia (4).

120 Helen Fickling: Design: Marie-Andrée Fortier, Art & Jardins, International Flora, Montreal, Canada (b); **Harpur Garden Library:** Jerry Harpur/Design: Vladimir Sitta (c).

120–121 Helen Fickling: Architect: Claude Cormier, International Flora, Montreal, Canada (t).

121 Marion Brenner: Design: Andrea Cochran Landscape Architect, San Francisco (c); **Harpur Garden Library:** Jerry Harpur/Design: Steve Martino (cr); **Steve Gunther:** Architect: Ricardo Legorreta/Landscape Architect: Mia Lehrer & Associates, LA (br); **Harpur Garden Library:** Jerry Harpur/Design: Peter Latz & Associates, Chaumont Festival, France (bl).

122 GAP Photos: Andrea Jones/Design: Adam Frost, Sponsor: Homebase (bl); **Clive Nichols:** Clive Nichols/Design: Wynniat-Husey Clarke (cl) (br).

123 Alamy Stock Photo: Ellen Rooney (t); **MMGI:** Marianne Majerus/Design: Nic Howard, *Aeon* sculpture: David Harber (bl); Marianne Majerus/Design: Nic Howard, sculpture: David Harber (br).

124 DK Images: Peter Anderson/Design: Robert Myers, RHS Chelsea 2011.

125 Richard Bloom: Richard Bloom/Design: Colm Joseph Gardens.

126 Alamy Stock Photo: Guy Bell/Alamy Live News/Design: Sarah Eberle.

127 GAP Photos: Paul Debois/Design: Antony Watkins.

128–129 The RHS Images Collection: RHS/Neil Hepworth, design: Andy Sturgeon, RHS Chelsea 2016.

130–131 GAP Photos: Joanna Kossak/Design: Chris Beardshaw.

132 Charles Mann.

133 MMGI: Marianne Majerus/Design: Sally Hull (crb).

136 MMGI: Marianne Majerus/Design: Julie Toll (bl).

137 DK Images: Design: Kate Frey, RHS Chelsea 2007 (t); **MMGI:** Marianne Majerus/Design: Wendy Booth & Leslie Howell (b).

138 MMGI: Marianne Majerus/Design: James Lee (l); Marianne Majerus/P & M Hargreaves, Grafton Cottage, Staffs (c); **GAP Photos:** Robert Mabic/Design: Tom Massey (r).

139 DK Images: Design: Jason Lock & Chris Deakin, RHS Chelsea 2008 (fbl); **GAP Photos:** Jerry Harpur/Design: Roberto Silva (cla); **The Garden Collection:** Derek St Romaine/Glen Chantry, Essex (fbr); Nicola Stocken (tr); **MMGI:** Marianne Majerus (cb); Marianne Majerus/Design: Charlotte Rowe (clb); **Photolibrary:** Ron Evans (crb).

140 The Garden Collection: Nicola Stocken (cl); **MMGI:** Marianne Majerus/Design: Anthony Paul Landscape Design (bl).

141 Nicola Browne: Design: Jinny Blom (c); **Jason Liske:** www.redwooddesign.com/Design: Bernard Trainor (bc); **Photolibrary:** Jerry Pavia (t).

142 GAP Photos: Nicola Stocken/Design: Andy Sturgeon.

143 GAP Photos: Jerry Harpur/Design: Scenic Blue, RHS Chelsea 2007 (cra).

144 GAP Photos: Mark Bolton, The Wild Kitchen Garden/Design: Ann Treneman (tr); Rob Whitworth/Design: Kate Gould, Sponsors: Kate Gould Gardens Ltd (br) (tl, bl).

145 GAP Photos: Maxine Adcock (tl); Thomas Alamy (bl); Jenny Lilly (tc); **GAP Photos:** Design: Luke Heydon, Sponsor: Thetford businesses and residents (bc); Anna Omiotek-Tott/Design: Andy Clayden, Dr Ross Cameron (tr) (br).

146 GAP Photos: Carole Drake/Design: Andrea Newill (tr, br); Clare Forbes/Design: Matthew Wilson, Sponsor: Royal Bank of Canada (c); Joanna Kossak/Design: Rosemary Coldstream (bl).

147 GAP Photos: Paul Debois/Design: Caro Garden Design (br); Rachel Warne (t, bl).

148 Alamy Images: CW Images (tl); **DK Images:** Alex Robinson (tr); **GAP Photos:** John Glover (cl); **DK Images:** Peter Anderson/Design: Kati Crome and Maggie Hughes, RHS Chelsea 2013 (cfr); **DK Images:** Jon Spaull (bl); **MMGI:** Marianne Majerus/Kingstone Cottages (br).

149 Jason Liske: www.redwooddesign.com/Design: Bernard Trainor (tr); **GAP Photos:** Elke Borkowski/Design: Adam Woolcott (cr); Clive Nichols (cl); **MMGI:** Marianne Majerus/Claire Mee Designs (fbr); Marianne Majerus/Design: Bunny Guinness (b).

150–151 The Garden Collection: Jonathan Buckley/Design: Diarmuid Gavin.

151 Design: Amanda Yorwerth.

152 The Garden Collection: Derek St Romaine/Design: Phil Nash (r); **MMGI:** Marianne Majerus/Design: Laara Copley-Smith (c); Marianne Majerus/Palazzo Cappello, Malipiero, Barnabo, Venice (l).

153 DK Images: Design: Sarah Eberle, RHS Chelsea 2007 (tl); **MMGI:** Marianne Majerus/Design: Lynne Marcus (cl).

154–155 Case study: Design: Fran Coulter, Owners: Jo & Paul Kelly.

155 The Garden Collection: Liz Eddison/Design: Kay Yamada, RHS Chelsea 2003 (br); **Harpur Garden Library:** Marcus Harpur/Design: Justin Greer (fbr); **MMGI:** Marianne Majerus/Design: Jessica Duncan (cr); Marianne Majerus/Design: Wendy Booth, Leslie Howell (ftr).

156 MMGI: Marianne Majerus/Claire Mee Design (t); Marianne Majerus/Design: Lynne Marcus, John Hall (b).

156–157 Marion Brenner: Design: Andrea Cochran Landscape Architecture.

157 Jason Liske: www.redwooddesign.com/Design: Bernard Trainor (tr).

158 Nicola Browne: Design: Jinny Blom (br); **DK Images:** Design: Graduates of the Pickard School of Garden Design (cl).

158–159 Harpur Garden Library: Jerry Harpur/Architect: Piet Boon, Planting Design: Piet Oudolf.

159 DK Images: Design: Paul Williams (bl); **The Garden Collection:** Gary Rogers/Chatsworth House (br); **Charles Hawes:** Designed & created by Tony Ridler, The Ridler Garden, Swansea, ammonite sculpture: Darren Yeadon (ca).

160 MMGI: Bennet Smith/Design: Ian Dexter, RHS Chelsea 2008 (c); Marianne Majerus/Design: Anthony Tuite (b).

160–161 The Garden Collection: Nicola Stocken.

161 GAP Photos: Nicola Stocken (tl); **MMGI:** Bennet Smith/Design: Thomas Hoblyn (br).

162 Garden Exposures Photo Library: Andrea Jones/Design: Dan Pearson & Steve Bradley (cl); **The Garden Collection:** Liz Eddison/Design: Alan Sargent, RHS Chelsea 1999 (bl).

162–163 The Garden Collection: Jonathan Buckley/Design: Joe Swift & Sam Joyce for the Plant Room.

163 Roger Foley: Scott Brinitzer Design Associates (br); **MMGI:** Marianne Majerus/Design: Paul Cooper (bc).

164 MMGI: Marianne Majerus/Design: Sara Jane Rothwell.

165 GAP Photos: Lynn Keddie (ca); Nicola Stocken (bl); **MMGI:** Marianne Majerus/Design: Charlotte Rowe (tl); Marianne Majerus/Design: Nicola Gammon, www.shootgardening.co.uk (tr); Marianne Majerus/Design: Fiona Lawrenson & Chris Moss (fbr); **Derek St Romaine:** Design: Koji Ninomiya, RHS Chelsea 2008 (br).

166 DK Images: Peter Anderson/RHS Hampton Court 2014 (tr); **MMGI:** Marianne Majerus/Fiveways Cottage (cla); Marianne Majerus/Design: Paul Dracott (bl); **B & P Perdereau:** Design: Yves Gosse de Gorre (crb).

167 The Garden Collection: Jonathan Buckley/Design: Diarmuid Gavin (bl); **MMGI:** Marianne Majerus/Design: Lynne Marcus (tl); Marianne Majerus/Design: Arabella Lennox-Boyd, RHS Chelsea 2008 (cra); Marianne Majerus/Design: Chris Perry, Claire Stuckey, Jill Crooks, & Roger Price, RHS Chelsea 2005 (br).

168 Harpur Garden Library: Jerry Harpur/Design: Made Wijaya & Priti Paul (bc); **Photolibrary:** Peter Anderson/Design: Martha Schwartz (br).

169 DK Images: Design: Marcus Barnett & Philip Nixon, RHS Chelsea 2007 (t); **The Garden Collection:** Derek Harris (c); **MMGI:** Marianne Majerus/Leonards Lee Gardens, West Sussex (b).

170 GAP Photos: Richard Bloom (cr); **MMGI:** Marianne Majerus/Design: Ali Ward (bc); **Photolibrary:** David Dixon (bl).

171 Peter Anderson (t); **GAP Photos:** Clive Nichols/Chenies Manor, Bucks (cl); **MMGI:** Andrew Lawson/Sticky Wicket, Dorset (bc); Marianne Majerus (bl) (br).

172 Helen Fickling: International Flora, Montreal (tr); **Harpur Garden Library:** Jerry Harpur/Design: Jimi Blake, Hunting Brook Gardens (cl); **MMGI:** Marianne Majerus/Design: Julie Toll (bl).

173 GAP Photos: J. S. Sira/Chenies Manor, Bucks (bc); **MMGI:** Andrew Lawson/Design: Philip Nash, RHS Chelsea 2008 (fbr); Bennet Smith/Paul Hensey with Knoll Garden, RHS Chelsea 2008 (tl); Marianne Majerus/Design: Piet Oudolf (cra); Marianne Majerus/Les Métiers du Paysage dans toute leur Excellence Jardins, Jardins aux Tuileries 2008, Christian Fournet (bl); **Clive Nichols:** Design: Wendy Smith & Fern Alder, RHS Hampton Court 2004 (cr); **Photolibrary:** Mark Bolton (tc).

174 (left to right): **DK Images; Clive Nichols:** Design: Fiona Lawrenson; **The Garden Collection:** Jonathan Buckley; **Forest Garden Ltd:** tel: 0844 248 9801 www.forestgarden.co.uk; **The Garden Collection:** Jonathan Buckley; **Photolibrary:** Roger Foley/Design: Raymond Jungles Landscape Architect (bc); **The Garden Collection:** Derek St Romaine/Design: Philip Nash (br); **Photolibrary:** Marie O'Hara/Design: Andrew Duff (bl).

175 GAP Photos: Rob Whitworth/Design: Mandy Buckland (Greencube Garden and Landscape Design), RHS Hampton Court 2010.

176 GAP Photos: Annie Green-Armytage/Design: Anna Dabrowska-Jaudi (c); Jacqui Hurst/Design: Tom Massey (tr); Joanna Kossak/Design: Tony Woods, Garden Club London (bl); Stephen Studd/Design: Kate Durr Garden Design (br).

177 GAP Photos: Leigh Clapp/Design: Tom Massey (bl); Caroline Mardon/Design: Karen Rogers www.krgardendesign.com (t); Nicola Stocken (cl); Suzie Gibbons/Design: Patricia Thirion & Janet Honour (cr); Anna Omiotek-Tott/Design: Anne Keenan (br).

178 GAP Photos: Richard Bloom/Design: Belderbos Landscapes Ltd (tr); Andrea Jones (l); Anna Omiotek-Tott/Design: Kate Gould (br).

179 GAP Photos: Thomas Alamy (br); J. S. Sira/Design: Jill Foxley, A Matter of Urgency, Hampton Court 2011 (tr); Jonathan Buckley/Design: Sarah Raven (tl); Leigh Clapp/Design: Acres Wild (bl); Richard Bloom (c).

180 GAP Photos: Clive Nichols/Design: Tony Heywood Conceptual Gardens.

181 The Garden Collection: Nicola Stocken (t).

182 The Garden Collection: Nicola Stocken (l); **MMGI:** Marianne Majerus/Design: Sara Jane Rothwell (t).

190–191 123RF.com: Dara-on Thongnoi (illo x 4).

191 TurboSquid: Agnes B. Jones (bc, crb).

192 DK Images: Design: Heidi Harvey & Fern Adler, RHS Hampton Court 2007 (bc); **MMGI:** Marianne Majerus/Leonardslee Gardens, West Sussex (br).

193 GAP Photos: Elke Borkowski (c); **MMGI:** Marianne Majerus/Coworth Garden Design (br).

194 DK Images: Design: Robert Myers, RHS Chelsea 2008 (tr); **The Garden Collection:** Nicola Stocken (b); **Charles Mann:** Sally Shoemaker, Phoenix AZ (cr); **MMGI:** Marianne Majerus/Scampston Hall, Yorks/Design: Piet Oudolf (tc); Marianne Majerus/Rectory Farm House, Orwell/Peter Reynolds (c).

195 DK Images: Design: Cleve West, RHS Chelsea 2008 (l).

196 DK Images: Design: Fran Coulter, Owners: Bob & Pat Ring (br); **GAP Photos:** Dave Zubraski (7); Sarah Cuttle (2); **Clive Nichols** (4).

197 DK Images: Design: Paul Williams (t); Design: Adam Frost (b); **GAP Photos:** Adrian Bloom (l/t); Richard Bloom (5/t) (5/b).

198 DK Images: Peter Anderson/Design: Adele Ford and Susan Willmott, RHS Hampton Court 2013.

199 GAP Photos: John Glover (crb).

200 GAP Photos: Jerry Harpur (tl); **MMGI:** Marianne Majerus (tc).

201 Brian North: (br); **Photolibrary:** Howard Rice/Cambridge Botanic Garden (cr).

202 GAP Photos: Elke Borkowski (bc); Jerry Harpur/Design: Julian & Isabel Bannerman (tr); **The Garden Collection:** Derek Harris (tc); **MMGI:** Marianne Majerus/Design: Bunny Guinness (cl).

203 Marion Brenner: Design: Mosaic Gardens, Eugene, Oregon.

204 The Garden Collection: Andrew Lawson (tc); Nicola Stocken (tr); **MMGI:** Marianne Majerus/Design: Susan Collier (bl); Marianne Majerus/RHS Wisley/Piet Oudolf (br).

205 The Garden Collection: Andrew Lawson (b); Derek St Romaine/Glen Chantry, Essex (cl); **MMGI:** Marianne Majerus/Woodpeckers, Warks (tr).

206 GAP Photos: Clive Nichols/Design: Duncan Heather (br); **MMGI:** Marianne Majerus (bc); Marianne Majerus/Design: Jill Billington & Barbara Hunt, "Flow" Garden, Weir House, Hants (bl).

207 DK Images: Steven Wooster/Design: Rebecca Phillips, Maria Ornberg, & Rebecca Heard, "Flow Glow" Garden, RHS Chelsea 2002 (r); **GAP Photos:** Elke Borkowski (l).

208 GAP Photos: Elke Borkowski (bl); John Glover (r).

209 DK Images: Design: Tom Stuart-Smith, RHS Chelsea 2008 (tr); **GAP Photos:** Elke Borkowski (br) (tl); J. S. Sira (cl); S & O (bc).

210 GAP Photos: Geoff du Feu (bl); Jerry Harpur/Design: Isabelle Van Groeningen & Gabriella Pape, RHS Chelsea 2007 (tc); **Clive Nichols:** RHS Wisley (tr).

210–211 GAP Photos: Mark Bolton.

211 GAP Photos: Elke Borkowski (tc) (cr); **Harpur Garden Library:** Jerry Harpur/Design: Beth Chatto (tr); Marcus Harpur/Writtle College (br).

212 GAP Photos: Jonathan Buckley/Design: John Massey, Ashwood Nurseries (cl); **MMGI:** Marianne Majerus/Mere House, Kent (tr); Marianne Majerus/Ashlie, Suffolk (bl).

213 GAP Photos: Clive Nichols (cl); Elke Borkowski (tl); Jonathan Buckley/Design: Wol & Sue Staines (panel right); **The Garden Collection:** Jonathan Buckley (bc).

215 MMGI: Marianne Majerus/Design: Declan Buckley (tl); Marianne Majerus/Design: Philip Nash, RHS Chelsea 2008 (tc); Marianne Majerus/Tanglefoot (bl); **Photolibrary:** Howard Rice (tr).

216 Alamy Stock Photo: Niall McDiarmid (tr); **Dorling Kindersley:** RHS Wisley (cb); **Dreamstime.com:** Jon Benito Iza (clb); Marinodenisenko (bc); **Getty Images/iStock:** annrapeepan (cla); **Shutterstock.com:** Peter Turner Photography (crb).

217 Dorling Kindersley: Mark Winwood/Downderry Nursery (ca); **Dreamstime.com:** Barmalini (tr); Makoto Hasegawa (cra); Wiertn (cb); Natalia Pavlova (bl); Iva Vagnerova (br); **Getty Images/iStock:** 2ndLookGraphics (clb).

218 Dreamstime.com: Alfotokunst (tr); Shihina (cla); Mykhailo Pavlenko (bc); **GAP Photos:** Visions (cra); **Getty Images/iStock:** ChamilleWhite (crb).

219 Alamy Stock Photo: blickwinkel/McPHOTO/HRM (cb); Marcus Harrison, plants (clb); **Dreamstime.com:** By Daphnusia (tr); Yavor Yanev (cra); Unique93 (ca); **Shutterstock.com:** Lesub (bl).

220 Alamy Stock Photo: Tim Gainey (tr); **Dreamstime.com:** John Caley (clb); Sielan (cra); Kamilpetran (bl); Tom Meaker (cb); **Getty Images:** Stockbyte/Maria Mosolova (crb); **Shutterstock.com:** jamesptrharris (cla).

221 Alamy Stock Photo: Botany vision (bl); John Richmond (tr); **Dreamstime.com:** Macsstock (crb); Michal Paulus (cla); Snowboy234 (cra); Seramo (cb).

222 DK Images: Peter Anderson/Design: Joe Swift, RHS Chelsea 2012.

223 DK Images: Design: Heidi Harvey & Fern Adler, RHS Hampton Court 2007 (cra); **GAP Photos:** J. S. Sira/Kent Design (crb).

224 Alamy Images: Mark Summerfield (bl); **DK Images:** Design: Phillippa Probert, RHS Tatton Park 2008 (br); **Harpur Garden Library:** Jerry Harpur/Design: University College Falmouth Students, RHS Chelsea 2007 (t); Jerry Harpur/East Ruston Old Vicarage, Norfolk (bc).

225 Harpur Garden Library: Jerry Harpur/Design: Julian & Isabel Bannerman (cl); Marcus Harpur/Design: Kate Gould, RHS Chelsea 2007 (cr); **MMGI:** Marianne Majerus (bl); Marianne Majerus/Design: Lynne Marcus & John Hall (bc); Marianne Majerus/Design: Michele Osborne (ca); **Photolibrary:** John Glover (tc); Stephen Wooster (cb).

226 Marion Brenner: Design: Shirley Watts, Alameda CA www.sawattsdesign.com (br); **GAP Photos:** Michael King/Ashwood Nurseries (bl); **MMGI:** Marianne Majerus/Design: Jonathan Baillie (bc); Anna Omiotek-Tott/Design: Joe Perkins (cr); **Clive Nichols:** Wingwell Nursery, Rutland (tr); **DK Images:** Design: Adam Frost, RHS Chelsea 2007 (c).

227 The Garden Collection: Jonathan Buckley/Design: Diarmuid Gavin (bc); **MMGI:** Marianne Majerus/Gardens of Gothenburg, Sweden 2008 (tr); **Photolibrary:** Botanica (br); Howard Rice (bl); **GAP Photos:** J. S. Sira, location: HCFS 2005/Design: Paul Hensey (cr).

228 DK Images: Design: Bob Latham, RHS Chelsea 2008 (bl); Design: Del Buono Gazerwitz, RHS Chelsea 2008 (br); Peter Anderson/Design: Harry & David Rich, RHS Chelsea 2013 (tl); **Harpur Garden Library:** Jerry Harpur/Design: Sam Martin, London (cr).

229 GAP Photos: Rob Whitworth/Design: Angela Potter & Ann Robinson (bc); **Harpur Garden Library:** Jerry Harpur/Design: Philip Nixon (tl); Marcus Harpur/Design: Growing Ambitions, RHS Chelsea 2008 (tr); **MMGI:** Marianne Majerus/Design: Jilayne Rickards (bl); Marianne Majerus/The Lyde Garden, The Manor House, Bledlow, Bucks (br).

230 DK Images: Design: Paul Dyer, RHS Tatton Park 2008 (br); **MMGI:** Marianne Majerus/Design: Peter Chan & Brenda Sacoor (cr).

232 DK Images: Design: Helen Derrrin, RHS Hampton Court 2008 (t); www.indian-ocean.co.uk (c); www.outer-eden.co.uk (b).

232–233 The RHS Images Collection: RHS/Neil Hepworth, design: Charlie Albone, RHS Chelsea 2016.

233 Nicola Browne: Design: Craig Bergman (tc); **GAP Photos:** Elke Borkowski (cr); **MMGI:** Marianne Majerus/Design: Diana Yakeley (br); www.wmstudio.co.uk (cl).

234 DK Images: Design: Francesca Cleary & Ian Lawrence, RHS Hampton Court 2007 (tr); Design: Noel Duffy, RHS Hampton Court 2008 (bl); James Merrell (tl); **GAP Photos:** Richard Bloom/Design: Katharina Nikl Landscapes (br).

235 DK Images: Brian North/Design: The Naturally Fashionable Garden Designer NDG+, RHS Chelsea 2010 (bl); Design: Philip Nash, RHS Chelsea 2008 (tc); **The Garden Collection:** Torie Chugg/Design: Sue Tymon, RHS Hampton Court 2005 (c); **The Interior Archive:** Fritz von der Schulenburg (tr); **Red Cover:** www.dylon.co.uk (br); **MMGI:** Marianne Majerus/Design: Sara Jane Rothwell, London Garden Designer (tl).

236 Nicola Browne: Design: Piet Oudolf (tr); **DK Images:** Design: Sadie May Stowell, RHS Hampton Court 2008 (tl); Design: Sim Flemons & John Warland, RHS Hampton Court 2008 (br); **GAP Photos:** Heather Edwards/Design: Mark Draper, Graduate Gardeners Ltd (bl).

237 The RHS Images Collection: RHS/Neil Hepworth, design: Chris Beardshaw, RHS Chelsea 2016 (t); **Helen Fickling:** Design: May & Watts, Loire Valley Wines, RHS Hampton Court 2003 (c); **MMGI:** Marianne Majerus/Design: Lynne Marcus (bl).

238 The Garden Collection: Marie O'Hara (br); Nicola Stocken (bc); Steven Wooster/Design: Anthony Paul (tl); **MMGI:** Marianne Majerus/Design: Charlotte Rowe (bl); Marianne Majerus/Design: Lucy Sommers (tr); **Clive Nichols:** Design: Mark Laurence (tc).

239 Nicola Browne: Design: Kristof Swinnen (tl); **The Garden Collection:** Liz Eddison/Design: David MacQueen, Orangebleu, RHS Chelsea 2005 (bc); **Harpur Garden Library:** Marcus Harpur/Design: Charlotte Rowe (br); **Clive Nichols:** Spidergarden.com/RHS Chelsea 2000 (c); **Red Cover:** Kim Sayer (bl); Mike Daines (cra).

240 www.janinepattison.com.

241 (left to right): **Clive Nichols:** Design: Charlotte Rowe; **Helen Fickling:** Claire Mee Designs; **GAP Photos:** Design: Cube 1994; Graham Strong; **Photolibrary:** Botanica (bl); **Red Cover:** Ken Hayden (bc); **Shutterstock** (br).

242–243 GAP Photos: Fiona McLeod/Design: Matt Keightley.

244 DK Images: Design: Sam Joyce (bc); **The Garden Collection:** Gary Rogers (br); **Shutterstock.com:** Ingo Bartussek (cl).

246 Alamy Stock Photo: Dave Burton (t); **GAP Photos:** Dave Bevan (bl); John Glover (br).

247 Alamy Stock Photo: ACORN1 (bl); Clare Gainey (tr); Carolyn Jenkins (br); **GAP Photos:** Graham Strong (tl).

248 Alamy Stock Photo: Paul Briden (c); Curtseyes (br); **GAP Photos:** Thomas Alamy (tr); Mel Watson (bl).

249 Alamy Stock Photo: ACORN1 (tl); Barbara Jean (tr); **GAP Photos:** (bl); Jo Whitworth/Design: Kate Frey, RHS Chelsea (br).

250 GAP Photos: Mel Watson (br); **Shutterstock.com:** Paul Brennan (tl).

ACKNOWLEDGMENTS

251 Alamy Stock Photo: Barrie Harwood (c); Graham Prentice (b); **GAP Photos:** Mel Watson (tr).

252 GAP Photos: Heather Edwards/Design: Charlotte Harris, Sponsor: RBC.

253 GAP Photos: Brent Wilson (t); **Clive Nichols** (b).

257 Dreamstime.com: Paul Maguire (bl).

261 GAP Photos: Matteo Carassale/Design: Cristina Mazzucchelli (br).

263 Dreamstime.com: Elena Vlasova (br).

265 The RHS Images Collection: RHS/Sarah Cuttle, design: Martin Royer, RHS Hampton Court 2016 (bl); **GAP Photos:** Fiona Lea (br); **MMGI:** Marianne Majerus/Design: Jill Billington & Barbara Hunt, "Flow" Garden, Weir House, Hants (cr).

270 Alamy Stock Photo: Gillian Pullinger (cr).

273 Getty Images/iStock: robertprzybysz (tr).

275 MMGI: Marianne Majerus (br).

276–277 Clive Nichols: Design: Helen Dillon.

279 GAP Photos: Neil Holmes (tr).

281 DK Images: Design: Xa Tollemache.

282 GAP Photos: Lee Avison (tr); Jonathan Buckley/Design: James Hitchmough (bl); Richard Bloom (br).

283 GAP Photos: Gary Smith (tr); Nicola Stocken (tl, bl).

284 Photoshot: Photos Horticultural (br).

285 Dreamstime.com: Daria Nipot (cb).

286–287 GAP Photos: Clive Nichols/Design: Andy Sturgeon, Sponsor: The Telegraph.

288 Dreamstime.com: Marinodenisenko (cl).

289 Dorling Kindersley: Peter Anderson (bl); **Shutterstock.com:** InfoFlowersPlants (cl).

290 Dreamstime.com: Tom Cardrick (bl).

291 Garden World Images: Paul Lane (tl); **Dorling Kindersley:** Mark Winwood/RHS Wisley (bc).

292 Garden World Images: Carolyn Jenkins (cl); **Dreamstime.com:** Alena Stalmashonak (cr).

293 The Garden Collection: Torie Chugg (c).

295 MMGI: Marianne Majerus/Design: Tom Stuart-Smith (bl).

296 Alamy Stock Photo: P. Tomlins (cl); **Shutterstock.com:** guentermanaus (bl).

297 Alamy Stock Photo: Zoltan Bagosi (tl); John Richmond (bc).

298 MMGI: Marianne Majerus/Saling Hall, Essex (bl).

299 Alamy Stock Photo: John Richmond (cl); **Dreamstime.com:** Apugach5 (bc); **Getty Images/iStock:** Iamnee (bl); **Shutterstock.com:** Simona Pavan (tl).

300 Shutterstock.com: Traveller70 (bl).

302 Alamy Stock Photo: garfotos (tl); **Dreamstime.com:** John Caley (bl).

303 Alamy Stock Photo: Martin Hughes-Jones (tl); **Dreamstime.com:** Wirestock (bl).

304 Dreamstime.com: Bat09mar (br).

305 Alamy Stock Photo: amomentintime (bl); Wiert Nieuman (c).

308 GAP Photos: Nova Photo Graphik (cl).

309 GAP Photos: Charles Hawes (cl).

310 Dreamstime.com: John Caley (br); Tom Cardrick (bl); **Shutterstock.com:** Peter Turner Photography (bc).

311 Shutterstock.com: Alex Manders (bl).

312 www.davidaustinroses.com (c).

313 GAP Photos: Visions Premium (tr).

314 Garden World Images: Martin Hughes-Jones (cl).

315 Alamy Stock Photo: Avalon.red/Photos Horticultural (tc).

316 Alamy Stock Photo: Chris Bosworth (br).

317 Shutterstock.com: John R Martin (tl).

318 Dreamstime.com: Artist Krolya (c).

319 DK Images: Roger Smith (tl); **The Garden Collection:** Nicola Stocken Tomkins (tc).

320 Alamy Stock Photo: Nick Higham (br); **Dreamstime.com:** Angelacottingham (c).

321 Alamy Stock Photo: RM Floral (tc); **Shutterstock.com:** AliScha (bc).

322 Getty Images/iStock: Martin Wahlborg (tr).

323 Alamy Stock Photo: Clare Gainey (bl); P. Tomlins (cr).

324 Dreamstime.com: Golden Shark (c); **GAP Photos:** Nicola Stocken (tr); **Shutterstock.com:** Holly Anne Cromer (tc).

325 Shutterstock.com: Alex Manders (cl).

326 GAP Photos: Modeste Herwig, White Cottage Daylilies; Nicholas Inigo Peirce (tc); **Shutterstock.com:** Sergey V Kalyakin (cl).

327 GAP Photos: Neil Holmes (cr); **The Garden Collection:** Andrew Lawson (c).

328 MMGI: Marianne Majerus (cl); **Shutterstock.com:** Walter Erhardt (tc).

329 Dreamstime.com: Anthony Baggett (cl); Leklek73 (c).

330 GAP Photos: FhF Greenmedia (cl); **Getty Images/iStock:** Olena Lialina (c).

331 Alamy Stock Photo: Avalon.red/Michael Warren (tl); Botanic World (tr); **Dreamstime.com:** Robert Adami (tc); **Shutterstock.com:** Flower_Garden (c).

332 Dreamstime.com: Imladris (bc); Natalia Pavlova (cr); Iva Vagnerova (br); **Getty Images/iStock:** fotomarekka (bl); **Shutterstock.com:** Lisa-Marie Brown (tc).

333 Alamy Stock Photo: Avalon.red/Photos Horticultural (bc); **Dreamstime.com:** Gabriela Beres

(tr); **GAP Photos:** Nova Photo Graphik (c); Visions Premium (tc); **Getty Images/iStock:** Steve Deming (bl); **Shutterstock.com:** Alex Manders (cl).

334 Alamy Stock Photo: Matthew Taylor (c); **Dreamstime.com:** Tom Cardrick (br); Orest Lyzhechka (tc); Wirestock (bl); **Getty Images/iStock:** magicflute002 (tr).

335 Dreamstime.com: Rob Lumen Captum (c); Meunierd (cl).

336 Photolibrary: Mark Bolton (c); **Shutterstock.com:** nnattalli (tr).

338 GAP Photos: Howard Rice (bc); **Photolibrary:** Mayer/Le Scanff (br).

339 Dreamstime.com: Tom Meaker (bl).

340 The Garden Collection: Andrew Lawson (bc).

341 GAP Photos: J. S. Sira (c).

342 Alamy Stock Photo: blickwinkel/McPHOTO/HRM (br); Dr Ian B Oldham (cl).

344 Garden World Images: (bl).

345 Shutterstock.com: Sergey V Kalyakin (cl); Ioana Rut (cr).

346 GAP Photos: Paul Debois (tl).

349 Shutterstock.com: crystal51 (cr).

350 GAP Photos: Elke Borkowski (bl); Jerry Harpur (br); www.stonemarket.co.uk (c) (bl).

351 www.stonemarket.co.uk (top row) (bl); www.bradstone.com/garden (c) (cr); www.organicstone.com (bc); **Dreamstime.com:** Thornton McLaughlin (bc).

352 Nuage Porcelain Paving by London Stone, www.londonstone.co.uk/Design & Build: Garden House Design, www.gardenhousedesign.co.uk: (tc); **DK Images:** Design: Martin Thornhill, RHS Tatton Park 2008 (cr); **Forest Garden Ltd,** tel: 0844 248 9801, www.forestgarden.co.uk (cl); Images supplied courtesy of Marshalls www.marshalls.co.uk/transform (bc); www.jcgardens.com (br).

353 DK Images: Design: Jane

Hudson & Erik de Maejer, RHS Chelsea 2004 (tc); Design: Jon Tilley, RHS Tatton Park 2008 (bl); Design: Martin Thornhill, RHS Tatton Park 2008 (br); **GAP Photos:** J. S. Sira (cl); Howard Rice (bc); www.specialistaggregates.com (cr); **Chris Young** (bc).

354 DK Images: Steven Wooster/Design: Claire Whitehouse, RHS Chelsea 2005 (c); Design: Geoff Whitten (br); **GAP Photos:** Elke Borkowski (bl); www.bradstone.com/garden (bc); **Images supplied courtesy of Marshalls** www.marshalls.co.uk/transform (tc).

355 DK Images: Design: Paul Hensey with Knoll Gardens, RHS Chelsea 2008 (c); Design: Toby & Stephanie Hickish, RHS Tatton Park 2008 (bc); Design: Niki Ludlow-Monk, RHS Hampton Court 2008 (br); Design: Ruth Holmes, RHS Hampton Court 2008 (cr); **GAP Photos:** Leigh Clapp/Design: David Baptiste (bl).

356 DK Images: Design: Helen Williams, RHS Hampton Court 2008 (cr); www.grangefencing.co.uk (tl); www.jacksons-fencing.co.uk (tr); **Forest Garden Ltd**, tel: 0844 248 9801 www.forestgarden.co.uk (cl) (c); www.kdm.co.uk (bc).

357 GAP Photos: Leigh Clapp (bc); **MMGI:** Marianne Majerus/Design: Hans Carlier (tr); **Forest Garden Ltd**, tel: 0844 248 9801 www.forestgarden.co.uk (tc) (bl); www.stonemarket.co.uk (br).

358 DK Images: Brian North/RHS Hampton Court 2010 (tl); Design: Mark Sparrow & Mark Hargreaves, RHS Tatton Park 2008 (bl); **Alamy Images:** Francisco Martinez (tc); **GAP Photos:** Jerry Harpur (tr); **Photolibrary:** John Glover/Design: Jonathan Baillie (c); www.breezehouse.co.uk (cl); www.cuprinol.co.uk (bc).

359 DK Images: Design: Jackie Knight Landscapes, RHS Tatton Park 2008 (tc); Design: Mark Gregory, RHS Chelsea 2008 (bc); www.garpa.co.uk (br); **MMGI:** Marianne Majerus/Design: Earl Hyde, Susan Bennett (cl); Marianne Majerus/Elton Hall, Herefordshire (c); www.jcgardens.com (cr); www.cuprinol.co.uk (tl) (bl).

360 DK Images: Design: David Gibson, RHS Tatton Park 2008 (cl); Design: Cleve West, RHS Chelsea 2008 (bl); **GAP Photos:** Elke Borkowski (cr); Jo Whitworth/Design: Tom Stuart-Smith, RHS Chelsea 2006 (br).

361 DK Images: Design: Tim Sharples, RHS Hampton Court 2008; **GAP Photos:** Tim Gainey (bl); **The Garden Collection:** Nicola Stocken Tomkins (tr); www.hayesgardenworld.co.uk (cr).

362 GAP Photos: Richard Bloom (bl); Nicola Stocken (tl); Heather Edwards/Design: Cleve West, The Daily Telegraph Garden, Gold Medal Winner (tc); Heather Edwards/Design: Peter Reader, Sponsor: Living Landscapes (tr); Martin Staffler (cl); Jenny Lilly (c); Fiona Lea/Design: Alison Galer (cr); Heather Edwards (bc); Stephen Studd/Design: Adam Frost, Sponsor: Homebase (br).

363 GAP Photos: Richard Bloom/The Swimming Pond Company (cr); Zara Napier/Design: Allison Armour-Wilson (tl); Rob Whitworth (tc); Nova Photo Graphik (tr); Annaick Guitteny (c); Richard Bloom/Design: Craig Reynolds (bl); Andrea Jones (bc); Mark Bolton (br); **MMGI:** Bennet Smith/Design: Thomas Hoblyn (cl).

364 GAP Photos: Mark Bolton (cr); Elke Borkowski (tl); GAP Photos/Design: Cube 1994 (cl); Nova Photo Graphik (c); Heather Edwards/Martin Young, Sitting Spiritually (bl); Paul Debois (bc); Clive Nichols/Design: Trevyn McDowell and Paul Thompson (br); **MMGI:** Marianne Majerus/Sculpture design: David Harber, Garden design: Nic Howard (tr); **The Crop Candle Company:** Duncan MacBrayne/Silver Cloud Photography (tc).

365 GAP Photos: Matteo Carassale/Design: Cristina Mazzucchelli (c); Anna Omiotek-Tott/Design: Will Williams (tl); Andrea Jones (tc); Nicola Stocken, Knolling with Daisies Garden/Design: Sue Kent (tr); Paul Debois (cl); Robert Mabic (cr); Joanna Kossak/Design: Tony Woods (bl); GAP Photos/Design: Cube 1994 (bc); **Clive Nichols:** Clive Nichols/Design: Wynniat-Husey Clarke (br).

366 Alamy Stock Photo: Cavan Images (br); guy harrop (cr); Paul Mogford (bc); **GAP Photos:** Richard Bloom/Design: Katharina Nikl Landscapes (cl); GAP Photos/Design: Rhiannon Williams, RHS Hampton Court 2018 (tl); Brent Wilson (tc); Howard Rice/Design: Cultivate Gardens, Cambridge (tr); Paul Debois (c); **Clive Nichols:** Clive Nichols/Design: Matt Keightley (bl).

367 Alamy Stock Photo: Tony Giammarino (t); **GAP Photos:** Elke Borkowski (cr); Joanna Kossak/Design: Tony Woods (tc); Nicola Stocken (c); Jerry Harpur (bl); J. S. Sira/Design: Martin Royer, Sponsors: Final5 (bc); Jenny Lilly (br); **MMGI:** Marianne Majerus/Design: Charlotte Rowe (tr); Marianne Majerus/Design: Dan Cooper (cl).

368–369 The RHS Images Collection: RHS/Neil Hepworth, design: Charlie Albone, RHS Chelsea 2016 (b).

Jacket Helen Elks-Smith

All other images:
© Dorling Kindersley

Thanks to the following people for allowing us to photograph and feature their gardens:

Zelda and Peter Blackadder, Jacqui Hobson, Jo and Paul Kelly, Bob and Pat Ring, Amanda Yorwerth.

Thanks to the following companies for their help on this project:

Blue Wave bluewave.dk

Brandon Hire brandonhirestation.com

Garpa garpa.co.uk

Marshalls marshalls.co.uk

Organicstone organicstone.com

Ormiston Wire ormiston-wire.co.uk

Stonemarket stonemarket.co.uk

Thanks to Naorem Anuja for editorial assistance, Nicola Powling for illustrations, John Friend for proofreading, and Vanessa Bird for indexing.

Thanks to the following DK staff for their work on the original edition of the book:

Senior Editor Zia Allaway
Senior Art Editor Joanne Doran
Airedale Publishing Ruth Prentice, David Murphy, Murdo Culver
Photographers Peter Anderson, Brian North
Illustrators Peter Bull Associates, Richard Lee, Peter Thomas
Plan Visualizers Joanne Doran, Vicky Read
Managing Editor Anna Kruger
Managing Art Editor Alison Donovan
Publisher Jonathan Metcalf
Associate Publisher Liz Wheeler
Art Director Bryn Walls

Index

A

Abelia 196
   A. × *grandiflora* 300
abstract sculpture 35, 122, 236, 237
Acacia dealbata 288
Acanthus 206
   A. spinosus 318
accent colours 36
access paths 153
Acer 59, 204, 211, 213
   A. campestre 288
   A. c. 'Schwerinii' 288
   A. griseum 292
   A. japonicum 'Vitifolium' 292
   A. negundo 101
   A. n. 'Variegatum' 290
   A. palmatum 98, 300
   A. p. 'Bloodgood' 292
   A. p. 'Fireglow' 54
   A. p. 'Ōsakazuki' 292
   A. p. 'Sango-kaku' 292
   A. p. var. *dissectum* 210
   A. platanoides 'Crimson King' 288
   A. rubrum 69
   A. r. 'October Glory' 288
Achillea 68, 80
   A. 'Lachsschönheit' (Galaxy Series) 322
   A. 'Moonshine' 54
   A. 'Taygetea' 322
acid soils 134, 200
Aconitum 213
   A. 'Spark's Variety' 318
Acorus
   A. calamus 56
   A. c. 'Argenteostriatus' 342
Actaea 213
   A. simplex Atropurpurea Group 'Brunette' 346
Actinidia kolomikta 314
Adiantum venustum 330
Aegopodium podagraria 'Variegatum' 209
aesthetics, alluring 26
Agapanthus 'Silver Baby' 330
Agastache 'Blackadder' 322
aggregates 231, 265, 353
Ajuga
   A. reptans 209, 330
   A. r. 'Catlin's Giant' 330
Akebia quinata 55, 314
Albizia julibrissin 292
Albone, Charlie 34
Alchemilla mollis 84, 330
alder see *Alnus*

Alhambra & Generalife (Spain) 38, 39
alkaline soils 134, 200
Allium 10, 94, 193, 196, 200, 206, 212
   A. cepa 111
   A. cristophii 336
   A. hollandicum 'Purple Sensation' 71, 336
   A. sphaerocephalon 55
   A. stipitatum 'Mount Everest' 336
allotments 90
   Manchester Allotment Society garden 110, 111
   small space gardens 90, 109
   vertical allotments 113
Alnus glutinosa (common alder) 85
Alnwick Garden (Northumberland) 65
alpine meadows 201
alpine strawberry 113
alpines 134, 179, 192, 277
aluminium 229, 231
   furniture 365
   panel walls 355
Amelanchier 176, 212
   A. lamarckii 218, 292
amenities 214
   identifying position of 136
Ampelopsis brevipedunculata 314
Anaphalis triplinervis 322
Anemanthele lessoniana 206, 342
Anemone 212
   A. blanda 'Pink Star' 336
   A. b. 'Radar' 336
   A. b. 'White Splendour' 336
   A. × *hybrida* 318
   A. sylvestris 330
Angelica archangelica 57
angel's fishing rod see *Dierama pulcherrimum*
Angiozanthos 129
annuals 57, 199, 200, 202, 249, 283, 284
Anthemis
   A. punctata subsp. *cupaniana* 330
   A. tinctoria 'E.C. Buxton' 207
apertures 227
   doors 227, 357
   gates 141, 227, 357
aphids 276
aquatic plants 200, 214, 230, 273, 347
Aquilegia
   A. chrysantha 54
   A. vulgaris 'William Guiness' 322
Aralia 122, 126
   A. elata 'Variegata' 296
arbours 48, 49, 52, 359
Arbutus unedo 128, 292

arches 48, 154, 358
architectural planting 36, 40, 90, 118, 210, 319
   see also sculptural plants
architecture 31, 33, 34, 158
art 31, 148, 166, 167, 223
   furniture as 235
   land art 117
   statement gardens 16, 118, 123
Art Deco 118
Artemisia 200
   A. arborescens 306
   A. ludoviciana 'Silver Queen' 322
   A. 'Powis Castle' 42, 100
Arts and Crafts Movement 47, 49
Arum italicum subsp. *italicum* 'Marmoratum' 330
arum lily see *Zantedeschia aethiopica*
Aruncus dioicus 'Kneiffii' 206, 346
Arundo donax var. *versicolor* 342
Asarum 209
aspect 134–35, 151, 171, 192
Asperula odorata 209
Asphodeline lutea 318
Aspidistra 41
Asplenium scolopendrium Crispum Group 322
assessing your garden 132–41
Astelia 119, 173
   A. chathamica 40, 43, 322
Aster 213
   A. amellus 'Veilchenkönigin' 330
Astilbe 192, 209, 215
   A. chinensis var. *taquetii* 'Purpurlanze' 56
   A. 'Fanal' 346
   A. 'Professor van der Wielen' 346
   A. 'Willie Buchanan' 346
Astrantia 85, 209
   A. major 'Hadspen Blood' 322
   A. m. 'Roma' 197
astroturf 25
asymmetry 36, 37, 41, 44, 45, 119
Athyrium
   A. filix-femina 323
   A. niponicum var. *pictum* 330
aubergines 111
Aubretia 47, 179
Aucuba japonica 'Crotonifolia' 218, 300
autumn
   autumn colour 202, 213, 295
   autumn-flowering shrubs 311
avenues 34
awnings 364, 367
axial layouts 38, 39
azaleas 200, 211
Azara microphylla 296

B

balance 33
Ballota 'All Hallows Green' 306
bamboo (as a material) 167
   fences 69
   screens 96, 155, 167, 173, 219, 357
bamboos (plants) 342–45
banana see *Musa*
Baptisia australis 318
barbecues 62, 366
barberry see *Berberis*
bare-root plants, planting 273, 274
bark 174
   as mulch 277, 279
   paths 161, 265, 352
   play areas 61, 71, 149, 352
Barnett, Marcus 44–45
Barragán, Luis (1902–88) 42
barrels 137, 215, 279, 361, 363
bars, outdoor 366
Barton, James 54, 57
basil see *Ocimum basilicum*
bay see *Laurus nobilis*
bean bags 364
bedding plants, summer 207
beds 139, 194
   see also borders
beech hedges 57, 204, 357
   see also *Fagus*
Beesia calthifolia 331
beetroot 25
Begonia 178
   B. 'Glowing Embers' 220
benches 43, 54, 91, 95, 161, 365
Berberis 200, 213, 219, 357
   B. darwinii 300
   B. julianae 300
   B. × *stenophylla* 306
   B. × *s.* 'Corallina Compacta' 306
   B. thunbergii 'Aurea' 306
   B. t. f. atropurpurea 'Atropurpurea Nana' 306
   B. t. f. a. 'Helmond Pillar' 306
Bergenia 206, 209
   B. 'Bressingham White' 218
   B. 'Pink Dragonfly' 331
Beta vulgaris subsp. *vulgaris* 111
Betonica officinalis 'Hummelo' 331
Betula (birch) 56, 72, 99, 173, 211, 213
   B. nigra 55, 288
   B. pendula (silver birch) 84, 99
   B. utilis var. *jacquemontii* 69, 99, 288
   B. u. var. *j.* 'Silver Shadow' 288
biennials 199, 200
bike storage 177
bindweed 247

bins 155, 177, 359
biodiversity 27, 28, 75, 217
birds 14, 77, 80, 105, 278, 281
blackthorn see *Prunus spinosa*
blades, water 362
bleeding heart see *Lamprocapnos spectabilis*
block planting 35, 37, 96, 194
blocks 254, 350
    cutting 255
    see also paving
Blom, Jinny 54, 55
bluebells see *Hyacinthoides non-scripta*
bog gardens 31, 133, 136
bog plants 214, 346–49
boggy soil, plants for 349
bold planting 95
boots, as planters 361
Borde Hill (West Sussex) 53
borders 20, 21, 34, 197
    cottage gardens 149
    country gardens 53, 79
    gravel borders 149, 264–65
Borinda papyrifera 342
boundaries 82, 140–41
    construction 245
    legal issues 133, 140, 141
    materials for 226–27
    narrow spaces between 151
    options 166–67
    and perspective 78
    softening 125, 178–79
    see also fencing; gates; hedges; walls, etc
bowls, water 362
box
    edging with 42, 159, 167
    hedges 34, 35, 57, 173
    parterres 31, 35, 38, 194, 199
    topiary 35, 39, 159, 210
    see also *Buxus*
Brahea armata 127
branches, removing 285
braziers 359
Breedon gravel 54, 265
bricks 47, 231
    designs in 33, 253
    edgings 42
    mowing strips 269
    paths 49, 73, 85, 104, 114, 179, 224, 254
    paving 47, 70, 350
    walls 170, 171, 226, 228, 258–59, 354
Briza maxima 342
Brockbank, Jane 70, 84, 85
Brogdale (Kent) 109
Brookes, John 23, 62, 91
broom see *Genista*
bubble diagrams/plans 143, 149, 182, 190, 194

bubble fountains/pools 154, 215, 360
bubble tubes 73
buckets 144
Buckley, Declan 40
Buddleja (butterfly bush)
    B. alternifolia 'Argentea' 296
    B. davidii 'Dartmoor' 300
    B. × *weyeriana* 'Sun Gold' 300
budgeting 138, 195, 244
bulbs 206, 208, 336–41
    foliage and texture 55
    naturalized 79, 213
    seasonal colour 193, 199
    spring bulbs 193, 199, 212, 213
    summer bulbs 212
    wet soil and 209
burdock 251
Buss, Nick 72–73
Butomus umbellatus 215, 346
butterfly bush see *Buddleja*
butyl liners 230, 270–71
Buxus 81
    B. sempervirens 40, 196
    B. s. 'Suffruticosa' 306
    see also box

C

cacti 118
CAD (computer-aided design) 181, 191
Calamagrostis × *acutiflora* 'Overdam' 342
Calamintha grandiflora 'Variegata' 331
Calendula 107, 109
Calibrachoa Million Bells Series 220
Calluna (heather)
    C. vulgaris 306
    C. v. 'Gold Haze' 306
    C. v. 'Spring Cream' 306
Caltha palustris 56, 215, 346
Calycanthus 'Aphrodite' 296
Camassia 283
Camden Children's Garden (London) 65
Camellia 135, 200, 218, 247
    C. japonica 'Bob's Tinsie' 300
    C. reticulata 'Leonard Messel' 296
Cameron, Alasdair 100–101
Camley Street Natural Park (London) 65
Campanula
    C. 'Burghaltii' 323
    C. glomerata 'Superba' 323
    C. 'Pink Octopus' 331
Campsis × *tagliabuana* 'Madame Galen' 314
candles 238, 241

Canna 127, 204, 213
    C. 'Durban' 336
    C. 'Striata' 336
canopies 43, 95, 140, 156, 165
cardoon see *Cynara cardunculus*
Carex (sedge) 200, 209, 215
    C. buchananii 342
    C. elata 'Aurea' 342
    C. oshimensis 'Evergold' 342
carpet stones 254–55, 350
Carpinus (hornbeam) 34, 38, 56, 90, 91, 176, 193
    C. betulus 57, 210
    C. b. 'Fastigiata' 290
carrots 111
Caryopteris × *clandonensis* 'Worcester Gold' 307
Casa Mirindiba (Brazil) 36–37
cascades 61, 90, 119
Catalpa bignonioides 'Aurea' 289, 290
catmint see *Nepeta*
cavolo nero 25
Ceanothus 179
    C. 'Concha' 300
    C × *delilianus* 'Gloire de Versailles' 307
    C. thyrsiflorus var. *repens* 307
Cedrus atlantica f. Glauca Group 289
Centaurea dealbata 'Steenbergii' 323
Centranthus 86, 124
    C. ruber 124
    C. r. 'Albus' 124
Cephalaria gigantea 318
ceramics 119, 174, 231
    ceramic pots 360
    see also mosaics
Ceratostigma willmottianum 307
Cercidiphyllum japonicum 289
Cercis
    C. canadensis 'Forest Pansy' 292
    C. siliquastrum 211, 293
Chaenomeles speciosa 'Moerloosei' ('Apple Blossom') 301
Chamaecyparis
    C. pisifera 'Filifera' 290
    C. p. 'Filifera Aurea' 290
Chamaerops humilis 127
Château de Villandry (France) 37, 109
Chaumont-sur-Loire (France) 121
cherry laurel see *Prunus laurocerasus*
cherry see *Prunus*
chestnut paling 356
chevron panel fences 356
chilli peppers 104, 107
chimeneas 240, 241

Chimonanthus praecox 'Grandiflorus' 296
chipped bark see bark
chives 113
Choisya × *dewitteana* 'Aztec Pearl' 301
Church, Thomas (1902–78) 62, 78
Cimicifuga see *Actaea*
circular shapes 150, 156, 160
Cirsium rivulare 'Atropurpureum' 70, 85, 318
Cistus
    C. × *dansereaui* 'Decumbens' 307
    C. × *purpureus* 307
city gardens see urban gardens
classical architecture 33, 34, 158
clay soils 134, 136, 200, 225, 280, 285
Clematis 177, 179, 201, 278, 279
    C. alpina 279
    C. armandii 314
    C. 'Bill MacKenzie' 314
    C. 'Étoile Violette' 314
    C. florida var. *florida* 'Sieboldiana' 314
    C. 'Huldine' 314
    C. integrifolia 323
    C. macropetala 279
    C. 'Markham's Pink' 315
    C. montana var. *montana* 'Tetrarose' 315
    C. 'Pink Fantasy' 196
Clerodendrum trichotomum var. *fargesii* 296
climate change 23, 27, 28, 146, 284
    plants for 216
climbers 199, 314–17
    care of 21
    how to plant 278–79
    roof gardens 92
    for screening 109, 140, 155, 166, 179, 193
    softening hard landscaping with 178, 179
    for spring and summer flowers 315
    supporting and training 178, 200, 201, 267, 278–79, 358, 359
    see also individual species
climbing roses 278, 279, 316–17, 359
cloud pruning 236
clump-forming plants 201
coastal gardens 124, 149, 201
cobbles 33, 48, 174, 254, 271, 353
cold frames 104
colour 168–69, 194, 199
    accent colours 36
    applying 172–73
    artificial colour 173

autumn colour 202, 213, 292, 293
colour effects 52, 170–71
colour-themed gardens 54, 55, 197
colour wheel 168–69, 170
combining colours 52, 169
effects of 16, 17, 143
family gardens 67, 68
flower and leaf colour 94, 199, 207
focal plants 211
furniture 235
hard landscaping 161
materials and 62, 223, 225
monochrome colours 173
neutral colours 173
properties of 171
relaxing colours 172
screens and boundaries 226
spring colour 212
statement gardens 118, 119, 120
summer colour 212, 309, 315, 339
vibrant colours 172
winter colour 205, 213
communal gardens 141
composite decking 352
compost (garden compost) 25, 76, 77, 133, 134, 145, 276, 277, 278
compost bins 25, 77, 111
computer-aided design (CAD) 181, 191
concept gardens 31, 117
see also statement gardens
concrete 174, 231
containers 360
cubes 121
kitchens 367
paving and paths 37, 105, 224, 350
polished 36
removing 249
rendered 228
screens 226, 227, 354
seating 120
sleepers 352
in small spaces 91
in statement gardens 120, 121
walls and panels 166, 174, 226, 228, 355
for water features 230
coneflower see *Echinacea*; *Rudbeckia*
conifers 166
conservatories 135
container-grown plants, planting 273, 274–75, 276
containers 201, 360–61
feeding 285
flexible 176
as focal points 154, 155, 156
grasses, sedges and bamboos for 343
growing climbers in 279
growing vegetables in 24, 104, 105, 106, 107, 115
perennials for 333
as screens 133, 139
seasonal 220
small spaces and 89, 92, 95
in statement gardens 118
trees in 176
in urban gardens 90, 91, 95
urns 35, 42, 197, 360
watering 20, 284
contemplation 17
contemporary gardens 19, 28, 156–57, 158, 236, 238
formal style 36
multi-level layouts 162
contractors 244, 245, 247, 248, 249, 253
Convallaria majalis 336
Convolvulus cneorum 206, 307
Cordyline 97, 204
C. *australis* 'Red Star' 296
Coreopsis verticillata 'Moonbeam' 331
corms 336–41
Cornerstone (California, USA) 121
cornflowers 45, 82
Cornus (dogwood) 211, 213
C. *alba* 'Aurea' 301
C. *a.* 'Sibirica' 301
C. *alternifolia* 210
C. *canadensis* 209
C. *controversa* 'Variegata' 293
C. *kousa* var. *chinensis* 'China Girl' 293
C. *mas* 296
C. *sericea* 'Flaviramea' 301
Corokia × *virgata* 129
Coronilla valentina subsp. *glauca* 307
corrugated iron 355
Corsican mint see *Mentha requienii*
Cortaderia (pampas grass) 210
C. *selloana* 'Aureolineata' 343
C. *s.* 'Pumila' 343
Corten steel 33, 34, 122, 125, 229, 360
Corylus (hazel) 213
C. *avellana* 'Contorta' 213, 293
C. *maxima* 'Purpurea' 297
Cosmos 110
C. *atrosanguineus* 221
Cotinus (smoke bush) 213
C. 'Grace' 297
Cotoneaster 213
C. *dammeri* 308
C. *franchetii* 219
C. *frigidus* 210
C. *f.* 'Cornubia' 297

C. *lacteus* 297
C. *salicifolius* 'Gnom' 69, 308
cottage gardens 28, 31, 47–59
blurring the edges 179
colour palette 202
furniture 232, 234
grouping plants 194
raised beds 228
sculpture in 236
Coulter, Fran 154–55, 196
country gardens 31, 47–59, 91, 160, 173, 234
courtyards 92
case study 114–15
climbers in 278
colour use 173
lighting 240
sculptures in 237
statement gardens 125
texture use in 174
crab apple see *Malus*
Crambe 124
C. *cordifolia* 318
cranesbill see *Geranium*
Crataegus (hawthorn) 213, 275
C. *laevigata* 217
C. *orientalis* 293
C. *persimilis* 'Prunifolia' 293
crazy paving 350
Crinum
C. × *powellii* 336
C. × *p.* 'Album' 336
Crocosmia 206
C. 'Firebird' 337
C. × *crocosmiiflora* 'Coleton Fishacre' 336
C. × *c.* 'Venus' 337
Crocus 212
C. *goulimyi* 337
C. *tommasinianus* 337
C. *t. f. albus* 337
crops 103
crop rotation 107
see also productive gardens; vegetables
crown imperial see *Fritillaria imperialis*
Cucurbita pepo (pumpkin) 111
Cupressus macrocarpa 'Goldcrest' (Monterey cypress) 293
curves 150, 160–61, 191
decking 224
paths 152, 153, 160, 161
cushions 68, 95, 235
Cycas revoluta 40
Cyclamen hederifolium 337
Cynara
C. *cardunculus* 318
C. *c.* Scolymus Group 206
Cyperus 215
C. *alternifolius* 215
cypress see *Cupressus macrocarpa*

D

daffodils see *Narcissus*
Dahlia 104, 109, 199, 213
D. 'Bishop of Llandaff' 337
D. 'David Howard' 337
D. 'Gay Princess' 337
daisies 82
damp conditions, perennials for 329
Daphne
D. *bholua* 'Jacqueline Postill' 301
D. *odora* 'Aureomarginata' 308
D. *transatlantica* 'Eternal Fragrance' ('Blafra') 308
Darmera 200
D. *peltata* 346
Daucus carota subsp. *sativus* 111
Davidia involucrata 290
Day-Glo colours 173
daybeds 365
daylily see *Hemerocallis*
deadheading 21, 276, 284
deciduous plants 199, 200, 204
trees 140, 176, 192, 205, 213
decking 156, 253, 282
composite 352
contemporary formal gardens 33, 36
creating texture with 174
curved 160, 224
family gardens 71
joist hangers 261
laying 260–61
making the deck frame 260–61
materials for 224, 225
paths 144
privacy and 140
putting up support posts 260–61
for slopes 133, 137, 162
small spaces 90, 96, 97
wood treatments 261
Delphinium Pacific Hybrids 319
dens 62, 64
Deschampsia
D. *cespitosa* 'Bronzeschleier' 99
D. *flexuosa* 'Tatra Gold' 343
design
assessing your garden 132–41
choosing materials 222–41
designing with plants 198–221
first principles 142–79
future proofing 146–47
gathering inspiration 148–49
how to design 130–241
inspiration 12–129
styles 28–129
design software 181, 191
Deutzia 204
diagonal layouts 150, 152, 157, 190, 191

Dial Park (Worcestershire) 48–49
Dianthus
   D. Allwoodii Group 'Bovey Belle' 331
   D. barbatus Festival Series 220
   D. cruentus 86
Diascia 'Hopleys' 323
Dicentra 'Bacchanal' 331
Dicksonia antarctica 293
Dierama pulcherrimum 319
Dig for Victory campaign 103
Digitalis (foxglove) 59
   D. 'Goldcrest' ('Waldigone') 323
   D. × *mertonensis* 323
   D. purpurea 85
   D. p. 'Alba' 54
dining areas 367
   aspect and 135
   family gardens 61, 62, 64–65, 66, 71
   furniture 232, 233
   positioning 154, 167, 223
   small spaces 92, 94, 98
   urban gardens 43
Dipelta floribunda 297
disabilities, gardeners with 105
diseases 75, 76, 103, 104, 107, 115, 279, 284, 285
divided gardens 196
DIY vs contractors 244
Djurovic, Vladimir 98
dog's-tooth violet *see Erythronium dens-canis*
dogwood *see Cornus*
domes, water 363
doors 227, 357
Doronicum 45
   D. 'Little Leo' 332
drainage 245, 249
   improving 106, 133, 134, 146, 280
   in retaining walls 228
   on slopes 136–37, 162, 163
   of surfaces 76, 136–37, 224, 225
   Sustainable Urban Drainage System (SuDS) 137
drifts, plant 52, 53, 54, 83, 76, 96, 194, 199, 204, 282, 283
   see also prairie-style planting
driftwood 236, 237
drives 353
drought 145, 146, 149
   drought-resistant plants 86, 124, 145, 192, 200, 216, 249
dry sites, plants for 200, 208, 299
dry stone walls 99, 174, 228, 354
Dryopteris 209
   D. affinis 'Cristata' 197
   D. erythrosora 324
   D. wallichiana 319

Dumbarton Oaks (USA) 39
Dunnett, Nigel 54, 56, 98, 99
Dutch ovens 366
dwarf French beans 112

E

easy-care gardens 15, 202
Eberle, Sarah 124, 126
Eccremocarpus scaber (Chilean glory vine) 315, 279
Echeveria elegens 220
Echinacea 282
   E. 'Glowing Dream' 324
   E. purpurea 'Virginia' 324
Echinops bannaticus 319
Echium pininana 100
eclectic influences 28, 118, 119
ecosystems 76, 217
edging 85, 152
   formal gardens 39, 42
   ideas for 225
   laying a path 254, 255
   softening 39, 179
   water features 230
edimental gardens 24
Edwardian style 49, 54
Elaeagnus
   E. × *ebbingei* 'Gilt Edge' 297
   E. 'Quicksilver'' 297
   E. × *submacrophylla* 219
elder *see Sambucus*
electrics
   electric heaters 240
   electric lights 241, 245
   electrical safety 223, 230, 238, 240, 241
elephant's ears *see Bergenia*
Elymus magellanicus 343
entertaining 14, 43, 59, 64–65, 66, 98, 146, 238, 359
environmental issues 23, 24–27, 28, 223, 233, 240, 351
Epimedium 209, 212
   E. 'Amber Queen' 332
   E. × *perralchicum* 332
equipment 245
Equisetum (horsetail) 215
   E. hyemale 215
   E. scirpoides 215
Eranthis hyemalis 338
Eremurus 68
   E. stenophyllus 324
Erica arborea var. *alpina* 301
ericaceous plants 134, 200
Eryngium 68
   E. agavifolium 319
   E. bourgatii 'Oxford Blue' 332
   E. b. 'Picos Blue' 324
Erythronium dens-canis 338
Eucomis bicolor 338
Euonymus

   E. alatus 'Compactus' 56
   E. fortunei 'Emerald Gaiety' 308
   E. japonicus 40
Eupatorium maculatum 'Altropurpureum Group' 346
Euphorbia
   E. characias subsp. *characia* 216
   E. c. subsp. *wulfenii* 'John Tomlinson' 308
   E. epithymoides 332
   E. griffithii 'Dixter' 324
   E. × *martini* 324
   E. myrsinites 332
   E. palustris 56
   E. × *pasteurii* 302
   E. polychroma 45
   E. schillingii 324
evening primrose *see Oenothera fruticosa*
evergreens 58, 62, 174, 199, 200, 202
   grasses, sedges and bamboos 345
   hedges 140, 204
   perennials 335
   shrubs 141, 206, 313
   trees 16, 41, 42, 97, 128, 289
   water features 215
   winter interest 205, 213
Exochorda × *macrantha* 'The Bride' 301
exotic plantings 28, 151

F

Fagus (beech) 81, 125
   F. sylvatica 57, 217
   F. s. (Atropurpurea Group) 'Riversii' 289
false perspectives 35, 39
family gardens 31, 60–73, 139, 154–55
   case study 72–73
   family garden plans 68–71
   family needs 14–15
   flexibility in 147
   interpreting the style 64–67
   what a family garden is 62–63
Fargesia
   F. murielae 343
   F. rufa 43
Farrand, Beatrix (1872–1959) 78
Fatsia 178
   F. japonica 40, 302
featheredge fences 356
fedges 357
feeding plants 285
Felicia amelloides 220
fences 166, 246
   bamboo 69
   climbers and 278, 279
   concreting the posts 262–63

fixing bolt-down supports 262
   height 141
   materials for 227, 355, 356–57
   metal spike supports 263
   multi-level layouts 163
   picket fences 47, 155, 227, 356
   putting up fence posts 262–63
   replacing old fence posts 262–63
   softening 178, 279
   staining 263
   statement gardens 122
fennel *see Foeniculum*
ferns 41, 84, 174, 176, 178, 209, 215
   see also individual species
fertilizer 276, 279, 280, 281, 285
Festival of Gardens (Chaumont-sur-Loire, France) 121
Festuca glauca 'Elijah Blue' 343
fibreglass pool liners 230
Ficus carica 'Brown Turkey' 294
Filipendula 215
   F. rubra 'Venusta' 347
fire
   fire pits 129, 146, 240, 364, 366
   fire tables 364
   fire and water features 364
   fireplaces 94, 240
firethorn *see Pyracantha*
fish 69
fishponds 362
flambeaux 43
flint 354
flooding 137
flooring kits 352
flowers
   colour 193, 207
   cutting flowers 106
   flowering period 192
focal points
   colour and focal plants 211
   containers as 154, 155, 156
   focal plants 199, 203, 210–11
   furniture as 48, 84, 232, 235
   natural style and 78
   sculpture as 33, 39, 90, 124, 159, 165, 237
   shrubs as 297
   trees as 291
Foeniculum (fennel)
   F. vulgare 206
   F. v. 'Purpureum' 319
Foerster, Karl (1874–1970) 54
foliage 28, 197, 207, 209
   climbers 317
   colour of 172, 192, 193
   formal gardens 33
   perennials 327
   shrubs 303
   texture and tone of 17, 94
follies 359

forest-edge statement gardens 126
Forest Stewardship Council see FSC
forget-me-not see *Myosotis*
form 10, 167, 205
formal gardens 23, 28, 32–45, 152, 172, 173
  case study 44–45
  contemporary style 33, 36
  formal garden plans 40–43
  interpreting the style 38–39
  multi-level layouts 162
  near the house 194
  plants 199, 210
  sculpture 90, 156, 165, 237
  seats 232
  shrubs for 297
  structures 191, 359
  symmetry and 28, 31, 143, 150
  trees for 291
  vegetable beds 106
  what formal style is 34–37
Fothergilla 211, 213
foundations 245, 248, 256
fountains 33, 35, 38, 39, 119, 215
  bubble fountains 154, 215, 360
  dancing 363
  tiered 363
  wall fountains 362
foxglove see *Digitalis*
fragrance see scent
frameworks, creating 204
framing views 154–55
Fritillaria (fritillary) 212
  F. imperialis 338
  F. i. 'Maxima Lutea' 338
  F. meleagris 338
front gardens 42, 173, 359
Frost, Adam 197
frost pockets 134
fruit 17, 48, 103, 105, 107, 269, 278
fruit trees 105, 107
FSC (Forest Stewardship Council) 223, 229, 233, 351
Fuchsia magellanica 213, 302
functional requirements of gardens 14–15, 31
  functional planting 202
funky gardens 19
furniture 149, 223
  architectural 37
  care of 233
  contemporary 37, 232, 235
  designing with 232–33, 253
  family gardens 66, 68
  as focal points 48, 84, 232, 235
  foldaway 232
  hardwoods 233
  informal gardens 59
  modernist 232, 234
  rain and UV damage 233
  rustic 49, 77, 173, 232, 233, 234
  sculptural 91, 232, 235
  small spaces 90, 96, 98, 101
  statement gardens 118, 120, 124
  as storage 233, 359
  storing 177
  styles 234–35
  types of 364–65
  see also seating and seats
fusion style 31
future proofing garden designs 146–47

G

gabions 354
Galanthus (snowdrop) 213
  G. 'Atkinsii' 338
Galium odoratum 332
Galtonia viridiflora 338
galvanized metal 361
Garden of Australian Dreams (Canberra, Australia) 121
garden compost 76, 77, 133, 134, 145, 276, 277, 278
garden offices/studios 229, 358
garden pavilions 365
garden projects 242–85
  assessing the site 251
  budgets 244
  building structures 252–71
  evaluating the garden 246
  materials 350–56
  pre-construction checklist 245
  preparations 244–45
  preparing new-build sites 250–51
  preparing overgrown sites 246–47
  rehabilitating soil 248–49
  site clearance 247
garden "rooms" (compartments) 48, 90
garden rooms (structures) 85, 99, 358
garden styles 18–19, 28–129
Gardens of Appeltern (The Netherlands) 97
Garrya 213
gas heaters 240
gates 141, 227, 357
Gazania 197
gazebos 118, 358, 367
Genista 208
geometric layouts 156–57, 224
  formal gardens 31, 33, 34, 36, 37
  how to use shapes 150–51
  modernist gardens 36, 45, 96
  productive gardens 104, 105, 111
  small space gardens 90, 98
  statement gardens 122
  symmetrical layouts 158, 159
geotextile membrane 260–61, 264, 265, 271
Geranium (cranesbill) 18, 84, 209
  G. 'Ann Folkard' 332
  G. 'Brookside' 324
  G. × *cantabrigiense* 'Biokovo' 332
  G. clarkei 'Kashmir White' 333
  G. endressii 209
  G. macrorrhizum 209, 325
  G. 'Nimbus' 325
  G. nodosum 'Silverwood' 333
  G. palmatum 40
  G. 'Patricia' ('Brempat') 55
  G. phaeum 325
  G. pratense (meadow cranesbill) 282
  G. 'Rozanne' ('Gerwat') 333
  G. sanguineum 42
  G. sylvaticum 56
Gesellschaft der Staudenfreunde 57
Geum 84
  G. 'Prinses Juliana' 99
  G. 'Totally Tangerine' 325
Ginkgo biloba 216
Gladiolus 193, 212
glass 231
  in formal gardens 37
  garden structures 229
  glass aggregates 353
  ground-glass 265
  panels 165, 355
  sculptural structures 167
  small spaces 91
  statement gardens 119
Gleditsia triacanthos f. *inermis* 'Sunburst' 290
globe thistle see *Echinops bannaticus*
Goldsworthy, Andy 117
Goodman, Will 118
Google Earth 189
gourds 109
gradients, measuring 185
granite 57, 225
  polished 174, 351
  setts 174, 225, 350
grape hyacinth see *Muscari*
grass lawns see lawns
grasses 199, 200, 206, 342–45
  for autumn colour 213
  country gardens 52, 53, 78–79
  formal gardens 37, 39, 158
  informal gardens 53, 55
  natural gardens 75, 76, 82, 84
  ornamental 85
  plants for senses 221
  small gardens 94, 95, 97
  statement gardens 119
  urban gardens 90, 91
  for winter interest 173
  see also prairie-planting and individual species
gravel 225, 353
  Breedon 54, 265
  family gardens 66, 70
  formal gardens 33, 34, 35, 36
  gravel borders 149, 264–65
  gravel gardens 192, 249
  informal gardens 47, 48, 49, 54
  as mulch 264, 277
  paths 34, 47, 49, 78, 80, 85, 105, 110, 173, 224, 253, 255
  resin-bound 353
  self-binding 54, 265, 353
  self-seeding gravel 55
  statement gardens 122
Gravetye Manor (West Sussex) 78
Great Dixter (East Sussex) 52, 53
green roofs 26, 56, 76, 77, 83, 96, 99, 177, 358
green walls 178
greenhouses 70, 104, 105, 110, 139, 358
Gresgarth Hall (Lancashire) 53
Grevillea rosmarinifolia 100
grey-leaved plants 200, 208
grey water 145, 284
grid-pattern gardens 39
grit 134
Groeningen, Isabelle Van 54
ground cover 56, 199, 203, 208–29
  shrubs for 305
grouping plants 194
growing conditions 133
Guinness, Bunny 110–11
Gunnera 204
  G. manicata 347
Gustafson, Kathryn 117
Gypsophila 206
  G. paniculata 'Bristol Fairy' 325

H

habitats 78, 138, 277
  natural gardens 57, 75, 76
  water features 31, 137, 215
Hakonechloa macra 'Aureola' 54, 83, 207, 343
Hall, Stephen 77
Hamamelis 193, 211, 213
  H. × *intermedia* 'Jelena' 297
  H. × *i.* 'Pallida' 297
hammocks 364
handkerchief tree see *Davidia involucrate*
hanging baskets 144
hard landscaping 76
  drainage issues 225
  family gardens 139

introducing colour with 161, 168
natural gardens 80
small spaces 89
softening 178–79
symmetrical designs 158
texture 174
using height and structure 164–65
see also decking; paths; walls, etc
Hardenbergia violacea 315
hardiness ratings 288, 370
hardwoods 229, 231, 233, 268, 351, 352, 356
hardy geraniums see *Geranium*
hawthorn see *Crataegus*
hazel see *Corylus*
health 17, 27
heather see *Calluna*
heating, outdoor 94, 98, 223, 240–41, 359
choosing 240–41
Hebe 193, 196, 206, 218
H. 'Emerald Gem' 179
H. 'Great Orme' 308
H. macrantha 308
H. 'Midsummer Beauty' 302
H. pinguifolia 206, 208
H. p. 'Pagei' 308
H. 'Red Edge' 309
Hedera (ivy) 179, 209, 279
H. colchica 'Dentata Variegata' 315
H. c. 'Sulphur Heart' ('Paddy's Pride') 315
H. helix 209, 219
H. h. 'Oro di Bogliasco' ('Goldheart') 315
H. h. 'Parsley Crested' 315
hedges 21, 166, 167, 202
beech 57, 204, 357
box 34, 35, 57, 173
country gardens 52, 53, 78, 79
deciduous 204
evergreen 35, 140, 204
formal gardens 33, 34, 35, 36, 357
informal gardens 47, 53, 56
low 155, 156
mixed 79, 274
planting 274–75
pleached 41, 110
and right to light 141
for structure 35, 52, 53, 78, 79, 204
tall 133, 141
topiary 36
yew 35, 53, 55, 79, 87, 193, 211, 237, 289
height, using 164–65, 192
Helenium 'Moerheim Beauty' 325
Helianthemum 206, 208

H. 'Wisley Primrose' 309
Helianthus 'Lemon Queen' 319
Helichrysum italicum subsp. *serotinum* 309
Helleborus (hellebore) 212
H. argutifolius 325
H. foetidus 325
H. orientalis 209
H. 'Walberton's Rosemary' (Walhero) 333
Hemerocallis (daylily)
H. 'Marion Vaughn' 326
H. 'Selma Longlegs' 326
herb gardens 159, 172
herbaceous perennials see perennials
herbs 17, 24
container-grown 144
ground cover 199
informal gardens 49
Mediterranean 134
productive gardens 103, 104, 106, 109, 111, 113, 115
raised beds 269
small spaces 31
Hestercombe (Somerset) 53
Heuchera
H. 'Chocolate Ruffles' 197
H. 'Pewter Moon' 333
H. 'Pink Pearls' 333
H. 'Plum Pudding' 207, 333
Hibiscus syriacus 'Diana' 302
Hidcote Manor (Gloucestershire) 48
Hidden Gardens 62
high-maintenance gardens 20–21, 202
highlights, creating 170
hillsides, natural 163
Hippophae rhamnoides 298
hit and miss fences 356
Hitchmough, James 52
holly see *Ilex*
holly oak/holm oak see *Quercus ilex*
honesty see *Lunaria*
honeysuckle see *Lonicera*
hop see *Humulus lupulus*
hornbeam see *Carpinus*
horsetail see *Equisetum*
Hosta 84, 154, 155, 173, 176, 206, 207, 209, 213, 215
H. 'Francee' 69
H. 'Krossa Regal' 197
H. 'Royal Standard' 54
H. sieboldiana var. *elegans* 326
H. 'Sum and Substance' 54, 326
H. Tardiana Group 'June' 333
hot, dry sites, shrubs for 299
houseleek see *Sempervivum tectorum*
hues 168, 170, 171, 172

human-made materials 91, 118, 119
Humulus lupulus 'Aureus' 315
hurdles 78, 174, 357
Hurst Garden (US) 118
hyacinth see *Hyacinthus*
Hyacinthoides non-scripta 338
Hyacinthus (hyacinth) 212
H. orientalis 'Blue Jacket' 338
Hydrangea 211, 279
H. anomala subsp. *petiolaris* 316
H. arborescens 'Annabelle' 302
H. aspera 'Hot Chocolate' 302
H. macrophylla 42
H. m. 'Mariesii Lilacina' 302
H. paniculata 'Unique' 298
H. quercifolia 'Snow Queen' ('Flemygea') 302
H. 'Runaway Bride Snow White' 309
Hylotelephium 'Matrona' 326

I

Iberis sempervirens 206
Ilex (holly) 141
I. aquifolium 'Silver Queen' 298
I. crenata 201, 210, 303
Imperata cylindrica 'Rubra' 343
Indigofera heterantha 303
infinity pools 99
informal gardens 46–59, 143, 152
case study 58–59
informal garden plans 54–57
informal ponds 253, 270–71
interpreting the style 50–53
planting 158
what informal style is 48–49
insects 14, 277, 279
insect hotels 24, 99, 283
natural gardens 77, 83
pollinating insects 24, 104, 106, 107, 112, 217, 283
water features and 214, 215
inspiration 18–19, 28, 143, 148–49, 192
intercropping 25
interlocking circles 160
International Garden Festival (Chaumont-sur-Loire, France) 97
Inula magnifica 319
Ipomoea (morning glory) 179, 279
Iris 97, 192, 208, 214, 215
I. 'Butter and Sugar' 347
I. 'Golden Alps' 339
I. laevigata 347
I. lazica 333
I. pallida 'Variegata' 339
I. pseudacorus 215
I. p. 'Variegata' 347
I. reticulata 208
I. sibirica 57

I. s. 'Perry's Blue' 347
I. 'Superstition' 339
I. versicolor 'Kermesina' 347
I. 'White Knight' 339
ironwork 355
irregularly shaped plots 186–87
drawing plans of 189
Islamic influence 35, 39, 158, 159
Isolepis cernua 215
Italian styles 41
Itea ilicifolia 298
ivy see *Hedera*
ivy-leaved geranium see *Pelargonium*

J

Japanese anemone see *Anemone × hybrida*
Japanese maple see *Acer*
Japanese-style 28, 232, 353
family gardens 69
small spaces 91, 92
statement gardens 118, 119
Jasione montana 124
Jasminum (jasmine) 72, 278, 279
J. nudiflorum 303
J. officinale 'Argenteovariegatum' 316
Jekyll, Gertrude (1843–1932) 47, 49, 53, 78
Jellicoe, Sir Geoffrey (1900–96) 54
Jensen, Jens (1860–1951) 76
jets 61, 90
Joseph, Colm 124, 125
Judycki, Maggie 69
Juncus 215
J. effusus f. *spiralis* 215
J. patens 'Carman's Gray' 215
jungle style 151
Juniperus
J. communis 'Hibernica' 298
J. × pfitzeriana 309
J. × p. 'Pfitzeriana Aurea' 309
J. procumbens 309
J. squamata 'Blue Carpet' 309

K

kale 25
Keightley, Matt 40, 41, 58–59
Kent, William (1685–1748) 53
Kiftsgate Court (Gloucestershire) 53
Kiley, Dan 42, 78
Kirengeshoma palmata 347
kitchen gardens 228
see also vegetable gardens
kitchens, outdoor 367
kitchenware, recycled as containers 361
Knautia macedonica 326

Kniphofia (red hot poker)
   K. 'Bees' Sunset' 326
   K. 'Percy's Pride' 326
knot gardens 33
Kogan, Marcio 36–37
Kolkwitzia amabilis 'Pink Cloud' 303

L

Laburnum 211
   L. × *watereri* 'Vossii' 294
lady's mantle see *Alchemilla mollis*
Lagurus ovatus 344
Lamium maculatum 'White Nancy' 334
Lamprocapnos spectabilis 'Alba' 326
land art 117
landscape, links to 78, 79, 151, 155
landscape fabric 277
large plants
   shrubs 296–99
   trees 288–89
late-flowering plants
   perennials 325
   shrubs 311
Lathyrus (sweet pea) 210
   L. odoratus 278, 279
   L. vernus 334
Laurus nobilis (bay) 115, 167, 176, 294
Lavandula (lavender) 18, 104, 167, 200, 208, 217, 237
   L. angustifolia 'Munstead' 309
   L. × *intermedia* 'Grosso' 221
   L. stoechas 310
Lavatera 201
lavender see *Lavandula*
lawns 147, 156, 284
   country gardens 78, 79
   family gardens 61, 66, 67
   formal gardens 33, 34, 35, 36, 37, 38, 41
   informal gardens 48, 53
   laying 273, 280–81
   mowing 268, 281
   natural gardens 82
   no-mow meadows 282, 283
   preparing the ground 380
   seeding 273, 281
   stripping turf 249
   weeding 285
   work involved 20–21
layering shapes 156–57
Le Nôtre, André (1613–1700) 35, 39
lead planters 361
leaf kale 106
leaves 200
   colour 207
   composting 145
   leafmould 277
   see also foliage
legal issues, boundaries 133, 140, 141
lemon balm see *Melissa officinalis*
lettuce 113
Leucanthemella serotina 320
Leucanthemum vulgare (ox-eye daisy) 83, 85, 282
Leucojum aestivum 'Gravetye Giant' 339
levels 156–57, 183
Liatris spicata 'Kobold' 327
lifestyle changes 28
light 38, 118, 143, 170, 192, 200
   colour and 171
   right to 141
lighting 141, 156
   choosing 240–41
   designing with 238–39
   family gardens 61
   formal gardens 36, 37, 38
   installations 223, 238, 240, 244, 245
   LEDs 119, 173, 238, 239, 240, 241
   light pollution 238
   lighting effects 239
   for mood 17, 239
   small scale gardens 98, 99
   solar 223, 241
   statement gardens 117, 118, 119
   types of 240
   urban gardens 91
Ligularia 206
   L. 'The Rocket' 348
Ligustrum
   L. delavayanum 42, 99, 210
   L. ovalifolium 219
   L. o. 'Aureum' 298
Lilium 212
   L. 'African Queen' 339
   L. 'Black Beauty' 339
   L. henryi 339
   L. martagon 339
   L. Pink Perfection Group 340
   L. regale 340
   L. 'Star Gazer' 340
lily see *Crinum*; *Lilium*
lily-of-the-valley see *Convallaria majalis*
lime see *Tilia*
limestone 36, 38, 41, 55, 81, 97, 120, 351
liquid feeds 285
Liquidambar styraciflua 'Worplesdon' 290
Liriodendron tulipifera 'Aureomarginatum' 289
Liriope muscari 334
living/green walls 113, 166, 177, 178
living willow 357, 358

Lloyd, Christopher (1921–2006) 52
loams 134, 136
local materials 75, 76, 77
lollipop trees 167, 210
London Garden Designer (garden design) 182–83
Long, Richard 117
long-term gardening 26
Lonicera (honeysuckle) 179, 213, 278, 279
   L. nitida 210
   L. n. 'Baggesen's Gold' 196
   L. periclymenum 217
   L. p. 'Scentsation' 221
   L. p. 'Serotina' 56, 316
   L. pileata 209, 310
Lost Gardens of Heligan (Cornwall) 109
Lotus hirsutus 206
Loudon, John Claudius (1783–1843) 91
low-maintenance gardens 21, 43, 91, 202
   see also prairie-style planting
Lunaria (honesty) 173
Lupinus 'Chandelier' 327
Lutyens, Edwin (1869–1944) 49, 53
Lychnis
   L. coronaria 327
   L. c. 'Alba' 327
   L. flos-cuculi 57, 83
Lysimachia 209, 214
   L. ephemerum 327
Lythrum 214
   L. salicaria 'Feuerkerze' 327

M

MacDonald, Catherine 68
machinery 245, 247
Macleaya microcarpa 'Kelway's Coral Plume' 320
Magnolia 212
   M. liliiflora 'Nigra' 303
   M. stellata 303
Mahonia 199, 213
   M. eurybracteata 'Soft Caress' 218
   M. japonica 303
   M. × *media* 'Charity' 298
maintenance 20–21, 284–85
mallow see *Lavatera*
Malus (apple)
   M. 'Evereste' 294
   M. 'Royalty' 294
   M. 'Rudolph' 125
Manchester Allotment Society 111
manure 134, 276, 277, 278, 285
maple see *Acer*
marble 351
marginal plants 77, 192, 214, 271, 347

marjoram see *Origanum*
Martino, Steve 118
Mason, Olive 48–49
mat-forming plants 201
materials 90, 350–56
   choosing 222–41, 245
   design materials checklist 231
   human-made 91, 118, 119
   local 75, 76, 77
   mixing 225
   modern 28, 119, 173
   natural 91
   paths 152, 153, 223, 350–53
   recycled 144
   reusing 145
   for screens and boundaries 226–27
   for slopes and structures 228–29
   statement gardens 118, 120, 124
   for surfaces 224–25
   for water features 230–31
matrix planting 54
Matteuccia struthiopteris 348
mature gardens, rejuvenating 138
meadows 21, 147, 249
   benefits of 283
   family gardens 70
   meadow plantings 47, 53, 75, 76, 282–83
   natural gardens 78, 79, 82, 85
measuring a plot 184, 185, 253
   irregular-shaped plots 186–87
   rectangular-shaped plots 184
Meconopsis betonicifolia 121
Mediterranean gardens 172, 208, 236
   case study 86–87
   Mediterranean plantings 118, 120, 124, 128–29, 277
Mediterranean herbs 134
medium-sized plants
   perennials 322–29
   shrubs 300–305
   trees 290–91
Mee, Claire 71
Melianthus 178
   M. major 320
Melissa officinalis (lemon balm) 221
Mentha requienii 20
metal 231
   containers 361
   grilles 352
   materials for slopes and structures 228, 229
   steel edging 39
   steps 229
   texture and 174
   use of in statement gardens 122, 123
   see also aluminium; steel

microclimates 28, 108, 140
midrange plants 199, 203, 204, 206–207
Millennium Park (Chicago, US) 65
minimalism 19, 28
mint see *Mentha*
Miscanthus 95, 204
  M. sinensis 213, 221
  M. s. 'Gracillimus' 344
  M. s. 'Kleine Silberspinne' 344
  M. s. 'Malepartus' 344
  M. s. 'Silberfeder' 344
  M. s. 'Zebrinus' 344
mixed planting
  borders 20–21
  hedges 79, 274
mock orange see *Philadelphus*
modern materials 28, 119, 173
modern schemes 28, 31, 194
  case study 44–45
  country gardens 58–59
  formal gardens 34, 36, 37, 41
  informal gardens 52
  furniture 232, 234
  materials for screens and boundaries 226–27
  materials for structures 228, 229
  materials for surfaces 224
  modernist influences 55
  natural gardens 228
  paths 152, 223
  roof gardens 100–101
  small spaces 89, 96, 98, 99
  statement gardens 118
  steps in 229
  sustainable 75, 76, 77
  walls and railings 354–55
  water features 230
moisture-loving plants 133, 136, 329
Molinia
  M. caerulea subsp. *arundinacea* 'Windspiel' 206
  M. c. subsp. *caerulea* 'Variegata' 344
Monarda didyma x *fistulosa* 'Oneida' 327
monastic gardens 103, 104
Mondrian, Piet 45
monochrome colours 173
monoculture plantings 52
"mood boards" 148
moods 16–17, 143, 151
Moorish influence 359
morning glory see *Ipomoea*
Morus nigra (mulberry) 290
mosaics 173, 225
  mosaic walls 354
mosses 215
movement 69, 78, 119, 156

mowing lawns 281
mowing strips 269
mulberry see *Morus*
mulches 145, 277, 284
  chipped bark 277, 279
  garden compost 25, 134, 277
  gravel 264, 277
  leafmould 277
  manure 134
  organic matter 276
multi-level layouts 162–63
multi-stemmed trees 44, 59
Munstead Wood (Surrey) 49
Musa basjoo (banana) 320
Muscari (grape hyacinth) 212
  M. armeniacum 'Blue Spike' 340
  M. latifolium 340
mycorrhizal connections 25
Myers, Robert 124
Myosotis (forget-me-not) 173, 215
  M. scorpioides 215, 348
mystery, sense of 161

N

Nandina
  N. domestica 303
  N. d. 'Fire Power' 303
Narcissus (daffodil) 212, 283
  N. 'Bridal Crown' 340
  N. poeticus var. *recurvus* 340
  N. 'Tête-à-Tête' 340
  N. 'Thalia' 340
Nasturtiums 106, 109, 112, 317
National Garden Scheme (NGS) 97
native plants 57, 215, 228, 238
natural forms 167
natural gardens 23, 28, 74–87, 179
  case study 86–87
  interpreting the style 80–83
  natural garden plans 84–85
  what natural style is 76–79
natural materials 91, 118, 228
naturalistic plantings 16, 70, 85, 99, 192, 194, 204
nature, benefits of 27
navigation 152–53
Nectaroscordum siculum subsp. *bulgaricum* 341
neighbours 140, 141, 245
  see also privacy
Nemesia 'Wisley Vanilla' 220
Nepeta 208
  N. grandiflora 'Dawn to Dusk' 327
  N. nervosa 196
  N. 'Six Hills Giant' 327
Nerine 193
  N. bowdenii 341

neutral colours 173
new-build sites 250–51
New Perennial Movement 75, 76, 282
new styles 31
Nixon, Philip 90–91
no dig approach 25
Nymphaea (waterlily) 94, 215, 271
  N. alba 57
  N. 'Darwin' 348
  N. 'Froebelii' 348
  N. 'Gonnère' 348
  N. 'Marliacea Chromatella' 348
  N. tetragona 215
Nyssa sinensis 290

O

oak see *Quercus*
obelisks 105, 279, 358
Ocimum basilicum (basil) 111
O'Connor, John 237
Oenothera fruticosa 'Fyrverkeri' 334
offsets (in surveying) 186
Olea (olive)
  O. europaea 71, 165, 294
Olearia macrodonta 298
olive see *Olea*
Olof, Clare 72–73
Omphalodes cappadocica 'Cherry Ingram' 334
Ophiopogon planiscapus 'Nigrescens' 344
options, assessing 138–39
organic layouts 143, 156, 160–61, 191, 224
organic matter 208, 209, 249, 251, 274, 276, 280, 285
orientation 171
  see also aspect
Origanum 59, 112
  O. 'Kent Beauty' 310
  O. laevigatum 'Herrenhausen' 328
  O. vulgare 217
  O. v. 'Aureum' 71
ornamental cherry see *Prunus*
ornamental ponds 362
ornamental vegetables 112
ornaments 34, 35
Orontium 214
  O. aquaticum 215
Osmanthus 44, 95
  O. × *burkwoodii* 304
Osmunda regalis 348
Oudolf, Piet 52, 53, 75
outbuildings 83
outdoor living 28, 36, 95, 98, 139, 146, 366–67
outdoor rooms 14, 36, 62, 90, 91, 139, 233, 238

ovens, open-air 366
overgrown sites, preparing 246–47
overlaid photographs 182, 187, 194
oxygenators 214

P

Pachysandra terminalis 209, 334
paddlestones 353
Paeonia (peony) 202, 213
  P. 'Bartzella' 328
  P. delavayi 304
  P. lactiflora 'Duchesse de Nemours' 54
paint 168, 351, 355, 358, 359
palettes, planting 202, 207
Paley Park (New York, US) 97
pallets 144, 355
palms 119, 120
  see also *Trachycarpus wagnerianus*
pampas grass see *Cortaderia*
Panicum virgatum 'Heavy Metal' 55, 344
pansy see *Viola*
Papaver 82
  P. Oriental Group 'Black and White' 328
Pape, Gabriella 54
papyrus see *Cyperus papyrus*
parasols 140
parterres 103, 159, 194, 199
  formal gardens 31, 33, 34, 35, 38
Parthenocissus
  P. henryana 316
  P. tricuspidata 'Veitchii' 316
Passiflora (passionflower) 179, 202, 278
  P. caerulea 316
paths 152–53, 167
  bark 161, 265, 352
  brick 49, 73, 85, 104, 114, 179, 224, 254
  cottage gardens 47, 48, 49
  curves 152, 153, 160, 161
  decking 144
  edgings 254
  grass 79
  gravel 34, 47, 49, 78, 80, 85, 104, 110, 173, 224, 253, 255
  informal gardens 57
  laying 254–55
  lighting 238
  limestone 81
  marking out 254–55
  materials for 152, 153, 223, 350–53
  natural gardens 78
  permeable 265
  planting 47, 48, 58, 153, 179, 225

in productive gardens 104, 114
routes of 143, 152–53, 157, 161
stone 54, 57, 80, 104, 224, 350–51
timber 122
width of 105, 253
patio ponds 360, 362
patios
aspect 135
construction 253, 352
family gardens 139
laying 249, 256–57
materials 225
patio kits 352
privacy 140, 141
productive gardens 103
site for fragrant plants 193
Paulownia tomentosa 291
pavilions, garden 83, 365
slatted side 367
paving
brick 70, 350
cottage gardens 48
crazy 350
cutting curves into 257
family gardens 70
formal gardens 33, 37, 38
geometric designs 156
irregular-shaped 128
laying paving slabs 256–57
materials for 224, 350–51
small spaces 90, 94
plants in 47, 48, 58, 153, 179, 225
statement gardens 118
stone 33, 35, 39, 86, 87, 98, 350–51
pear see *Pyrus*
pebbles 173, 192, 225, 265
Pelargonium 18
Pennisetum 95
P. alopecuroides 201, 345
Penstemon
P. 'Alice Hindley' 328
P. 'Andenken an Friedrich Hahn' 328
peony see *Paeonia*
perennials 24, 139, 199, 200, 202, 203, 204
for architectural interest 319
for attracting wildlife 321
for autumn colour 213
for containers 333
for damp conditions 329
deadheading 284
early-flowering 323
evergreen 335
for foliage interest 327
formal gardens 37
informal gardens 52, 53, 56
late-flowering 325
medium-sized 322–39

natural gardens 76
perennial plantings 282–83
planting 276–77
shade-tolerant 331
small 330–35
for summer colour 193
tall 318–21
weeds 273, 276, 280, 285
for winter interest 173
woodland 79
see also drifts, planting; New Perennial Movement
perfume see scent
pergolas 154, 156, 165
awnings 367
building 253, 266–67
constructing the roof 266–67
erecting the arches 266–67
family gardens 71
informal gardens 52
kits 253, 266, 359
making the arches 266–67
materials for 228, 229, 253, 266–67, 359
for privacy 90, 140, 196
productive gardens 106
for shade 359
periwinkle see *Vinca*
permeable hard surfaces 76
Persicaria
P. amplexicaulis 'Firetail' 328
P. bistorta 'Superba' 197, 328
perspective 164, 165, 186–87, 210
Perspex 117, 119, 120, 173, 231
pesticides 75
pests 49, 75, 76, 103, 104, 107, 115, 276
Petersen bricks 70, 85
pH of soils 134, 200, 248
Phaseolus coccineus 111
Philadelphus 'Belle Étoile' 304
Phlomis 52
P. fruticosa 310
P. russeliana 328
Phlox
P. paniculata 'Blue Paradise' 320
P. p. 'Norah Leigh' 328
Phormium 193, 199, 210
P. cookianum subsp. *hookeri* 'Tricolor' 329
P. tenax Purpureum Group 320
Photinia × *fraseri* 'Red Robin' 298
photographs, designing with 182, 187, 192, 194
Phuopsis stylosa 334
Phygelius × *rectus* 'African Queen' 310
Phyllostachys
P. aureosulcata f. *aureocaulis* 219
P. nigra 40, 345
P. sulphurea f. *viridis* 167
P. vivax f. *aureocaulis* 345

physic gardens 104
Physocarpus opulifolius 'Diabolo' 304
Picea (spruce)
P. breweriana 291
P. pungens (Glauca Group) 'Koster' 291
picket fences 47, 155, 227, 356
Pieris 200
P. japonica 304
P. j. 'Blush' 304
pine see *Pinus*
pine cone mulch 145
pink see *Dianthus*
Pinus (pine)
P. mugo 218
P. m. 'Mops' 310
P. sylvestris (Aurea Group) 'Aurea' 291
P. wallichiana 289
Pittosporum
P. tenuifolium 'Golf Ball' 310
P. t. 'Silver Queen' 299
P. t. 'Tom Thumb' 310
P. tobira 'Nanum' 310
pizza ovens 366
planning controls and permission 140, 141, 245
plans 10, 143
bubble diagrams 143, 149, 182, 190, 194
creating 180–97
creating site plans 184–85
cross-sections 183
experimenting with 190–91
finished plans 183
overhead plans 183
planting plans 181, 183, 188, 192–96
scale plans 185, 186, 187, 188–89, 190–91, 194, 195
symbols 182, 183
understanding plans 182–83
working plans 182
see also planting plans; site plans
plantain 251
planters 96, 109, 144, 174, 360, 361
see also containers
planting and plantings 245
bare-root plants 273, 274
between paving 47, 48, 58, 153, 179, 225
blocks planting 35, 37, 96, 194
climbers 278–79
container-grown plants 273, 274–75, 276
cottage gardens 47, 48–9
formal plantings 194
hedges 274–75
informal plantings 158
naturalistic 16, 56–57, 70, 85, 99, 192, 194, 204
perennials 276–77
planting density 194, 195
planting palette 192, 202, 207
seasonal plantings 158
shrubs 276–77
structural plantings 167
techniques 272–85
trees 274–75
water features 214–15
see also Japanese-style; Mediterranean-style gardens; planting plans, etc
planting plans 183, 192–96
cottage gardens 54–55
drawing up 194–95
foliage gardens 40–41
formal gardens 42–43
Japanese-style gardens 69
natural gardens 56–57
scale for 188
sustainable gardens 56–57
urban gardens 43
plants
assessing existing plants 247
designing with 198–221
growth habits 201
increasing numbers of 24
plant guide 288–349
selecting 192–93, 202–203
sourcing 24
spread of 194, 204, 273
types and design uses 203
understanding 200–201
plastic, reducing 25
platforms 137, 162
play areas 14, 28, 139, 149, 155, 359
family gardens 61, 62–63, 64–65, 66, 71, 73, 160
pleaching
hedges 41, 110
trees 33, 34, 38, 90, 91, 176
polished granite 174, 351
pollarded trees 291
pollinating insects 24, 104, 106, 107, 112, 217, 283
polyanthus see *Primula*
Polypodium × *mantoniae* 'Cornubiense' 334
polyrattan furniture 365
ponds and pools 16, 133, 137
country gardens 79
family gardens 61, 62, 64, 65, 66
fishponds 362
formal 34, 35, 39, 43
informal 253, 270–71
lighting 239
lining and edging 270–71
Japanese gardens 69
making 270–71
materials for 230

natural swimming ponds 363
ornamental 362
patio ponds 360
plants for 347
raised pools 230
safety 61, 64, 139
siting 214
small spaces 94, 97
swimming pools and ponds 36–37, 64, 93, 174, 363
for wildlife 31, 62, 65, 66, 75, 77, 137, 215, 230, 362, 363
Pontederia 214
poppy see *Papaver*
porcelain tiles 351
postmodernism 118
potagers 37, 103, 104–105
Potentilla 208
  P. fruticosa 'Abbotswood' 311
  P. f. 'Dart's Golddigger' 208
  P. f. 'Goldfinger' 311
pots 25, 144, 174, 360, 361
  informal gardens 59
  statement gardens 118
  terracotta 360
  see also containers; planters
powder-coated metal 361
powdery mildew 279
"power" gardening 34
prairie-style planting 28, 194, 282–83
  natural gardens 75, 76, 78, 82
Pratia angulate 'Treadwellii' 125
pressure-treated timber 229, 253, 268
Primula 212, 214, 215
  P. alpicola 348
  P. beesiana 349
  P. 'Inverewe' 349
  P. vialii 215
privacy 133, 141, 155, 246
  family gardens 71
  formal gardens 40, 42, 43
  pergolas for 90, 91, 140, 196
  productive gardens 109
  screens for 40, 141, 165, 202
  small spaces 95, 101
productive gardens 23, 31, 102–15
  case study 114–15
  design influences 103, 105
  formal vs relaxed 108
  interpreting the style 106–109
  productive garden plans 110–13
  what productive style is 104–105
professionals 244, 245, 247, 248, 249, 253
  see also surveyors
proportion 36, 158, 186
pruning 21, 141, 210, 236, 276, 285
Prunus 212, 213

P. laurocerasus 197, 219
P. l. 'Zabeliana' 311
P. 'Mount Fuji' 294
P. padus 'Watereri' 291
P. serrula 93, 294
P. spinosa 141
P. 'Spire' 294
P. × *subhirtella* 'Autumnalis' 213
P. × *s.* 'Autumnalis Rosea' 295
Pseudosasa japonica 40
Pulmonaria
  P. 'Blue Ensign' 334
  P. 'Diana Clare' 335
pumpkin 111
pumps, pond 230, 271
PVC liner 270–71
Pyracantha 141, 179, 357
  P. 'Saphyr Jaune' ('Cadaune') 304
Pyrus
  P. calleryana 'Chanticleer' 216
  P. salicifolia 'Pendula' 295

Q

Quercus ilex (holm/holly oak) 91, 128, 289
quince, ornamental see *Chaenomeles speciosa*

R

railings 122, 163, 355
rain chains 363
rain shadow 219
rainbow chard 106
rainwater 87, 146
  capture of 26, 76, 77, 103, 137, 145, 177, 284
  drainage 76, 225, 265, 351
raised beds 146, 147, 173, 228
  family gardens 70
  making 253, 268–69
  measuring up the base 268
  productive gardens 105, 106, 107, 112, 115
  small spaces 94
  statement gardens 120
rambling roses 278, 316
ramps 162, 163
random planting 194
Ranunculus
  R. aquatilis 214
  R. flammula 215
raspberries 112
rattan furniture 233, 234
reclaimed wood 57, 173, 229, 232, 233, 268
rectangular plots 37
  measuring 184
  site plans 185
rectilinear layouts 44, 118, 157, 190, 196, 236

recycling 144, 359
  natural gardens 75, 76, 77
  recycled wood 57, 173, 229, 233, 268
  recycling cupboards 359
  water 137, 284
red chard 111
red hot poker see *Kniphofia*
reflections 93
  in water 35, 37, 43, 239
rejuvenating a mature garden 138, 139
relaxation 15, 31, 61, 98, 147, 149, 238
religious influences 28
Renaissance styles 35, 37
rendered walls 164, 174, 226, 228, 231, 237, 354
  Mediterranean gardens 86
  modernist gardens 97
repetition 28, 33, 165
  in planting 118, 122, 158–59, 164, 204
resin-bound gravel 353
retaining walls 162, 228, 248
  building 258–59
Rhamnus alaternus 'Argenteovariegata' 299
Rheum (rhubarb) 200
  R. palmatum 'Atrosanguineum' 349
Rhinanthus minor (yellow rattle) 283
Rhodanthemum hosmariense 335
Rhododendron 200, 212, 247
  R. 'Golden Torch' 311
  R. 'Kure-no-yuki' ('Kurume') 311
  R. luteum 299
RHS Chelsea Flower Show (London) 54, 97
RHS Flower Show Tatton Park 111
RHS Garden Wisley (Surrey) 109
RHS Hampton Court Palace Garden Festival (Surrey) 121
RHS (Royal Horticultural Society) 219
rhubarb see *Rheum*
Rhus 176, 213
  R. typhina 295
Ribes
  R. sanguineum 217
  R. s. 'Pulborough Scarlet' 304
right-angled shapes 150
rills 122, 145, 146, 174, 362
  formal gardens 34, 35, 39
  making a rill 270–71
  natural gardens 86
Robinson, William (1838–1935) 76, 78
rock gardens 201
rock rose see *Cistus*; *Helianthemum*

Rockcliffe Gardens (Gloucestershire) 110
rocks 120, 126
Rodgersia 126, 200, 206, 209
  R. pinnata 'Superba' 349
role of the garden 14–15, 31
Romneya coulteri 'White Cloud' 320
roof gardens
  case study 100–101
  formal gardens 43
  productive gardens 113
  small spaces 92, 96
roofs, green 56, 76, 77, 83, 96, 99, 177, 358
"rooms" in a garden 48, 90
Rosa (rose)
  R. 'Anna Ford' ('Harpiccolo') 311
  R. 'Compassion' 316
  R. 'The Fairy' (Poly) 312
  R. 'Félicité Perpétue' 316
  R. 'For Your Eyes Only' ('Cheweyesup') 311
  R. 'The Generous Gardener' ('Ausdrawn') 317
  R. 'Geranium' *moyesii* hybrid 304
  R. 'Golden Wings' 311
  R. 'Madame Alfred Carrière' 316
  R. 'New Dawn' 196
  R. 'Pearl Drift' ('Leggab') 312
  R. rugosa 217, 357
  R. 'Souvenir du Docteur Jamain' 197
  R. 'Wildeve' ('Ausbonny') 312
  R. xanthina 'Canary Bird' 304
rosemary 41
roses 21, 179, 200, 316
  climbing 278, 279, 316–17, 359
  formal gardens 49
  natural gardens 83
  rambling 278, 316
  see also *Rosa* (rose)
Rothwell, Sara Jane 182
Rousham Park House (Oxfordshire) 53
routes and navigation 143, 152–53, 154
Rowe, Charlotte 40, 42
rubber 117, 353, 359
Rudbeckia 80, 82, 213
  R. fulgida var. *sullivantii* 'Goldsturm' 329
  R. laciniata 'Goldquelle' 329
runner beans 104, 106, 111, 113, 279
rustic furniture 49, 77, 173, 232, 233, 234
Ruta graveolens 312

S

S-shaped designs 160
Sackville-West, Vita (1892–1962) 54, 78
safety 163, 244
    electrical 223, 230, 238, 240, 241
    play areas 149, 359
    water features 61, 64, 139, 214
sage see *Salvia*
Sagittaria sagittifolia 214
sails 43, 95, 140, 165, 364
salad crops 104, 109, 113
Salix (willow) 73, 211
    S. alba var. *sericea* 291
    S. elaeagnos subsp. *angustifolia* 197
    S. × *sepulcralis* var. *chrysocoma* 291
Salvia (sage) 124
    S. argentea 221
    S. 'Blue Spire' 312
    S. 'Hot Lips' 329
    S. microphylla 312
    S. nemorosa 335
    S. officinalis 208
    S. o. 'Purpurascens' 312
    S. o. 'Tricolor' 312
    S. rosmarinus 312
    S. uliginosa 320
Sambucus
    S. nigra f. *porphyrophylla* 'Eva' 305
    S. racemosa 'Plumosa Aurea' 305
sand pits 62, 149
sandstone 120, 174, 351
sandy soils 134, 136, 200, 276, 285
Sanguisorba
    S. canadensis 349
    S. 'Tanna' 329
Santolina
    S. chamaecyparissus 208
    S. pinnata subsp. *neapolitana* 'Sulphurea' 313
Sarcococca 213
    S. hookeriana var. *digyna* 313
    S. h. var. *humilis* 69
    S. h. 'Winter Gem' 313
Sassafras albidum 69
Saururus 214
scale 33, 34, 36, 118, 150, 158
    choosing a scale 188
    scale plans 185, 186, 187, 188–89, 190–91, 194, 195
Scampston Hall (North Yorkshire) 53
Scarpa, Carlo (1906–78) 55
scent 17, 193, 199, 202, 208, 341
    informal gardens 58
    bulbs, corms and tubers for 341

Schefflera 178
    S. taiwaniana 299
Schizophragma integrifolium 317
Schoenoplectus lacustris subsp. *tabernaemontani* 'Albescens' 215
Schultz, Michael 118
Schwartz, Martha 117
Scilla siberica 341
screens 156, 223, 226, 278, 356–57
    bamboo 96, 155, 167, 173, 357
    bin storage 155, 177
    containers as 133
    formal gardens 41
    internal 17, 166
    materials for 226–27, 278, 356–57
    metal 129
    natural gardens 78, 79
    for play areas 155
    pleached trees 90, 91
    for privacy 40, 42, 141, 165, 202
    small spaces 92, 94
    screen walls 354
    temporary 140, 165
    transparent 164, 165, 166
    and views 154, 155, 223
sculptural furniture 91, 232, 235
sculptural plants 18, 90, 98, 118, 119, 129, 210
    see also architectural plants
sculptural structures 167
sculpture 19, 31
    choosing 236
    commissioning 237
    as focal point 33, 39, 90, 124, 159, 165, 237
    formal gardens 33, 34, 39
    integrating into a design 236–37
    lighting 239
    living sculptures 93
    materials and cost 237
    natural gardens 78, 87
    positioning 237
    scale and proportion 237
    statement gardens 117, 118, 123
sea holly see *Eryngium*
sea kale 110
seaside-themed gardens 149, 201
seasonal interest 176, 199, 202, 203, 204, 212–13, 220
seating and seats 172, 202, 233, 234, 235, 364–65
    arbour seats 359
    benches 43, 54, 95
    built-in 90, 226, 235
    cottage gardens 48, 49
    family gardens 62, 66, 68, 69
    focal points 48, 232
    informal gardens 59
    natural gardens 81, 83, 84

seating areas 53, 57, 135, 140, 172, 191, 193, 253
    small spaces 94
    statement gardens 120, 124
    swing seats 155
    temporary 235
    walls as 164
    see also arbours
security 141, 237, 239, 358
sedges 200, 342–45
Sedum 53, 213
    green roofs 76, 77, 96, 358
    Sedum rupestre 'Angelina' 335
    S. takesimense 'Atlantis' ('Nonsitnal') 335
    S. 'Vera Jameson' 335
seedheads 52, 55, 174, 221, 283
seeds and seedlings
    growing seedlings 25
    meadow seed 282, 283
    seeding a lawn 281
    self-seeding 47, 55, 194
self-binding gravel 265, 353
Selinum wallichianum 68, 320
Sempervivum tectorum 335
senses, plants for the 221
services 214
    identifying position of 136
Sesleria
    S. autumnalis 345
    S. nitida 124
setts 48, 49, 54, 73, 174, 225, 253, 350
shade 135, 141, 146, 155, 176, 192, 202
    canopies for 43, 140, 156, 165, 364
    colour and 171
    family gardens 72
    formal gardens 38
    natural gardens 87
    perennials for 331
    pergolas for 359
    plants for 72, 197, 200, 209, 301, 331
    shrubs for 301
    small spaces 93
shades (colour) 168, 170, 171
shadow 118, 170
shapes and spaces 17, 150–51, 192, 199, 206
sheds 154, 155, 177, 228, 229, 358
shells 265, 353
    shell mosaic walls 354
shiplap fences 356
shrubs 21, 134, 199, 200
    for autumn colour 213, 311
    in containers 176
    evergreen 141, 206, 313
    for focal points 297
    for foliage interest 303
    for ground cover 305

    for hot, dry sites 299
    how to plant 276–77
    large 296–99
    medium-sized 300–305
    positioning 194
    for shade 301
    small 203, 206, 306–13
    for spring interest 307
    for summer colour 309
    wall shrubs 179, 278
silver-leafed plants 124, 200, 208, 216, 217
sinks 367
Sissinghurst Castle (Kent) 48
Sisyrinchium
    S. striatum 71
    S. s. 'Aunt May' 329
site clearance 247
site plans 192
    creating 184–85
sketches 194
Skimmia × *confusa* 'Kew Green' 305
slate 167, 225, 351
    chips 145, 265, 271, 353
slatted side pavilion 367
slatted wood fences 356
sleepers 146, 162, 268, 352
sloping gardens 133, 134, 136–37, 146, 162–63, 183
    materials for 228, 229
    measuring 185
slugs 109, 276
small plants
    perennials 330–35
    shrubs 306–13
    trees 292–95
small spaces 88–101, 151
    case study 100–101
    interpreting the style 92–97
    small garden solutions 176–77
    small space garden plans 98–99
    what small space gardens are 90–91
smoke bush see *Cotinus*
smokers 366
snails 109, 276
snake's head fritillary see *Fritillaria meleagris*
Sneesby, Richard 190, 191
snowdrop see *Galanthus*
snowy mespilus see *Amelanchier lamarckii*
sofas, outdoor 365
software packages 181, 191
softwood 229, 231, 351, 352
soil
    assessing 133, 134
    improving 76, 133, 134, 248–49, 277, 285
    no dig approach 25
    pH 134, 200, 248

plants for different soils 200
removing soil 248
soil types 134, 136, 181, 192
waterlogged 146
Solanum
  S. crispum 317
  S. c. 'Glasnevin' 317
  S. laxum 317
  S. l. 'Album' 317
  S. melongena 111
solar lighting 223, 241
Soleirolia soleirolii 58, 225
Sorbus 205, 213
  S. aria 'Lutescens' 295
  S. commixta 295
  S. c. 'Embley' 295
sound 73, 86, 95, 119
spaces, shapes and 150–51
Spartium 124
  S. junceum 101
specimen trees 36
Spiraea nipponica 'Snowmound' 305
spouts, wall 362
spring interest 212
  bulbs, corms, tubers for 337
  shrubs for 307
  trees for 293
spruce see *Picea*
squash plants 106
Stachys
  S. byzantina 208
  S. b. 'Big Ears' 216
staking trees 274–75
statement gardens 31, 116–29
  case study 128–29
  interpreting the style 120–23
  statement garden plans 124–27
  what statement style is 118–19
statuary 35, 36, 204, 236, 239
steel 37, 91, 229, 231
  Corten steel 33, 34, 122, 125, 229, 360
  edgings 39
  steel screens 129
  statement gardens 118, 119
  weathering steel 229, 360
stepping stones 244, 253, 352
steps 162, 163, 235
  lighting 238
  materials for 229
Stewartia sinensis 295
Stipa 97
  S. arundinacea 68
  S. gigantea 99, 216, 345
  S. ichu 345
  S. tenuissima 221, 345
stone 164, 225, 230, 231, 237, 244
  chippings 174
  family gardens 61, 66
  informal gardens 47

natural gardens 79
paths 54, 57, 80, 104, 224, 350–51
paving 33, 35, 39, 86, 87, 98, 350–51
walls 36–37, 45, 47, 97, 174, 226, 228, 354
stone urns 360
Stopherd, Chuck 62
storage 176, 177, 233, 358, 359
  furniture 233, 234
strawberries 107, 109, 113, 144
  strawberry pots 360
strawberry tree see *Arbutus unedo*
streams 79, 230, 363
  edging and lining 230
structure 139, 164–65
  structural plants 167, 193, 199, 203, 204–205, 210
structures 143, 166–67, 358–59
  building 252–71
  materials for 228, 229
Sturgeon, Andy 40, 43, 78, 128–29
Styphnolobium japonicum (pagoda tree) 87
subsoil 250, 251
subtropical gardens 127
succulents 178, 192, 220
summer colour 193, 207, 212
  bulbs, corms, tubers for 339
  climbers for 315
  shrubs for 309
sunflower see *Helianthus*
sunny sites 135, 200, 208
surfaces
  drainage 76, 136–37, 224, 225
  materials for 224–5, 350–3
surveyors 181, 183, 184, 185, 186
sustainability 23, 24–27, 28
  sustainable choices 144–45
  sustainable gardens 74–87
  sustainable materials 28, 75, 76, 77
Sustainable Urban Drainage System (SuDS) 137
sweet peas see *Lathyrus odoratus*
swimming pools and ponds 36–37, 64, 93, 174, 363
swings 71, 149, 155, 364
symbols, for planting plans 182, 183
symmetry 31, 158, 215
  formal gardens 28, 31, 33, 34, 35, 36, 37, 38–39, 143, 150, 194
  symmetrical layouts 143, 158–59
Symphyotrichum
  S. ericoides 'White Heather' 329
  S. novae-angliae 'Andenken an Alma Pötschke' 329

S. 'Ochtendgloren' 321
Syringa vulgaris 'Mrs Edward Harding' 299

T

tables 365
  fire tables 364
Tagetes 111
tall perennials 318–21
Tamarix (tamarisk)
  T. ramosissima 299
  T. r. 'Pink Cascade' 299
Taxus (yew) 59
  T. baccata 197, 289
  T. b. 'Fastigiata' 295
  T. b. 'Standishii' 289
tents and tepees 62
terraces 223
terracing 137, 146, 162–3
terracotta
  pots 360
  tiles 350
terrazzo 174, 351, 360
Teucrium chamaedrys 206
texture 16, 97, 143, 174, 225
  combining 44, 174
  foliage 94
  incorporating into a design 174–75
  materials 91, 174–75
  plants 38, 174–75, 192, 206
  types of 174
Thalia dealbata 349
Thalictrum 84
  T. delavayi 206
  T. 'Elin' 321
  T. flavum subsp. *glaucum* 321
thatched gazebos 367
Thompson, Jo 84
Thuja plicata 210
Thymus (thyme) 81, 112, 113, 179, 208
tiles 73, 231
  decking tiles 352
  glazed 350
  porcelain 351
  terracotta 350
  wood-effect 352
Tilia (lime) 38, 91, 126
timber 174, 229, 231, 237
  paths 122
  pressure-treated 229, 253, 268
  reclaimed 57, 173, 229, 232, 233, 268
  sustainable 76, 77
  walls 228
  see also FSC; wood
time to devote to the garden 20–21
tints 168, 170, 171
tomatoes 104, 108, 109, 111, 115

tones 168, 171
  tonal accents 36
topiary 26, 159, 166, 205, 210, 211, 236
  formal gardens 34, 35, 36, 39, 42
  informal gardens 47
topsoil 245, 248, 249, 250, 251
Trachelospermum jasminoides 58, 127, 196
Trachycarpus wagnerianus 127
Trachystemon orientalis 209
training plants 210, 278
trampolines 67
transparent screens 164, 165, 166
travertine 351
tree ferns 293
tree houses 65
trees 21, 24, 146, 199, 200, 205
  for autumn colour 213, 295
  deciduous 140, 176, 192, 205, 213
  evergreens 16, 41, 42, 97, 128, 289
  family gardens 65, 73
  as focal points 291
  fruit trees 105, 107
  growing climbers on 278
  how to plant 274–75
  large 288–89
  "lollipop" trees 167, 210
  mature trees 139
  medium-sized 290–91
  multi-stemmed 44, 59, 176
  pleached 33, 34, 38, 90, 91, 176
  pollarded 291
  positioning 194, 204
  and privacy 101, 246
  removing 139
  and right to light 141
  small 292–95
  specimen trees 36
  tree preservation orders 139
  for spring interest 293
  using in a small space 176
trellis 140, 164, 165, 166, 177, 178, 179, 193, 227, 278, 279, 356
trends, garden design 22–27
triangulation 186, 187
Trillium
  T. grandiflorum 341
  T. g. 'Flore Pleno' 341
Triteleia 212
Trithrinax campestris 299
Tropaeolum speciosum (nasturtium) 106, 109, 112, 317
troughs 215, 360, 361, 363
Tsuga canadensis 'Aurea' 295
tubers 336–41
Tulipa 45, 173, 212
  T. 'Flaming Parrot' 341
  T. kaufmanniana 208

T. linifolia Batalinii Group 208
T. 'Prinses Irene' 341
T. 'Queen of Night' 207, 341
T. 'Spring Green' 341
turf 35, 61, 251
   laying 273, 280–81
   stripping 249
   wildflower 283
   see also lawns
Typha minima 349
tyre planters 361

U

Uncinia rubra 345
United States (US) 76
University of Sheffield 56, 76
upcycled containers 144
upright plants 201
urban gardens 31, 173, 197
   family gardens 71
   formal gardens 38, 40–1, 42
   kitchen gardens 109
   microclimates 28, 40, 134
   plants for 219
   small spaces 90, 91
   statement gardens 127
urns 35, 42, 197, 360

V

Valeriana phu 'Aurea' 321
Vaux le Vicomte (France) 35, 39
vegetable beds 48, 49, 228, 269
vegetable gardens 31, 105, 159
   see also potagers; productive gardens
vegetables 24, 31, 49, 103, 104, 114, 228, 269
   growing on walls 177
   planting in rows 105
   see also individual types of vegetable
Verbascum 124, 201
   V. 'Cotswold Queen' 321
Verbena 80, 112
   V. bonariensis 55, 216, 321
Veronica 196
   V. gentianoides 335
   V. 'Shirley Blue' 54
Veronicastrum
   V. virginicum 321
   V. v. 'Fascination' 321
Versailles (France) 34, 35, 39
Versailles planters 361
vertical growing 95, 113, 166, 177, 178
Viburnum 212, 213
   V. × *bodnantense* 'Dawn' 305
   V. × *bodnantense* 'Deben' 305
   V. × *burkwoodii* 'Anne Russell' 313

   V. carlesii 'Aurora' 305
   V. davidii 313
   V. opulus 299
   V. o. 'Roseum' 216
   V. plicatum f. *tomentosum* 'Kilimanjaro' ('Jwwl') 305
   V. p. f. t. 'Mariesii' 210, 211, 305
views 154–55, 160, 193, 227
   borrowed views 155, 213
   country gardens 78, 79
   formal gardens 33, 34, 35, 39
   from the house 194
Villa Gamberaia (Italy) 39
Vinca (periwinkle) 209
   V. major 209
   V. minor 209
   V. m. 'La Grave' ('Bowles Blue') 313
vine see *Vitis*
Viola 220
vistas see views and vistas
visualization technique 192, 194
Vitis (vine) 111, 278, 359
   V. coignetiae 317
   V. vinifera 'Purpurea' 196, 317

W

Wade, Charles (1883–1956) 54
walkways 224, 235
wall shrubs 135
walls 90, 163, 164, 166, 246, 354–55
   brick 170, 171, 226, 228, 354
   concrete 174, 228, 355
   coping 226
   dry stone walls 99, 174, 228, 354
   enhancing 226
   faced 118, 119
   feature 161
   living/green walls 95, 113, 166, 177, 178
   materials for 223, 226, 354–55
   planting climbers on 278
   planting in 226
   rendered 86, 97, 164, 174, 226, 228, 231, 237, 354
   retaining 162, 228, 248, 258–59
   softening 178
   stone 36–37, 47, 97, 174, 226, 228, 354
   timber 228
   wall fountains 362
Washingtonia robusta 120
water 16, 174
   grey water 145
   management of 26
   recycling 137, 284
   reflections 35, 36–37, 43, 93, 239
   water usage 145
water butts 25, 76, 77, 137, 145, 177, 284

water features 16, 19, 26, 174, 223, 230, 244, 364
   containers for 360
   formal gardens 33, 34, 37
   lighting 238, 239
   materials for 230–31
   Mediterranean gardens 86
   modern 215
   planting 214–15
   positioning 214
   safety 61, 64, 139, 214
   small spaces 89, 93
   sound 73, 86, 95, 119
   statement gardens 119, 121, 125
   types of 362–63
   see also cascades; fountains; ponds and pools; rills; waterfalls
water plants 346–49
waterfalls 17, 215, 230
watering
   automatic irrigation 113, 284
   climbers 279
   lawns 273
   when and how to 284
waterlily see *Nymphaea*
Watkins, Antony 124, 127
weedkillers 285
weeds and weeding 251
   clearing 20, 21, 246, 247, 285
   hand weeding 285
   perennial weeds 273, 276, 280, 285
   spot weeding lawns 281
   weed suppressants 265, 277
weekend gardeners 20
Weigela
   W. florida 'Foliis Purpureis' 313
   W. NAOMI CAMPBELL ('Bokrashine') 196
Weisse, Rosemary 76
wellbeing 27
Wendy houses 359
West, Cleve 86–87
West Dean (West Sussex) 109
wildflowers 179, 249, 282–83
   natural gardens 79, 83, 85
   productive gardens 111
   wildflower gardens 236, 321
wildlife 14, 23, 24, 27, 28, 62, 67, 79, 147
   hedges and 274, 275
   meadows and 283
   natural gardens 57, 75, 76–77
   plants for 217
   ponds and water features for 31, 61, 62, 65, 66, 75, 77, 133, 137, 214, 215, 230, 362, 363
   wildlife walls 355
   see also birds; habitats; insects
Williams, Paul 197
Williams-Ellis, Nick 114–15

willow
   hurdles and screens 357
   living willow 357, 358
   tree seats 234
   see also *Salix*
Wilson, Andrew 98, 99
window boxes 18
windy sites 134, 135
   windbreaks 78, 110, 134, 193, 202, 227
winter interest 173, 193, 202, 205, 213
   winter-flowering shrubs 311
Wirtz, Jacques 52
Wisteria 179, 278, 359
   W. floribunda 317
   W. f. 'Multijuga' 317
wood 96, 174, 351
   containers 361
   pressure-treated 229, 253, 268
   recycled 57, 173, 229, 232, 233, 268
   wooden block walls 355
   wooden pallets as walls 355
   see also FSC; timber
wood preservative and stain 229, 351, 358
woodland-style gardens 125, 126, 135, 201, 212, 229, 236, 275, 277
   informal gardens 56
   natural gardens 84
   statement gardens 123
working plans 182, 189
workload 20–21

Y

yarrow 179
year-round interest 193, 205, 213
   plants for 218
yew 59, 122
   hedges 35, 53, 55, 79, 87, 193, 211, 237, 289
   topiary 159, 210, 289
   see also *Taxus baccata*
Yorkstone 42, 79, 351
Yucca 119, 210
   Y. aloifolia 99
   Y. filamentosa 'Bright Edge' 313

Z

Zantedeschia 178, 215
   Z. aethiopica 349
Zelkova carpinifolia 289

About the contributors

Editor-in-Chief

Chris Young is a landscape designer, horticultural content specialist, and garden business consultant. Formerly, he was Head of Editorial for the Royal Horticultural Society, where he oversaw all the publishing across books, websites, and magazines. He has been an awards judge for the Society of Garden Designers, Garden Media Guild, and was chair of the RHS Photographic Competition for 10 years. He originally studied landscape architecture at the University of Gloucestershire, England, and has won two Garden Media Guild awards for his writing. Other books Chris has authored include *Take Chelsea Home*.

Authors

Philip Clayton is a horticultural author, plant expert, and gardening enthusiast. His book *RHS A Plant for Every Day of the Year* (DK) charts the myriad plants on offer throughout the seasons. He trained at RHS Garden Wisley, was deputy editor on RHS *The Garden* magazine for more than 10 years, and is one of the leading writers on plants in the UK.

Andi Clevely has worked in gardening for over 50 years and is the best-selling author of *The Allotment Book*, as well as over 20 other titles. He also writes for magazines and has twice been awarded Practical Journalist of the Year by the Garden Media Guild. He lives in mid-Wales, where he tends a wild garden and allotment.

Jenny Hendy has a degree in botany and is an author, garden designer, teacher, and presenter. She has written books on a wide range of subjects, including design, planting techniques, and topiary, and writes for the gardening press. She is a regular contributor to BBC local radio and runs gardening workshops in North Wales.

Richard Sneesby is a landscape architect, garden designer, and lecturer based in Cornwall, with over 25 years' experience in the design of private and public landscapes and gardens. He has presented a number of television series, writes regularly for the garden press, and runs workshops for garden and landscape designers.

Paul Williams has spent a lifetime in horticulture, working and designing with plants. Trained at Pershore College of Horticulture, he has used his passion for plants and gardens to build a thriving horticultural consultancy and design practice. He has written several books on plants and gardening, and lectures on gardening.

Andrew Wilson is a multi-award-winning garden designer, director of garden design studies at the London College of Garden Design, co-director of design practice Wilson McWilliam Studio, and a lecturer and author. Together with his design partner, Gavin McWilliam, he has won a string of awards for his show gardens, both in the UK and internationally. He is also a fellow and former chairman of the Society of Garden Designers.

REVISED EDITION

Project Editor Dawn Titmus
Project Designer Christine Keilty
Picture Researcher Jackie Swanson
Consultant Gardening Publisher Chris Young

DK UK
Senior Editor Alastair Laing
Senior Designer Barbara Zuniga
Jacket Coordinator Jasmin Lennie
Production Editor David Almond
Producer Rebecca Parton
Editorial Manager Ruth O'Rourke
Art Director Maxine Pedliham
Publishing Director Katie Cowan

DK INDIA
Editor Ankita Gupta
Managing Editor Soma B. Chowdhury
DTP Coordinator Pushpak Tyagi
DTP Designer Manish Upreti
Assistant Picture Researcher Geetam Biswas
Pre-production Manager Balwant Singh
Creative Head Malavika Talukder

ROYAL HORTICULTURAL SOCIETY
Book Publishing Manager Helen Griffin
Editor Simon Maughan

This edition published in Great Britain in 2023 in association with The Royal Horticultural Society by Dorling Kindersley Limited, One Embassy Gardens, 8 Viaduct Gardens, London, SW11 7BW
First published in Great Britain in 2009

The authorized representative in the EEA is Dorling Kindersley Verlag GmbH. Arnulfstr. 124, 80636 Munich, Germany

Copyright © 2009, 2013, 2017, 2023
Dorling Kindersley Limited
Text copyright © 2009, 2013, 2017, 2023 Royal Horticultural Society and Dorling Kindersley Limited
A Penguin Random House company
10 9 8 7 6 5 4 3
005-333489-Sep/2023

All rights reserved.
No part of this publication may be reproduced, stored in or introduced into a retrieval system, or transmitted, in any form, or by any means (electronic, mechanical, photocopying, recording, or otherwise), without the prior written permission of the copyright owner.

A CIP catalogue record for this book is available from the British Library
ISBN 978-0-2415-9338-7

Printed and bound in China

For the curious
www.dk.com

This book was made with Forest Stewardship Council™ certified paper - one small step in DK's commitment to a sustainable future. For more information go to www.dk.com/our-green-pledge